Fourth Edition **INDUSTRIAL ARTS FOR THE GENERAL SHOP**

A STUDY OF AMERICAN TECHNOLOGY IN THE LABORATORY OF INDUSTRIES

Industrial Arts for the General Shop

FOURTH EDITION

DELMAR W. OLSON

Professor of Industrial and Technical Education
and Coordinator of Graduate Studies for Industrial Arts
North Carolina State University
Raleigh, North Carolina

PRENTICE-HALL, INC. Englewood Cliffs, New Jersey

PRENTICE-HALL INDUSTRIAL ARTS SERIES

Basic Electricity, by Abraham Marcus

Basic Electronics, by Abraham Marcus

Elements of Radio, by Abraham Marcus and William Marcus

Industrial Arts for the General Shop, by Delmar W. Olson

Mechanical Drafting Essentials, by Francis T. McCabe, Charles W. Keith, and Walter E. Farnham

Woods and Woodworking for Industrial Arts, by Delmar W. Olson

Architectural Drafting, by William J. Hornung

INDUSTRIAL ARTS FOR THE GENERAL SHOP, Fourth Edition, by Delmar W. Olson

ISBN 0-13-459131-3

10   9   8   7

Designed by Gary Schuermann

Prentice-Hall International, Inc., *London*
Prentice-Hall of Australia, Pty. Ltd., *Sydney*
Prentice-Hall of Canada, Ltd., *Toronto*
Prentice-Hall of India Private Ltd., *New Delhi*
Prentice-Hall of Japan, Inc., *Tokyo*

dedication

To my Jonny and Marene, and to every Jimmy and Mary, Joey and Martha, Jerry and Millie, and all of their friends now enrolled in industrial arts. May your lives be enriched and your futures brightened because you are in it.

And to every industrial arts teacher who believes as I do that what the chunk of clay and piece of wood do to the student is more important than what the student does to them. But let us not forget that what the young person does to the clay or the wood is a good measure of what it is doing to him.

DELMAR W. OLSON

Special credit is given to:

Institute of American Indian Arts, Santa Fe, N.M., and in particular the students Joyce Sisneros and Frances Makil, photos of whose work are included herein, and their photographer, Kay V. Wiest.

Department of Education, County of Burlington, N.J., for photos illustrating the work of Industrial Arts students on a gyrocopter. This project was arranged by William J. Howard, Jr., education consultant.

a word with the teacher

As I leaf through the earlier editions of this book, especially the first, I am reminded of a decision I had to make in planning it. It was my firm conviction, and still is, that if a text is to assist the teacher in accomplishing the purposes of industrial arts it should least of all be a book of recipes or a series of how-to-do-it, by-the-numbers exercises. Student input is a vital part of the learning experience. This explains why, instead of presenting projects with all of the thinking done in them, including step-by-step instructions, the book still concentrates on pointing the way or suggesting possibilities. It explains, too, the great variety of project ideas, as well as the encouragement to experiment with a variety of materials in the search for better ideas.

Through the several editions I have gradually increased the sophistication of the content, in order to keep pace with the increasing sophistication of the student. This edition includes new sections on industrial arts recreation, industrial arts and environment, occupations, and the consumer. Each of these is a base for the launching of studies that you can give as much timeliness and importance as you see fit.

Emphasis on the study of the technology is greater than before. This reflects not only my increasing concern for industrial arts as *interpreter of technology* for the American school, but the growing national awareness of technology as a socio-cultural force affecting both man's immediate environment and his entire planet.

As interpreter of technology, industrial arts has three primary purposes that distinguish it from all other educational disciplines and identify it at any grade level, kindergarten through the university. The first: to acquaint the student with the nature of the technology and its impact on and consequences for man, his culture, and his environment. The second: to assist the student in the discovery, development, realization, and release of his own talent potential within the context of the technology. The third: to assist the student in coping with a culture continually in change because of the continually advancing technology.

At this point I offer my assistance. If you have questions or problems you would like to pass on to me as you accept the challenge of industrial arts as the interpreter of technology for the American school, I shall be happy to share with you my ideas and recommendations.

Best wishes, DELMAR W. OLSON

contents

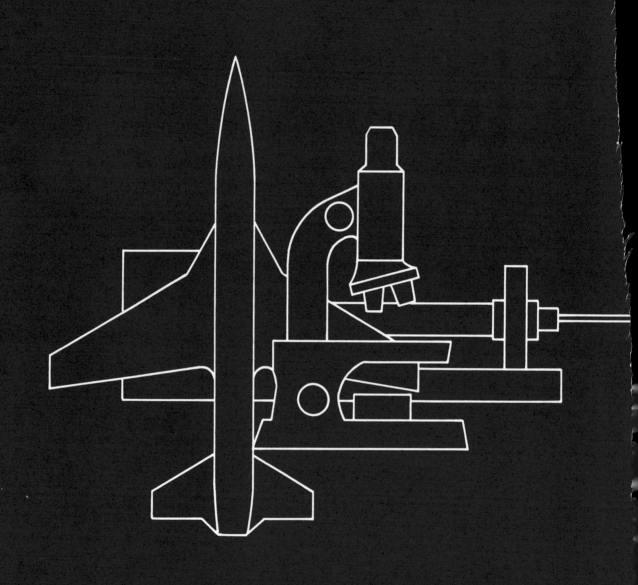

Industrial Arts, Technology, and You

1

Industrial arts is an opportunity for each boy and girl. It is first of all an opportunity for dreaming; then out of this dreaming can develop ideas in materials with the aid of tools and machines.

Tools and machines—they are by now almost second nature to modern man. They are the basis of today's *technology*. Let's look at this thing called technology. It is the story of what man has done in the past and can do tomorrow with materials.

The Beginnings of Technology

Technology began with man, or even before. Some **anthropologists** (an-thro-*pol*-o-gists), scientists who study the origin and development of man, believe that a preman creature used *pebble* tools. These were small stones that could crack mussel shells and nuts for food. The earlier theory was that the creature became man before he used tools. In either case, tools are the beginning of technology. Because man had hands that could grasp and manipulate things, he could be a *tool user*.

Pounding is thought to have been the first tool process. This was done with a rock or a club. Eventually the rock was tied to a stick. This increased man's ability

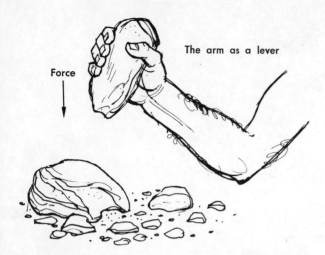

Force

The arm as a lever

Fig. 1–1. The discovery of the *lever* was probably the first step toward machines. The first tool, a stone held in the hand for pounding, gave early man an advantage over Nature. He learned to use stones to break mussel shells.

Fig. 1–2. (top) A rock tied to a stick gave early man a better tool for pounding. During the Old Stone Age he discovered how to flake stone to make sharp edges. Then he had an axe for chopping. The stick gave added leverage. **Fig. 1–3.** (bottom) The use of the lever enabled early man to multiply his strength in moving objects.

to pound. Then came scraping and cutting with the edges of shells and broken animal bones.

During the **Paleolithic** (pay-lee-o-*lith*-ic) or **Old Stone Age** man became a tool maker, a designer. He discovered that a certain rock could be chipped or flaked when struck with another rock. This enabled him to produce sharper and more durable cutting edges than he got with shells and broken bones.

The Old Stone Age gave way to the New Stone Age, the **Neolithic Age** (nee-o-*lith*-ic). With imagination added to his toolmaking, man was able to produce new tools and improve old ones. For example, he saw that the flakes chipped from rock were sharper than the rock itself. Setting flakes in grooves in small tree branches made saws. These cut by means of what we call teeth. By this time, man had made tools for pounding, scraping, cutting, boring, chopping, and sawing. With them he was gaining an advantage over Nature. Animals could not do this.

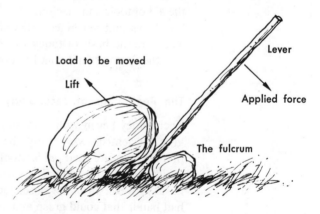

Load to be moved

Lift

Lever

Applied force

The fulcrum

Fig. 1–4. The idea of the *wheel* probably came from the use of logs as rollers to move heavy objects. We can imagine that this man was moving a heavy log. Where he could hollow it out for a canoe, he would have used fire to char the wood. With a stone axe he chopped out the burned wood.

Fig. 1–5. Some people have said that a potter's wheel was the discovery of the wheel. A flat stone turned on another stone made it easier to shape moist clay into pots. Rotation of clay for shaping on a wheel could have introduced the idea of machine processes.

Man Became a Potter

When man learned to plant seeds and harvest the grain, he needed containers to store it. But Nature did not provide these. Perhaps watching birds build nests, which are containers, man figured out how to make baskets. Eventually, he noticed that his footprints in the soft slippery material at the edge of the river held water. He found that he could coat the inside of a basket with the material we know today as clay. The basket would then hold water. This led to learning how to cook food. He dropped hot stones into the water to heat it. One day a claylined basket may have caught fire. When it cooled, a hardened shell remained. Here, Mr. Early Man may have become a potter.

About fifty centuries ago the Egyptians were able to shape clay on a flat rock, which was turned while it rested on another rock. Some **archaeolo-** gists (ark-ee-*olo*-gists) say that this was the discovery of the wheel. The stories of early civilizations are often pieced together by archæologists from remnants of fired clay pottery. These pieces are called potsherds.

Man Became a Metalworker

Man became a metalworker at about the same time that he became a potter. It is said that in the ashes of a campfire he probably discovered some small, shiny, heavy nuggets. This was his first metal. It could have happened when the fire had been built on an outcropping of ore. This first man-made metal was bronze. It is an alloy of copper and tin. The discovery ushered in the Bronze Age. The metal was used for tools, utensils, nails, pins, coins, jewelry, and sculpture, and in temple construction. The famous Parthenon,

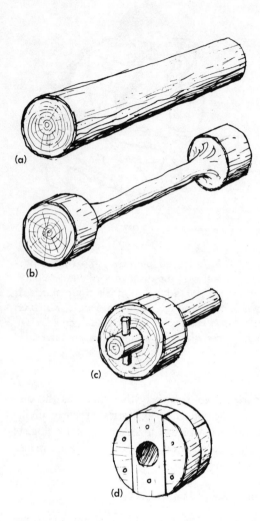

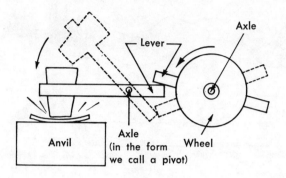

Fig. 1–7. *Mechanization.* The wheel, axle, and lever were used to mechanize hammering. This was probably done during the Bronze Age. Copper can be formed by hammering. This principle is used today in shaping certain metals. In industry it is called *forging,* or *drop forging.*

built during the Golden Age of Greece, is constructed of huge blocks of cut marble. These are held in place with bronze dowels or pegs fitted into holes in the blocks.

The Iron Age

The know-how of the Bronze Age led to the Iron Age. The Persians are said to have smelted iron as early as 1200 B.C. Invention of the bellows made it possible to reach higher temperatures than required to melt bronze. This knowledge spread around the Mediterranean Sea and up into Europe in the next 2000 years. By the year A.D. 1700 the skills, tools, and machines necessary to work iron had been highly developed. Experiments with iron water pumps and steam engines were being carried on in England. One of the key tools in this progress was the file. The key machine tool was the metal turning lathe. Both of these are used today in industrial arts to shape metals.

Fig. 1–6. *The axle.* (a) While log rollers could be used to move heavy objects their use was limited. (b) The same log with the center cut away rolled more easily. It was used for the first carts. We call this the discovery of the axle. Could it be so named because it was hewn with an axe? (c) But the most significant idea was that of a wheel free to turn on an axle. (d) The wheel itself has gone through many developments. For example, compare a bicycle wheel with this one made of boards.

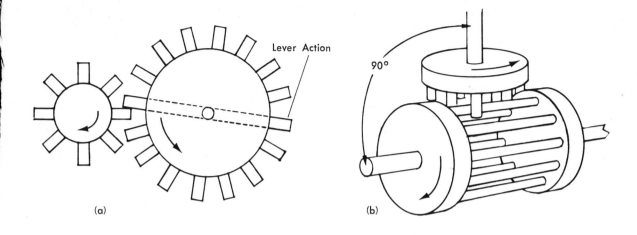

Fig. 1–8. *The Gear.* (a) Gears are wheels with levers over their rims. These levers are called teeth. The first gears were made of wood. Pegs were driven into holes in the rim. The speeds of a pair of gears vary as the number of teeth. For example, when one gear has twice as many teeth as the other it turns at half the revolutions of the other. (b) The cage gear was an early idea for changing the direction of motion.

The Age of Steel

The refinement of iron into steel brought the Age of Steel with improved tools and products. Steel was being made in Europe by A.D. 1800. To this point in history the actual production of things was done by craftsmen. They spent their lifetimes mastering their materials and the known processes. There was the metalsmith, the tinner, the leather-worker, the joiner (carpenter), the cabinet maker, the weaver, and others. Originally each made the complete article, whatever it was. A boy who would become a metalsmith, for example, was *apprenticed* to a master craftsman for several years. He lived with the master. Here he learned to read, write, and calculate as well as the "mysteries" of the trade. At the time of the founding of our country, apprenticeship was common practice. Public education for such skills came later.

AMERICAN TECHNOLOGY

The Indian in America at the time of the Pilgrims had his own technology. For example, he flaked stone to make arrowheads, axes, and tomahawks. He made clay pots and baskets as well as bows, arrows, and traps. He cured and sewed animal skins and fabricated canoes from bark and skins. Of course, all tribes did not have the same tools and skills. Can you figure out or find out why?

The early settlers from Europe had a more advanced technology than they found among the Indians. On the *Mayflower,* the passengers brought only such simple tools as hammers, axes, saws, hoes, shovels, and needles. Nevertheless among these early settlers there were craftsmen in wood, metal, clay, leather, glass, weaving, and printing. They got American technology underway. It was necessary for their survival.

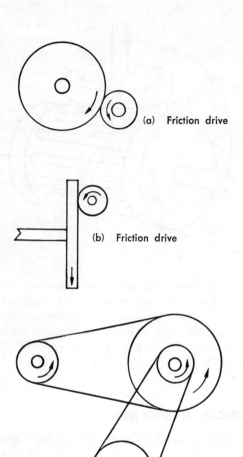

(a) **Friction drive**

(b) **Friction drive**

(c)

Fig. 1–9. *Drives.* There are several types of gear drives. Here are some other types. (a) *Friction Drive.* Contact between wheel rims causes the motion. (b) Friction Drive at 90 degrees. (c) *Belt Drive.* Flat belts and pulleys were the first types. Today V-Belts and pulleys are common. Note that pulleys turn in the same direction. The ratio of a pair of diameters is the ratio of their speeds. Study the machines in your industrial arts laboratory. How many kinds of gears and drives are there?

From the landing at Plymouth Rock to the War of 1812, the colonists depended largely on England for the manufactured products they needed for living. They exported raw materials and imported finished goods. The break in dependence on England in 1812 was accompanied by the arrival of the Industrial Revolution from Europe. From this point on in our history, machines began to replace hands and muscles. By 1900, steam engines were powering trains, ships, factories, electric power plants, and farm tractors. At the same time the internal combustion engine and the stationary electric motor were threatening to make steam power obsolete.

Technological Advance Brings Changes

The period since 1900 has seen a rapid succession of developments in technology. Each has brought new ideas and better ways of doing things. Each has been amazing in one way or another to us. The successful flights of the Wright Brothers at Kitty Hawk, North Carolina, must have been a sensation to the nation at that time. The whole world watched on television the first step of our man on the moon.

Technological advance is a parade of innovations. Each new idea makes obsolete whatever it replaces. As it advances, technology changes things and it changes our ways of living, of thinking, of doing. It even changes us as individuals. How would a youth of your age, living a hundred years ago, be different from you because of the difference in the technology of the times? Consider skills, goals, opportunities, attitude toward life.

Advances in technology serve us in many ways. They eliminate drudgery from work as machine power replaces muscles. They lessen work as machines take over. Along with this comes an increasing amount of leisure time. Today the work week is approximately 37 hours. A century ago it was nearly twice that. It has become increasingly easy to go where we wish on land and sea and in

Within the present century, technology has become a powerful force in doing things for us as well as against us. We must remember, however, that technology itself is inert. It is man who creates it, uses it, and causes it to abuse Nature and himself. Perhaps we should assume that *what man creates in his technology he should be able to understand, to use wisely, to control, and to change.* Industrial arts aims to help you to understand this great force and to find your way around in the technological culture. This is why it offers you an opportunity to create, invent, and develop your ideas with materials, tools, and machines. You get in on the ground floor of technology, where understanding begins.

Fig. I–I0. A student has designed and constructed his project. Here he is installing the lift engine that drives the propeller. Following tests, the experimental project will be evaluated, and the student will make such modifications as are needed to improve its efficiency.

AMERICAN INDUSTRY

Had you been living at the turn of this century, you could not have gone to the store and purchased what you can today. There were no radios, television sets, miniature cameras, comic books, or any of thousands of other products that are available today. They hadn't been invented or developed yet. Products change so frequently that it is almost impossible to predict what will be common in the year 2000. You might want to start a hobby. Make your predictions over the next year; then be sure to keep your list in a safe place until you can verify them years from now.

air and space. We can be the best informed people if we take advantage of the various communications media available. We can create about any kind of environment we wish, from homes to cities and from suburbs to parks. We can get better food, clothing, medical care, and all of the many consumer items.

At the same time, technological advance keeps making obsolete our houses, cities, highways, jobs, skills, and consumer products. It congests the space we have on land and in air with buildings and vehicles. It pollutes the air, water supply, lakes, rivers, and seas. It surrounds us with noise from which we can't escape. It destroys the natural environment and upsets the balance of Nature. It gives us more leisure than we can wisely use.

Industry today is the organization, the economic institution, that provides the quantities of goods and services we use in living. It is owned by stockholders for whom it must earn a profit. Technology existed long before industry was born and we would have a technology even if today's industry were to disappear.

Today's industries can be grouped as manufacturing, power, transportation, communications, construction, electrical-electronic, services, and others. Our story here focuses largely on manufacturing. This begins with the production and refining of raw materials. It extends into the pro-

Stockholders

President

Treasurer	Personnel Manager	Factory Manager	Chief Engineer	Chief Designer	Director of Research and Development	Director of Sales

Purchasing	Manufacturing	Plant Engineer	Safety Engineer
Material Control	General Superintendent	Plant Maintenance Protection Construction	Safety, Health

Parts Manufacturing Superintendent	Assembly Superintendent	Quality Control Superintendent	Tools Superintendent	Shipping Superintendent

Foundry Foremen	Machining Foremen	Sub-Assembly Foremen	Final Assembly Foremen	Inspection and Testing Foremen	Tool Room Foremen	Shipping Foremen

Group Leaders	Group Leaders	Group Leaders	Group Leaders	Group Leaders	Group Leaders	Group Leaders
Workers	Workers	Workers	Workers	Inspectors	Workers	Workers

duction of products to be used by the individual, the family, the community, business, industry, the military, and the medical. Any one industry in this group produces great quantities of identical items, from bolts and nuts to breakfast cereal, to automobiles and aircraft. This is called *mass production*. Today's manufacturing industries are not geared to producing one-of-a-kind products, except for experimental work. In industrial arts, however, we are mostly concerned with learning how to express our own ideas in materials with tools and machines. So we produce mostly one-of-a-kind items.

Interchangeable Parts

When a part breaks on your bicycle or car, you expect to be able to get a replacement part to correct the trouble. This part must be interchangeable with the original. When an auto manufacturer produces, say, 100,000 engines of the same model, he also turns out the required quantity of replacement parts to supply the dealers. A typical eight-cylinder engine would require eight identical

Fig. 1–12. (top) Bicycles are mass produced in order to meet the demand at reasonable prices. Here is the final assembly line in a bike factory. **Fig. 1–13.** (bottom) For every bike there are two wheels. Replacement wheels are also necessary. This machine forms a perfectly true rim automatically.

Fig. 1–11 *(opposite page)*. Flow chart showing the personnel organization of a typical manufacturing industry. The stockholders are the owners of the company. They have invested their money to buy stock. They elect a board of directors who appoint a president. The president is responsible for operating the plant. Now look at the bottom of the chart. You will see small groups of workers. These are supervised by group leaders who are responsible to foremen, who in turn are responsible to their superintendents, and so on up the line. In large industries where there may be fifty or more separate departments, each with a specific type of work, all these thousands of employees must have their tasks accurately coordinated so that the products flow off the production line on schedule.

pistons and sixteen valves, for example. In the factory the many parts flow to the assembly line where the engines are put together. The production of interchangeable parts is the practice that makes the mass production of assembled products possible. They are produced in the least time, at the lowest cost, and in the necessary quantities. It also facilitates the servicing. Eli Whitney is credited with introducing the concept of interchangeability into American manufacturing in

Fig. 1–14. A bike sprocket is an example of a simple gear driven by a chain. Here several sprockets are lifted from a chrome plating bath.

Owned by the Henry Ford Museum

Fig. 1–15. (top) Henry Ford and his quadricycle. This was his first automobile built by him as a young man. It was first driven on June 4, 1896. Automotive engineering got its start with inventors such as Ford. They dreamed, reasoned, designed, experimented, constructed, tested, and improved their ideas largely by trial and error. They depended much on "rule of thumb." Fig. 1–16. (bottom) Henry Ford's famous "999" racer set a speed record of 91.37 MPH in 1903 on the ice of Lake St. Clair in Michigan. Notice the size of the engine and the steering bar. All working parts were exposed. The driver probably got an oil bath. Compare the quadricycle and "999" with today's cars.

Owned by the Henry Ford Museum

Fig. 1–17. Ideas for new automobiles are created in a design studio. Artists and engineers team up here in designing the bodies.

Fig. 1–18. (top) Auto engine parts are made and assembled largely by automated processes. The complete engine is installed in the chassis in a matter of minutes. Fig. 1–19. (bottom) Auto bodies are assembled of parts formed in huge presses. The parts are loaded into assembly fixtures. Automatic welding fastens them together on a moving conveyor line.

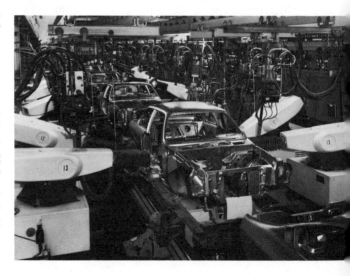

1799. He was able to demonstrate before a Congressional Committee that he could assemble muskets from boxes of identical parts.

Identical parts are made in molds, dies, jigs, and fixtures. A *mold* is a device in which or over which a material is formed to the desired shape. It is usually made of metal, wood, or plaster, depending on the material in the part to be formed. A *die* is a device for forming, stamping, cutting, or shaping identical parts. An automobile fender, for example, is pressed from sheet steel. It is first cut or stamped to the correct flat shape in stamping dies. This piece is fed into a pair of forming or pressing dies that produce the final shape. A *jig* holds a piece of material or a part for processing, such as drilling, bending, welding. Whitney made the jigs for his musket parts. A *fixture* is a holding device for parts while they are being assembled, as in an airplane wing.

Industrial Arts, Technology, and You

rect operation. Such machines are called automatic; semiautomatic machines are only partially automatic.

Automation

Completely automatic production is called *automation*. The goal here is machines and systems of machines to take over the complete responsibility of production. To make this possible, systems of precise control are necessary. They include such systems as electrical-electronic, hydraulic, mechanical, and pneumatic. The completely automated factory is one in which the product specifications are recorded, as on tape. The tape activates the factory. It feeds the material, processes it, evaluates its accuracy, and makes the necessary corrections. It moves the parts along to

Fig. 1–20. New autos are loaded by machine into special railroad cars. Note that they are standing on end in the cars. The story of mass production of autos is similar to that of any product. Machines do most of the work. Man provides the thinking.

Fig. 1–21. This is the bicycle used by the Wright Brothers for air pressure tests in the design and development of their airplane flown at Kitty Hawk, North Carolina. This is an example of what we call research and development. There were no books on aeronautical engineering for the Wrights to study. They reasoned, experimented, and often by trial and error found what would work.

Owned by the Henry Ford Museum

Mechanization

Mechanization is the conversion of manufacturing from hand processes to machine. This was the goal of the first century of the Industrial Revolution. In a mechanized factory each machine was operated by one man, called a machine operator. Such machines were controlled by hand, although driven by motors or engines.

Automatics

At about the turn of this century improvements in mechanization led to the development of machines that could run and control themselves. The machine operator could now supervise several machines. He turned them on and off, supplied the materials for processing, and checked on the corthe next station, and assembles them into the finished products in one continuous flow. It even packages them if required and moves them out.

Very few technicians are required to operate such a factory. Several, who are highly skilled in electronics, hydraulics, mechanics, and pneumatics, may be required to keep it running. This is essential because of the tremendous investment required to set up such a plant. It must run continuously to earn a profit.

What Next?

From hand-made goods to machine made, to automated factory is the story of industrial development. We can wonder what is next. Persons who are studying this question tell us that we are already in a *post-industrial* society. As industry in all of its forms becomes more automated, we move ahead into a society, a culture, unlike any we have ever known. This post-industrial society is one in which work becomes obsolete and leisure takes over. Instead of man finding his reason for existence in his work, as has been customary for hundreds of years, he now must find it in leisure. If you will plot a curve showing the declining length of the work week for the past century and then project it to the year 2000, you find that shortly after that date the work week has a length of zero. This is theoretical of course, but the trend is real. Comparatively few persons will be required to do what work is necessary. The post-industrial society also refers to the changing nature of work as well as to the work week. New and higher skills are required to make the machines. Greater knowledge is necessary to keep them running.

In this period automatics and automation will come into full development. They will increasingly be applied in industry, the home, transportation, communication, medicine, the military. You may eventually live in a home where the meals for the week are programmed into a computer, which at the right time activates the appliances to prepare and to serve them automatically.

Fig. 1–22. Man uses heat obtained from coal and gas to melt metal ores that he digs from the earth. He uses the metals to make millions of things. Here he is tapping a blast furnace.

Research and Development

Research and development is known as the "industry of discovery." It supplies manufacturers with new materials, processes, and products. Originally industry depended on inventors to come up with ideas. Major industries today generally have their own "R and D" departments. There mathematicians, scientists, engineers, and technicians work together searching for better ideas and better ways. Industrial arts offers you research and development experiences. Here you can invent, design, and demonstrate new ideas.

Fig. 1-23. (top left) In industrial arts many interesting things are done. This gyrocopter was built by industrial arts students in the high school at Burlington, New Jersey. Here it is being test flown. **Fig. 1-24.** (right) The parts of the "flying machine" are laid out ready for assembly. The students and their teacher put it together in 150 man-hours. **Fig. 1-25.** (bottom left) The gyrocopter is loaded on a trailer and towed to the air field from the school. You can see that it is not a toy.

TECHNOLOGY, MAN, AND CULTURE

Technology today is much more than nuts and bolts and hammers and nails. It has many faces. Since it grows out of ideas with materials, however, we can call it our *material culture*. This means that it is the sum total of all that man knows about and does with materials. Of course, one can't do much with materials without tools and machines. But these are made of materials, too.

Technology may be seen as *man's means to gaining advantage over and control of Nature*. He uses Nature's materials to make things, to build houses, automobiles, aircraft, highways, and such. Nature does not provide these for him, but he uses Nature to get what he wants. She provides the materials.

Technology is also seen as *man creating his own environment*. He changes the landscape at will. He moves mountains, levels hills, and builds lakes and cities. He can change his environment as he moves from place to place in his vehicles. Technology is the man-made environment.

In all of this, technology can be seen as *man expressing himself*. By nature he is a builder, a creator. Using materials, tools, machines, and energies he builds what he likes. This is a satisfying and a pleasurable experience for him. Can you explain why you enjoy expressing your ideas with materials? It is as natural for man to express himself with materials as it is with words. You are showing and telling about yourself as you express your ideas in industrial arts projects.

We can think of technology also as *man making himself*. The thought is that as man develops new technology it in turn acts on him and changes him. This is an interesting theory. You can do some exploring here. For example, ask several of the long-experienced teachers in your school for their opinions on how the automobile, airplane, and television have changed people. Perhaps your industrial arts teacher can help you design a questionnaire to get a large number of opinions on

such questions. Do you suppose that the understanding gap between youth and adults may be due at least in part to their technology gaps? The older one grows, the greater are the changes he has witnessed.

With technology having so many faces and capabilities, perhaps you will agree, as was suggested earlier, that what man creates in his technology he should be able to understand, to use wisely, to control, and to change. Your study and work in industrial arts is intended to help you do this.

TECHNOLOGY AND ENVIRONMENT

The preservation and restoration of our natural environment has national attention, as you well know. Perhaps a course in environmental education is offered in your school. As earlier pointed out, technology is man creating his own environment and gaining advantage over Nature. It is here that he has been causing trouble for himself. There are two basic kinds of environment: the natural as Nature provides it, and the man-made. The two do not necessarily live in harmony. Experience shows us that when man with his technology makes his own environment, he does one or more of the following:

1. He consumes Nature's materials faster than she can replace them. Some are not replaceable. It takes 50 to 75 years to grow a black walnut tree to maturity. How long does it take to cut one down? The minerals that man uses so freely required millions of years to form.

2. He destroys Nature as he moves mountains and fills lakes and valleys to build highways, pipelines, cities, and other bits of his environment. He replaces the natural with what he thinks is better. He rearranges the natural to suit his own temporary plans. He puts parks and industries, lakes and air terminals where he chooses.

3. He produces unnatural noise, waste, litter, junk, and excess heat as he converts natural materials into his usable things. He then dumps his

Fig. 1–26. An open pit copper mine. Note how man changes the face of Nature to get the natural minerals he wants.

discards back onto Nature, assuming that she can take care of them. But he is finding that even huge lakes like Lake Erie and Lake Michigan cannot accept an endless supply of garbage and sewage. The pollution kills off fish and wildlife, eliminates swimming, and spoils beaches. Nature doesn't enjoy being a cesspool.

4. He pollutes the natural environment by contaminating the air and the water. Smog filters out the sunlight. Insecticides poison wildlife, and the Balance of Nature is upset.

Technology and Nature in Harmony

At the beginning of this chapter we said that man is by nature a maker, a builder, a creator with materials. Because he is, he has developed his tremendous technology. Now he is discovering that he has been carried away by his ability to make things and his concern for profit in doing so. He has been building recklessly. He is finding that the supply of materials on this planet is limited. He

has mass-produced more junk and waste materials than Nature can accept.

Correcting this situation will help man to live in harmony with his natural environment. He knows he must do this if he is to survive. At this very time laws and regulations are being made and enforced to control pollution. Industry is recycling more materials than ever before. Less frequent changes in models of new products are being planned. Out of this is coming a new technology. It will enable man to live in a balance with Nature. Technology is the man-made environment. The greatest challenge of all history is likely to be to develop this new technology. With it, man can preserve and restore the natural environment. Watch for news of new developments on this front as you study this book.

YOU AND INDUSTRIAL ARTS

For our purposes in this book, industrial arts offers to acquaint you with materials, tools, machines, products, ideas, principles, and practices.

Fig. 1–27. Trips to industrial plants with your shop teacher can sometimes help you choose work you would like to do.

With this experience you gain technical knowledge and skills. Industrial arts also helps you:

(1) to discover career possibilities among the technical fields

(2) to find and try out hobby interests

(3) to be a wise consumer as you select, use, care for, and evaluate your purchases of products

(4) to understand and to appreciate the technology, its impact on culture and environment, and its promises and problems for you and for all of society

(5) to better know yourself as you discover, develop, release, and realize your talents for thinking, creating, and developing ideas in materials

With this operation you are invited to come up with your own goals for what you wish to learn, and for projects to develop. Discuss these with your teacher. He can help you achieve your ambitions.

The Environment for Industrial Arts

Your school has a center for its industrial arts activity. It is a laboratory containing materials, tools, machines, supplies, and other equipment that you can learn to use. Sometimes it is called a shop, or a general shop. It is the school environment for your study about technology. But the whole school, the city, and even the world can be your environment. Books, magazines, TV, visits, will push out the walls for you and enlarge your laboratory.

The Industrial Arts Project

When you see the laboratory with all of its materials and equipment you can begin to dream up ideas for things to do and to make. Materials, tools, and machines are intended for this. A project begins with an idea. As you and your teacher accept the idea it becomes your project, *your thing*. Since you spend most of your class time on the project, it ought to be a good idea to

Fig. 1–28. A corner in a home workshop equipped for working with wood. What kind of a workshop would you like? This man needs some advice on safety. What will you suggest?

begin with. The final chapter in this book is full of project ideas. You can add many more.

To Get the Most from Your Industrial Arts

Note these tips to help you get the greatest satisfaction from your experience in industrial arts:

1. Keep telling yourself that learning new things is your best way to grow.

2. Your brain is your idea machine. As you feed it with ideas and information, it produces other and different ideas. Keep this machine working. It is a necessary part of the equipment in industrial arts.

3. Tools and machines are designed to be used for certain things and in certain ways. For best results and less chance of injury, follow your teacher's instructions carefully. You'll be glad you did.

4. Carry your full share of the load for the smooth operation of your class. When each of you does this, your teacher can spend full time helping you with your projects.

YOUR RESEARCH AND DEVELOPMENT

A research and development project begins with a problem. You are attempting to solve it or come up with a better idea. The research includes reading, study, imagination, and plenty of good logical thinking. It takes research to find out if the problem has been solved before and how it was done. It involves making numerous proposed solutions in drawings and often in scale models. It requires evaluation of the ideas and the final selection of one for development.

Development takes the selected idea and produces a prototype. It studies, criticizes, tests, and modifies this original. It then produces the finished item. Along the way, experts are consulted for their reactions and suggestions. These are such professionals as engineers, chemists, electronics specialists, and technicians—often in local industries. Many times correspondence is required with experts in other parts of the country. We recall Richard, who was interested in World War I military aircraft. He wanted to make a collection of scale models. To get the necessary data he corresponded with aeronautical engineers in England, France, Germany, and Russia, and even had his school librarian get books for him from libraries in these countries.

If you are told that your idea won't work, you have several choices of action. You can drop it. You can modify it to eliminate the weaknesses. You can move ahead with it because you honestly feel, and your research supports you, that it has to work. An R and D (Research and Development) project, you see, is much more than making something. You are exploring, pioneering, and creating something of significance. Now and then a student hits on an idea that even the professionals haven't thought of. Wow!

MASS PRODUCTION IN INDUSTRIAL ARTS

You can carry on a mass production project in industrial arts. The entire class or just a group of students may be involved. A product in graphic arts is a good choice because printing is a mass production process in itself. In ceramics, the processes of slip casting and jiggering with clay are for producing like items. In the other areas you will need to adapt processes to mass production. There are certain logical steps in any mass production project you undertake.

1. Project Planning. Planning the project involves identifying problems and questions about the project and coming up with the best possible answers. The enthusiastic support of all the participants is required. The whole class or group functions as idea men and engineers. These are typical points to consider in the planning stage:

a. Why put on a mass production project? Will each person's time be as well spent as when he works on his own project?

b. Where can it bog down or fail along the way?

c. How and by whom are the important decisions to be made?

d. How will it be financed? Must it produce a profit to be worthwhile? What factors should be included in the per item cost?

e. What about "goofing-off" and absenteeism?

f. How can safety practices be enforced?

g. How will personnel be selected?

h. Should a company be organized?

2. Product Selection and Design. The idea for the product should, of course, be a good one. It should be worth producing in quantity. It should bring credit to industrial arts and to those who produced it. It must be so designed that it can be mass produced with the facilities and the time available. Several prototypes should be developed and evaluated. Make a *product analysis* from the reactions of a number of persons not involved in the project. Has a student already developed an idea that would be appropriate?

3. Engineering Analysis. The selected product is analyzed for materials, parts, subassemblies, fastenings, finishes. Each process to be included is identified. The function of the item is carefully evaluated. Appropriate structure and construction is important.

4. Production Planning. The best sequence of processing is worked out. A work schedule for each work station is charted and the manpower needs determined. Materials are ordered (in these days of concern for the environment, you might consider the possibility of using discarded materials).

5. Tooling Up. The necessary jigs, fixtures, molds, dies are designed, constructed, and arranged in the proper sequence. Conveyors and parts transfer systems are set up.

6. Personnel Assignments. From the manpower needs already determined, students are assigned to the tasks for which they are best qualified. Applications may be taken for the positions.

7. Trial Run. A trial run is made to check on the accuracy and fit of the parts. Any design changes are made along with any changes in the tooling.

8. Production. Preparation for the run takes considerable time, perhaps weeks. The actual manufacture may take but a few class periods. Dismantling the "factory" will also be necessary. All of the time consumed in these several steps should be included in calculating the cost per item. You can see that the more items produced the lower the individual cost. This is the economic objective in mass production.

9. Evaluation. The project should be evaluated in terms of the value you received for the amount of time invested. What were the bottlenecks? Why were they? Was the overall planning complete and efficient? What about the product quality? What were the toughest problems encountered? Were they technical, organizational, personnel? What are your recommendations to the next class that may wish to try a project in mass production.

CAREER CHOICES

Your experiences in industrial arts can help you to see yourself in many possible careers. At this time, what are your preferences? If you aren't sure, if you find it difficult to make choices, it's understandable. Had you been born a century earlier the choice would have been much less of a problem. There were fewer job possibilities then, and the son's tendency to follow his father was common. Today you can choose from among more than 50,000 recognized different occupations. The list is continually changing as some items become obsolete and new ones are added. But of the following you can be sure as you consider your future:

1. A person can be successful in many different fields of work. One's talents are so many that as he finds and develops them, they open up opportunities for careers. As you get into different activities in industrial arts, some will no doubt "turn you on." These may be clues to talents.

2. You can look forward to changing your occupation completely several times during your lifetime, especially if you will be working in industry or business. Consequently, it is well to have several career interests.

3. The ongoing best qualification for success in all careers is continuing study. Engineers, technicians, operators, for example, must keep studying to keep up-to-date with changes. To like to study is becoming increasingly important in the world of careers.

4. Your best preparation for careers while you are in school is likely to be all-around study and experience to help you most fully discover and develop your interests and talents. It is with these that your choices begin. This is why industrial arts emphasizes self-discovery within the technology.

5. Tryout experience is valuable in making career choices. In some cases you may be able to get part-time work while attending school. All of the subject areas in industrial arts represent fields

of career opportunities. You can get real experience in these, even though industrial arts as a study is not intended for specific job or trade training.

6. The most salable skill of all is the ability to think, especially to think creatively. This is demanded in all fields of work, and premium rewards are paid for it. Your projects show how creatively you think. Are you giving them your best efforts?

7. Generally a job will be as glamorous or as dull, as pleasurable or as boring, as one wishes to make it. Attitude toward work is largely the measure of a person's satisfaction in it. And success on a job often depends on one's ability to get along with people.

8. The mechanically repetitive jobs in industry are usually the first to be replaced by machines. Those jobs requiring no special training, education, or experience are the first to be made obsolete.

9. Hobby interests and skills often lead to desirable careers. Keep this possibility in mind, especially as we move ahead into the post-industrial society.

10. You can also get ideas for careers from persons in them, from field trips, movies, and TV programs, books, and magazines. Be alert to possibilities. It's your future with which you and we are concerned.

Your Own Business

Many students, while in industrial arts, have developed interests and skills with which they have opened part-time businesses at home. Some have earned their ways through college with these. Others have found business careers in them. Any industrial arts activities can have such possibilities. For example, there is custom printing, silk-screened signs, ornamental metalwork, ceramics, jewelry, weaving, furniture, refinishing, photography, and many others. With imagination and hustle along with your skills you can be in business. Both custom work for individuals and the production of items for sale have potential. There is also repairing and servicing to consider. Get acquainted with several "small business" men. They will know about the qualifications, problems, and rewards. If there is a Junior Achievement Program in your school, you can get real business experience there.

The Industrial Arts Teacher

Some of your classmates should eventually become teachers of industrial arts. Perhaps you, too, may look forward to teaching. It needs men and women who have wide interests and talents in technical fields along with a desire to work with boys and girls. A college degree in industrial arts teacher education is necessary. Ask your teacher to discuss the nature, qualifications, problems, and rewards in this field.

Success and Failure

Each of us is interested in a successful career. Yet we know that failures happen in all fields. As you prepare to choose your career, consider what makes for success and for failure in it. For example, ask employers in industry and business why people fail. You can do this on a field trip, by inviting a business man to speak to your class, or even by letter. He may tell you that a person rarely fails because he cannot do the work satisfactorily. It's more likely that he has a faulty attitude. He may be unwilling to follow directions. Perhaps he can't get along with fellow workers. He may want to set his own hours and time for coming to work. Success is made up of many factors. But one who is willing to learn and gets along well with others has the first qualifications. A healthy attitude toward work you must develop by yourself. You don't get it with a college degree or in a technical training course.

INDUSTRIAL ARTS RECREATION

Designing and making things is naturally fun. It is exciting, satisfying, and refreshing. Because of this, industrial arts makes a real contribution to recreation. The experience you get with materials, tools, and machines can be used at home in your leisure time. Some schools offer industrial arts recreation as separate courses. In these there are no examinations, required assignments, and no grades. The purposes are solely recreational. Students, parents, and teachers attend these courses. There are opportunities for them to "do their things." Perhaps you could develop a hobby from some activity you particularly enjoy in industrial arts. It not only can give you something to do when you have time of your own, but it might lead to a career.

A Home Workshop. Each of the chapters in this book is a source of ideas and activities. They have possibilities for both school and home workshop use. To be used at home a workshop is likely to be necessary. Space and cost are often problems. Let's discuss them.

How Much Space? Most of us can't have all of the space we'd like for our home workshop. But whether we can have an entire basement, a tiny closet, or a corner of the bedroom, a workshop is possible. Of course the things we can make in a basement shop will be different from those we make in a corner.

Suppose you have no space at all in which to set up your workshop. You have two choices. Do those things which require no fixed space. There is drawing, designing, collecting, model making and others, for example. Or you can use portable equipment. This can be stored away when not in use. Such equipment can be used for printing and photography, ceramics with clay and enamels, woodcarving, model making, radio, and many other possibilities.

What Will it Cost? You can spend a lot or little for your home workshop. You can build

items of equipment in your industrial arts classes. Make a list of what you need. Then look through this book for examples of machines and equipment that can be constructed. Magazines such as *Popular Science, Popular Mechanics,* and *Mechanics Illustrated* have articles on building equipment. Perhaps you can interest other members of your family in home workshop activities. Any cost then might seem less prohibitive. You may be able to sell some of the things you make. We know of two boys who while in school made some excellent model airplanes. Soon friends wanted kits, and eventually the two boys formed a company to make and sell them.

A Home Workshop Club. If there is a home workshop club in your neighborhood you may find it helpful in the pursuit of your hobby. The members get together to watch demonstrations, to study processes, see movies, and plan exhibits. In some schools the industrial arts students have such clubs. Talk to your teacher about the possibility of having such a club.

Craftsmanship. Craftsmanship begins when you care enough to do your very best with your materials, tools, and machines. It shows up in the things you make. They exhibit good construction, the right materials, proper finishes, and clever, imaginative designs. As you gain control over the materials and the tools with which you work, so that they behave as you intend, you are becoming a craftsman. Without concern for this type of excellence, industrial arts projects as well as industry's products are no more than junk. Your teacher knows that it takes time, experience, and a strong desire to become a good craftsman. He plans his instruction to help you achieve this goal. Ask him to discuss it with your class. He can give you some tips.

As you move on through this book we wish you many exciting and satisfying experiences. We would like very much to know about your projects or other activity. How about a letter or a snapshot?

TO BE A WISE CONSUMER

Someone has said that spending what one earns is just as important as earning. Do you agree? The thought may be that it is not so much what one earns as what one spends that is important. It may also suggest that spending usually means purchasing goods or services. Here one can easily make poor choices. You are probably aware of the current national concern for the consumer. Both state and Federal governments are attempting to protect him by passing consumer protection laws. Labels on containers must now accurately describe the nature of the contents as well as the quantity. Automobile safety regulations to protect the occupants and to cut down pollution have been passed. These are but a few examples. You can find many more if you search around.

Each of us is a consumer. We have to buy ready-made most of the things we use in living. A century ago the people were much less dependent on industry for these things. Many were produced in the home. For example, there were home-canned fruits, vegetables, and meats as well as home-made clothing, bedding, and toys. Farmers were even more self-sufficient.

Industrial arts for many years has included the study of consumership. It believes that a wise consumer knows how to select, purchase, use, and evaluate the things he buys. From your study of the following chapters you will learn about many materials and how they are changed into useful products. You have actually used materials, tools, and machines, and made things with them. You are now in a position to know when a product is well designed. You can tell if it is functional and durable. You can be intelligently critical of its form, beauty, color, and decoration. You can more easily tell if it is safe or dangerous to use. (We suggest you try to do this with children's toys.) And you are better qualified to use a product wisely.

There is another question that each of us as a consumer must answer. What effect will my choice have on the environment? If it is to be used once and disposed of, will it be biodegradable? (This means will Nature be able to make it decompose?) Will it cause pollution of air or water? Does it use a minimum of materials so that none are wasted? Is it made from recyclable materials? Such questions keep us mindful that the only source of materials we have on this planet is from the planet itself. They must also be returned to the planet sooner or later. When they are, can Nature reclaim them? Such questions have not been much considered in times past. Today we must consider them and act accordingly. Why?

The following checklist can help you to become a wiser consumer. Why not show it to your parents and discuss it with them? See also the checklist on good design in Chapter 2. Much of it applies here.

A CHECKLIST FOR CONSUMER PRODUCTS

The items in this checklist are typical of the questions that you and I as consumers should be asking about products we are about to purchase. You can add many more. To use this checklist, select the items that apply to the product. Some may not. Then use an appropriate rating scale. If, for example, three weights or scores are to be used, try 0, 5, and 10. If you prefer more choices, try 0, 2, 4, 6, and 8. Evaluate the item. Add up the total scores. Divide this by the total possible to get the percentage score. Put the scale opposite each item as shown in the first item below.

Function

1. How well will it do what it is designed to do? (0 2 4 6 8 10)
2. Is it simple to operate?
3. Is the structure sound, sensible?
4. Is it as durable as it should be?
5. Is there a good choice of mechanisms, circuits, systems?
6. Are the materials appropriate?

Economics

1. Is the price right?
2. Will it have a reasonable depreciation?
3. Is a trade-in possible?
4. Do I really need it?

Maintenance

1. Is it easily maintained, serviced?
2. Are parts easily available?
3. Can service be done conveniently?
4. Can the owner service it?

Safety

1. Is it shockproof?
2. Can I operate it safely?
3. Is it fireproof?
4. Can it be safely operated in the presence of others?

Environment

1. Is it made of recyclable or biodegradable materials?
2. Is it clean—no pollution of air, water, land?
3. Does it use a minimum of materials?
4. Will it operate quietly?
5. Will it occupy a minimum of space?

Cultural

1. Does it effectively reflect the present?
2. Does it contribute to improving the culture?
3. Does it display fine craftsmanship, pride of workmanship?
4. Is it an important invention for mankind?

Aesthetic

1. Is it really beautiful?
2. Is the beauty a part of the design (or is it added on)?
3. Is it of lasting beauty (not faddish)?
4. Is the product design imaginative and fresh?

Personal-Social

1. Does its ownership reflect favorably on the owner?
2. Does it really contribute to a finer living?
3. Is its possession acceptable in the family, neighborhood?
4. Is its possession legal, proper?

Use

1. Is it "fool" proof?
2. Is special instruction or skill needed?
3. Does it require a minimum of special care?
4. Does one's ability to use it contribute to his own growth and development?

Total Score _____
Total Possible _____
Percentage Score _____

FOR MORE IDEAS AND INFORMATION

Books

1. Donovan, Frank. *Wheels for a Nation*. New York: Thomas Y. Crowell Co., 1965.
2. Esterer, Arnulf K. *Tools! Shapers of Civilization*. New York: Julian Messner, 1967.
3. Fontana, John M. *Mankind's Greatest Invention*. Brooklyn: John M. Fontana. 1964.
4. Forbes, R. J. *Man the Maker*. New York: Abelard-Schuman Ltd., 1958.
5. Gies, Joseph. *Wonders of the Modern World*. New York: Thomas Y. Crowell Co., 1966.

6. Johnson, H. D., editor. *No Deposit—No Return*. Reading, Mass.: Addison-Wesley Publishing Co., 1970.
7. Keller, A. G. *A Theater of Machines*. New York: The Macmillan Co., 1965.
8. Oakley, K. P. *Man the Tool-Maker*. Chicago, Ill.: The University of Chicago Press, 1964.

Booklets

1. *Aerospace Highlights*
2. *Helicopters at Work*
3. *Jets*
4. *Wright Brothers, Highlights from Skylights*
   National Aerospace Education Council, 611 Shoreham Bldg., 806 15th Street, N. W., Washington, D. C. 20005
5. *Age of Flight*
   United Aircraft Corporation, Public Relations Dept., 400 Main St., E. Hartford, Conn. 06108
6. *Atomic Energy*
   Consolidated Edison, 4 Irving Place, New York, N.Y. 10003
7. *Automation: Life with a Little Black Box*
8. *Theory of the Leisure Masses,* Vol. 24, No. 19
   Kaiser Aluminum and Chemical Corp., Technical Publications, 300 Lakeside Drive, Oakland, Cal. 94604
9. *Count Down to Survival*
10. *The Glory Trail*
    National Wildlife Federation, Educational Services Section, 1412 16th St., N. W., Washington, D. C. 20036
11. *Designing a Better Tomorrow*
12. *Checklist for Cities* (guide for improving design)
    American Institute of Architects, 1735 New York Ave., N. W., Washington, D. C. 20006
13. *Educational Aids*—a list of booklets, charts, films
    General Motors Corp., Public Relations Staff, Room 1-101. G. M. Bldg, Detroit, Mich., 48202
14. *Getting the Right Job*
    Glidden-Durkee Division, Scoville Manufacturing Corp., 900 Union Commerce Bldg., Cleveland, Ohio 44115
15. *How Living Improves*
16. *Keep America Great* (series of brochures)
    DoAll Co., 254 N. Laurel Ave., Desplaines, Ill. 60016
17. *Teaching Aids Catalog*
    Teachers Publishing Corp., 23 Leroy Ave., Darien, Conn. 06802
18. *Teachers Resource Reference*

American Petroleum Inst., Committee of Public Affairs, 1271 Avenue of the Americas, New York, N.Y. 10020

19. *Time Telling* (History of)
    Hamilton Watch Co., Public Relations Dept., Lancaster, Penn. 17604
20. *Using Your Money Wisely*
    American Bankers Asso., Banking Education Committee, 90 Park Avenue, New York, N.Y. 10016 (or see your local bank)

Charts

1. *Productivity*
2. *Why Living Improves in America*
3. *Civilization Through Tools*
4. *This Is the Industrial Revolution*
   DoAll Company, 254 N. Laurel Ave., Desplaines, Ill. 60016
5. *The World Makes an Automobile*
   Automobile Manufacturers Asso. Educational Services, 320 New Center Bldg., Detroit, Mich. 48202

Films, Filmstrips

1. *Automation—What Is It?* (13 m. free)
2. *American Industry—Past and Present* (13 m. free)
   National Association of Manufacturers, 2 East 48th St., New York, N.Y. 10017
3. *Catalog of Free Films*
   Modern Talking Picture Service, Inc., 3 East 54th St., New York, N.Y. 10022
4. *Catalog of Free Films*
   U. S. Atomic Energy Commission, Educational Materials Section, Division of Technical Information, P.O. Box 62, Oak Ridge, Tenn. 37831
5. *Checklist of Films*
   American Institute of Architects, 1735 New York Ave., N. W., Washington, D. C. 20006
6. *Colonial America* (List of booklets, films)
   Colonial Williamsburg, Inc., Box C, Williamsburg, Va. 23185
7. *One Hoe for Kalabo* (27 m. free)
8. *The Littlest Giant* (13 m. free)
   Modern Talking Picture Service, Inc., 3 E. 54th St., New York, N.Y. 10022
9. *Science, Technology and Society* (filmstrips, free)
10. *The Cradle of American Industry* (filmstrip, free)
    American Iron and Steel Institute, Public Relations, 150 E. 42nd St., New York, N.Y. 10017

11. *Skills for Progress* (27 m. free)
   U. S. Department of Labor, 819 N. 7th St., Milwaukee, Wisc. 53233

Technical Drawing and Design 2

Words alone do not adequately describe technical ideas. For these we usually need drawings. Engineers, architects, technicians, draftsmen, designers, decorators, contractors, and craftsmen all use drawings. They use drawings of different types. With them they can be sure of accurately communicating with everyone concerned. An accurately made technical drawing has but one meaning to those who can read it. But a single word can have several different meanings. (Look up the word "draw," for example.)

As you learn to draw you will find that it is useful in many ways. Your idea for a project can be clearly described so that your teacher can know what you are thinking. Drawing your ideas will help you to figure out details and structure. It helps you to decide on the best procedure for constructing the project. On a drawing, "bugs" are easily spotted before you cut the materials. From a drawing you can accurately estimate the cost of the project. Drawing is a means of thinking on paper. It aids the flow of ideas. You'll find it mighty useful around the home, too.

TYPES OF TECHNICAL DRAWINGS

There are actually more than a hundred types of drawings in common use. The following types are used by engineers and designers.

Design Drawings. Drawings by an industrial designer usually begin as freehand sketches. They are ideas to be studied, proposed, improved on, and often discarded. The accepted idea will later be drawn as a picture and rendered in color.

Fig. 2–1. An automobile design studio is a constantly busy place, with various members of the design team working on different phases of the design process. In the foreground of this scene a designer works out a detail idea in sketches. In the background, the shape of the experimental car is taking form in a full-sized clay model.

Engineering Design Drawings. Design engineers use the industrial designer's drawing to work out engineering details. Specifications for shape, size, materials, finishes, fasteners, construction, costs, and such are included on the drawings.

Tool Engineering Drawings. From the design engineer's drawings, the tool engineer designs and draws the molds, dies, jigs, tools, and machines needed to manufacture the product.

Production Engineering Drawings. From the engineering and tool engineering drawings, the production engineer draws plans for the sequence and flow of the manufacturing production.

Working Drawings. These are prints of drawings that give the machinist, welder, pattern-maker the information they need to make parts, tools, molds, dies.

Drawings by Machine. Engineering design specifications can now be recorded on tape. The tape controls the machines in the manufacturing processes. There are automatic drafting (engineering drawing) machines into which tapes feed the specifications. The machine converts these into engineering drawings faster than a draftsman can do it. While new developments such as these make many drawing practices obsolete, probably drawings in one form or another always will be needed. For creating ideas visually, for communicating certain kinds of information, drawings are immensely helpful. And besides drawing is much fun. The better you can do it, the more fun.

Drawing Fields. Each technical or engineering field has its own drawings, drawing style, and symbols. For example, there are architectural, civil engineering, mechanical engineering, electrical, electronics, marine, and aerospace drawings, to mention but a few. They have much in common, however. When you can draw one kind you can easily learn another.

Architectural Drawing. The architect develops the overall design for the building. He employs engineers to design the structure, the heating and cooling installations, the electrical systems,

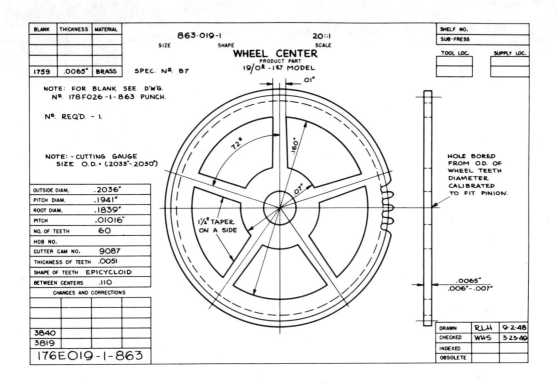

Fig. 2–2. A working drawing for the center wheel in a wrist watch. The drawing was made twenty times as large as the wheel. Why not actual size?

and the plumbing. He makes floor plans, elevations, and pictorial drawings.

Machine Drawings. These are drawings as used in mechanical engineering. They are often called mechanical drawings in schools. Different views of an object are used to show shape and construction. Two or more such views are usually necessary. Pictorial or picture drawings may be added.

Electrical Drawings. Such drawings commonly include *circuit* drawings, showing current flow; *schematic* diagrams, showing symbols of components; and *layouts*. They are adapted to building construction, electrical equipment, and electronic circuits.

DRAWING TOOLS

There are special tools for drawing just as there are for all crafts. The professional draftsman has a number of special tools. For instance, he uses drafting machines and electric erasers. But there are available to a beginner many tools that both he and the professional may share.

Drawing Board. The drawing board is usually made of basswood or white pine and is available in many sizes.

T Square. The T square enables you to draw horizontal lines that are always parallel. The head of the square must be kept firmly against the left side of the board.

Fig. 2–3. (top) Photographs tell much about an object. However, they show only what the eye can see. The right photo was taken with the camera above the airplane. The left, from ground level. **Fig. 2–4.** (bottom) This is a three-view drawing of the airplane giving overall dimensions. Note that the dimensions are in feet and inches.

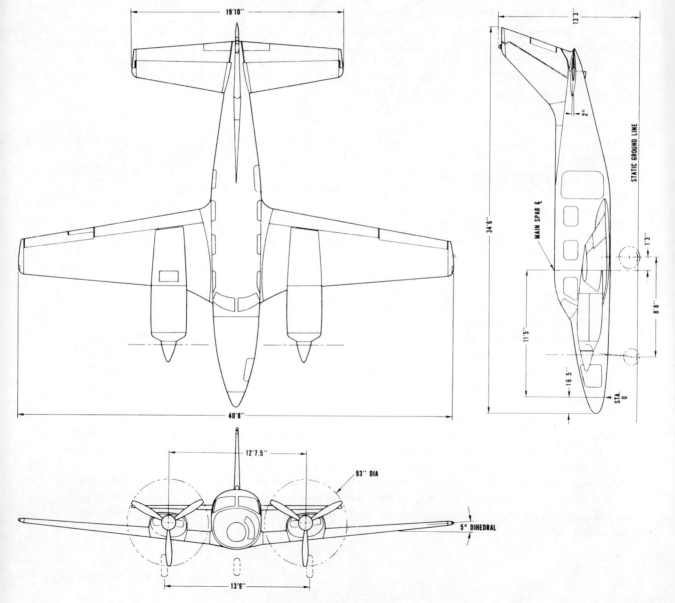

Triangles. With triangles you can draw angular lines and vertical lines. The 45-degree triangle is used for lines at that angle. The 30-degree and 60-degree triangles draw the respective angles. Each is used on the T square. They can be added together to make other angles; for example, 45 degrees plus 30 degrees equals 75 degrees.

Architects' Scale. The architects' scale is used for measuring on the drawing. It has a foot marked off in inches and sixteenths. A foot is squeezed down to 3 inches, to 1½, 1, ¾, ½, ⅜, $\frac{3}{16}$, and ⅛ inch. With these you can make true, accurate drawings smaller than the full size of the object to such proportions as $3'' = 1' - 0''$; $1'' = 1' - 0''$; and $\frac{1}{4}'' = 1' - 0''$.

Compass. This is the tool for drawing circles and parts of circles. It also steps off distances. For the latter use, the pencil lead is replaced by a steel point.

Fig. 2-6. (top). A draftsman uses a drafting machine as he makes this drawing of an automobile tire. Note the two blades of the machine. They are at right angles and are marked off for use as scales.

Fig. 2-5. (bottom) A set of drawing tools.

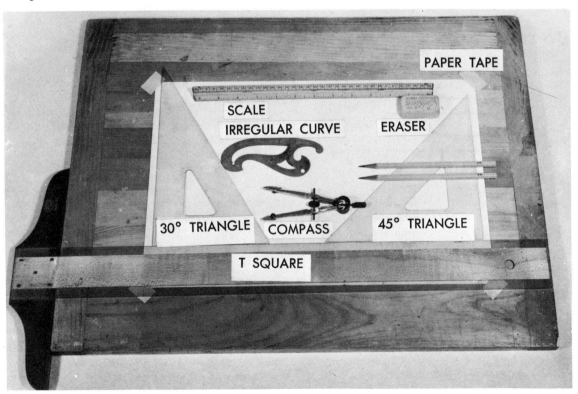

PAPER TAPE

SCALE

IRREGULAR CURVE   ERASER

30° TRIANGLE   COMPASS   45° TRIANGLE

T SQUARE

LINES

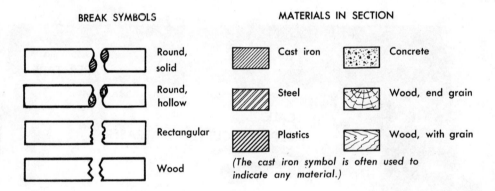

Object Line	*Thick* ———————	Bold, outstanding
Section Line	*Thin* ———————	Spaced evenly at 45°
Hidden Line	*Medium* - - - - - - -	Dash about 1/8"; spaces 1/16"
Center Line	*Thin* —— · —— · —— ·	Long and short dashes
Dimension,	*Thin* |←————— 2 —————→|	
Extension line	|—— —— —— ——|	These do not touch the object
Cutting Plane	*Thick* —— ·· —— ·· —— ··	One long, two short dashes
Break Line	*Thick* ~~~~~~~	Freehand

BREAK SYMBOLS

- Round, solid
- Round, hollow
- Rectangular
- Wood

MATERIALS IN SECTION

- Cast iron
- Steel
- Plastics
- Concrete
- Wood, end grain
- Wood, with grain

(The cast iron symbol is often used to indicate any material.)

Fig. 2–7. The alphabet of lines and drawing symbols. This is not a complete alphabet but it should serve your needs. Different companies often have different symbols.

Irregular Curve. This is a plastic guide for drawing curves that do not have fixed centers as do compass curves.

Paper. There are special papers for technical drawing. You should have one that is tough and quite hard. The 8½ by 11 inch size, or multiples of it, is the standard.

Pencils. Drawing pencils are graded by the lead. Those from 2H to 9H are hard. The H, F, HB, and B are medium. Those from 2B to 6B are soft. The more H's, the harder; the more B's, the softer. A 2H is recommended for drawing and an F for sketching and lettering.

Erasers. A ruby-type eraser is good for removing lines. An art gum or similar type is for cleaning.

Paper Tape. Paper tape is recommended for holding the drawing paper in place on the board. When it is removed it leaves no stickiness and the drawing is clean.

The Alphabet of Drawing

The language of drawing has an alphabet. You will learn it as you draw. There are certain types of lines that carry the same meaning in all technical drawing. Many symbols are used for convenience in drawing and ease in reading. (See Fig. 2–7.)

Fig. 2–8(a). (right) Some symbols used in electrical drawings. (See page 210 for electronics communications symbols.) Fig. 2–8(b). (bottom) In electrical drawings symbols are used for the actual parts.

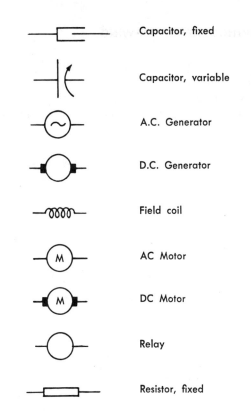

Capacitor, fixed

Capacitor, variable

A.C. Generator

D.C. Generator

Field coil

AC Motor

DC Motor

Relay

Resistor, fixed

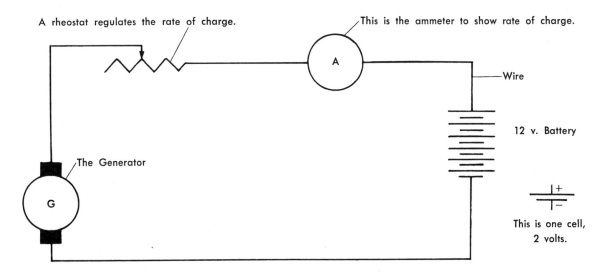

A rheostat regulates the rate of charge.

This is the ammeter to show rate of charge.

Wire

12 v. Battery

This is one cell, 2 volts.

The Generator

This is a circuit diagram for charging a 12-volt storage battery.

ORTHOGRAPHIC DRAWING

An orthographic drawing shows the object in its true shape and size by means of several different views or separate drawings. Two, three, or more views are needed to describe the object completely. The word *orthographic* comes from the Greek and means "true line." Each view is projected perpendicular to the drawing surface.

Hidden Lines. When details of the object cannot actually be seen in any view, they are shown by hidden lines. (See the Alphabet of

Fig. 2–9. The origin of orthographic views. Follow the steps to see how orthographic views are determined.

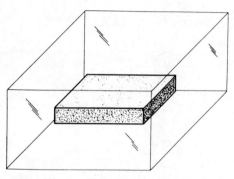

1. Imagine an object, in this case a block, suspended inside a clear transparent plastic box, with its longest side to the front.

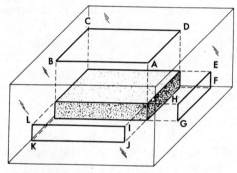

3. When you look at the block from the right end, the end view, EFGH, is seen and similarly, the front view, IJKL, is obtained.

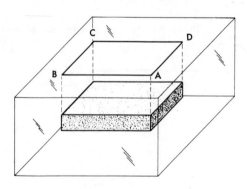

2. As you look down at the top of the box, you see the outline of the top of the block. This outline is projected up to the top of the box and is drawn on it, ABCD.

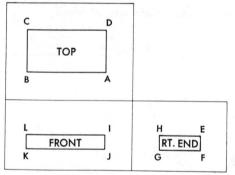

4. When the top, front, and right end of the box are opened and laid flat, these three views appear in their positions in orthographic drawing. The top view is directly above the front view. The right end view is to the right of, and in line with, the front view. These positions are sometimes changed, but only for exceptional cases.

Lines, Fig. 2–7.) If we make a cut in this same block, as is shown, note how hidden lines account for it. (See Fig. 2–16.)

Fig. 2–10. How to make an orthographic drawing.

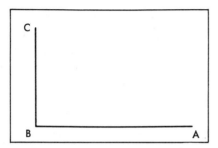

1. Lightly draw the base line, AB and BC, at right angles.

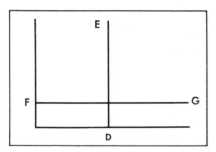

2. Mark off the length and height of the object. Add lines DE and FG.

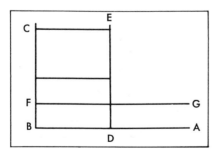

3. Mark off the width of the object in the top view. Add lines to show it.

How to Make an Orthographic Drawing

You don't need the plastic box when you make these drawings, but you must keep in mind the origin of the views. In Fig. 2–10 you can learn

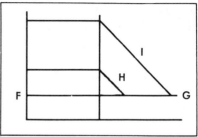

4. Draw 2 lines with a 45° triangle from the right hand corners of the top view to line F G. These determine the width of the end view.

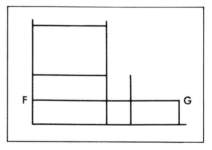

5. Drop these points to complete the end view.

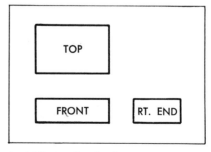

6. Add any details, then erase construction lines and retrace the object lines to make them bold.

what the steps are in making an orthographic drawing. Follow the steps carefully.

ISOMETRIC DRAWING

In isometric drawings, the true shape of the object cannot be shown. A square is not a true square, nor is a circle a true circle. Actually it is an oval or an ellipse. Isometric drawings are pictures made mechanically, so you can do them easily. The word *isometric* means in Greek "equal measure." It refers to the equal angles between the three axes as shown in step 1 of Fig. 2–12.

How to Make an Isometric Drawing

1. Using a T square and a 30 to 60 degree triangle, lay out the nearest corner of the object, lines *A, B* and *C*.

2. Measure off the length, width, and height on these lines. Add parallel lines to complete the object.

3. Add any details to the object. Remember to measure along the angular and vertical lines. Never measure horizontally. Each dimension is placed in the same plane as the face that includes it. The captions will help you.

Fig. 2–11. Four views of a model car. A dotted line in one view means that this line is not visible in this view. It appears as a solid line in the view in which it can be seen. The long dash lines show how points are projected from one view to another. They are called extension lines in the alphabet. Try this drawing. Add details to a view and then account for them in the other views.

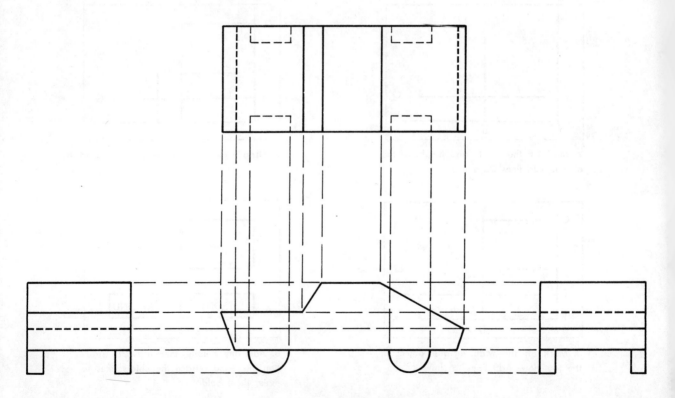

Isometric Circles. Circles are not shown in their true shape in isometric drawings. They are ovals or ellipses.

As a circle fits into a square, so an oval fits into a diamond. The curve touches the midpoint of each side. It is *tangent* at each midpoint. First construct the isometric square. Then find the centers, *I* and *J*. Swing the four arcs from points *A*, *C*, *I*, and *J*.

OBLIQUE DRAWING

Oblique drawing is a fast, simple form of pictorial or picture drawing. It has an advantage over isometric drawing because one face of the object is always shown in its true shape. Try this method for your freehand sketching. Follow the steps in Fig. 2–16.

Dimensioning. The front face is dimensioned just as in orthographic drawing. Do the

Fig. 2–12. How to make an isometric drawing. This is a type of picture drawing with no views shown in true shape.

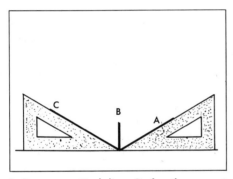

1. Draw a vertical line, B, for the nearest corner. With 30° triangle on T square, draw lines A and C.

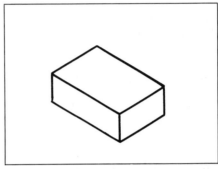

3. Draw the parallel lines to complete the view. Add any details and retrace the object lines.

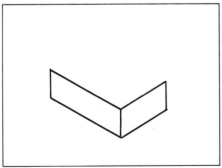

2. Mark off length, width, and height on lines C, A, and B. Add parallel lines to complete the two sides.

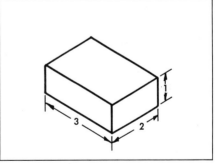

4. Dimensions are done isometrically. Hidden lines are not shown. Any isometric drawing starts with a box.

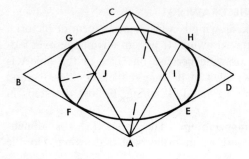

1. Draw an isometric square, ABCD.

2. Draw lines CE, CF, AH, and AG, bisecting opposite sides.

3. With points I and J as centers, swing the small arcs.

4. With points A and C as centers, swing the large arcs.

Fig. 2–13. The isometric circle.

others as in isometric drawing. Always use standard dimensioning practice.

SECTIONS

Sections are used to show the inside of an object. They show how parts are put together, and the materials of which they are made. When you slice through a layer cake you can see the thickness of the frosting and the kind of cake. The cake is sliced with a knife, but in a drawing it is done with a cutting plane. This cut is only imaginary. You draw what you know is there but really can't see. When the cake or the object is cut clear across, we call it a *full section*. When cut half through, it is a *half section*. A partial cut is *partial section*.

Sectioning is indicated by a symbol consisting of light, uniformly spaced lines at 45 degrees to the right or left. For your drawings this spacing will usually be from $\frac{1}{16}''$ to $\frac{1}{8}''$. This effect is *crosshatching*. (See Fig. 2–18.)

DIMENSIONING

Dimensions specify size and give locations of holes, cuts, parts, and such. All dimensions up to and including 72 inches should be given in inches, and when all dimensions in the drawing are in inches, no inch symbols (″) are used.

Location of Dimensions. Dimensions should be so located that there can be no question as to what or where they apply. Try not to crowd them and keep them clear and simple. Follow these tips:

1. Put the dimensions on that view which most clearly shows the part being dimensioned.

2. Dimensions should be readable from the bottom or the right side of the drawing.

3. Do not dimension from or to hidden lines. A center line should not be used as a dimension line.

4. It is generally better to place the dimensions outside the object, between the views. Can you tell why?

5. Arrange consecutive dimensions in a straight line, with the overall dimension outside of them.

6. Start extension lines about $\frac{1}{16}''$ from the object.

DRAWING TO SCALE

For greatest accuracy, working drawings are made to scale; freehand sketches are usually not. The full-size drawing, made to the actual size of the object, is preferred whenever it is practicable. It would be quite a task, however, to make full-size drawings of the *SS United States* or of the parts of a wrist watch. One is too large and the other is too small. Accurate drawings of the ship can be made smaller than full size. For example, one inch on the drawing might equal fifty feet on the ship. Watch parts are drawn larger than actual size so that the details can be more easily included. (See Fig. 2–2.)

Make your drawings to as large a scale as possible and be sure to indicate the scale on each. Notice the scale given for the drawing for Figure 2–2. Some common scales are: ¼″ = 1′ − 0″, meaning, ¼″ on the drawing equals 1′ − 0″ on the object; 1″ = 1′ − 0″; Half Size; and Full Size. Patterns are always full size.

Making Tracings. In industry the original drawings do not go out into the shop. Tracings are made and, from these, the prints. Each department needing prints can then have them. To make a tracing, fasten a piece of transparent tracing paper over the drawing. Trace the drawing with an H or an F pencil. Make the lines dark and bold. If the light in the printing machine cannot

Fig. 2–14. The relationship between orthographic views and a pictorial (isometric) drawing.

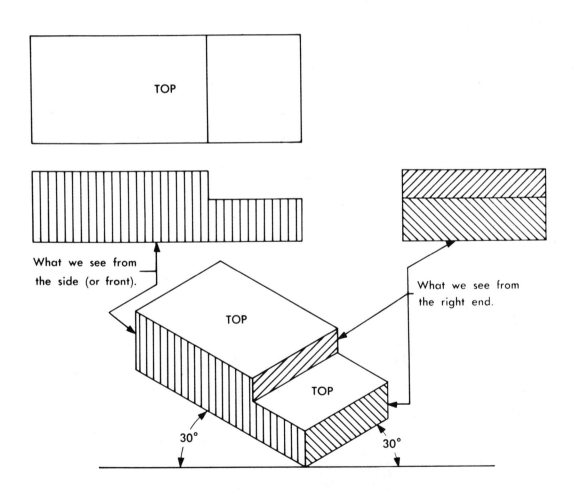

go through your lines, they will be sharp and clear on the print.

Sketching. Sketching is freehand drawing. It is done without drafting tools. Only paper and pencil are needed. Coordinate paper helps to get lines, angles, and measurements. Use loose, free arm movement in making the strokes with a soft pencil. Don't draw the lines so that they are perfect, sketch them.

THE ARCHITECT AND DRAWINGS

Houses, schools, city buildings, and factory buildings are planned by architects. Their drawings and specifications completely describe a building. The client is the one for whom the building is to be constructed. He decides whether or not he wants it on the basis of these plans. The building contractor with his carpenters, bricklayers, and steel

Fig. 2–15. The use of cross sections to show shape and interior details. This is a tractor toy for mass production. Only enough details are given here to show you the idea. Before you begin to build it, make a full-size working drawing showing dimensions. Make all parts, and smooth and paint them before assembling. Can you design a trailer or an implement to fit the tractor?

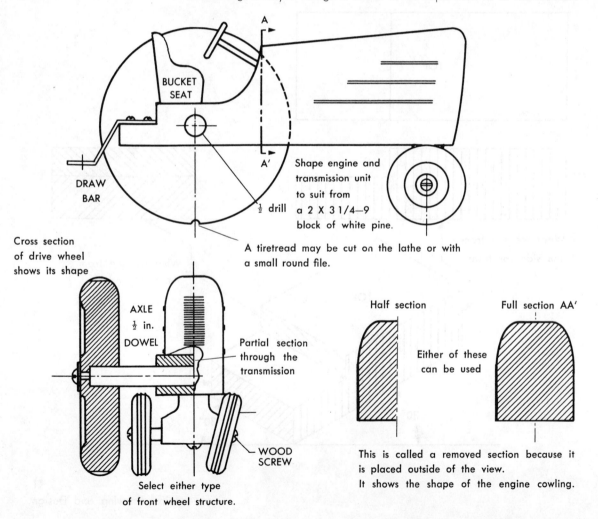

A

BUCKET SEAT

A'

DRAW BAR

½ drill

Shape engine and transmission unit to suit from a 2 X 3 1/4—9 block of white pine.

Cross section of drive wheel shows its shape

A tiretread may be cut on the lathe or with a small round file.

AXLE ½ in. DOWEL

Partial section through the transmission

WOOD SCREW

Select either type of front wheel structure.

Half section

Full section AA'

Either of these can be used

This is called a removed section because it is placed outside of the view.
It shows the shape of the engine cowling.

workers constructs according to the plans. The plumbing, electrical, painting, heating, and ventilating contractors all use them for their particular function. As the work progresses, the architect checks it against the plans. Can you imagine the troubles all these people might have if there were no drawings to guide them?

DESIGN AND DESIGNING

Doing your own designing is "doing your own thing." It involves dreaming, imagining, creating, inventing, planning, experimenting, testing, evaluating—all of which you have a capability for. In industrial arts you can develop these talents. As

Fig. 2–16. Pictorial drawings are easily made in *oblique*. Dimensioning is partly orthographic, partly isometric.

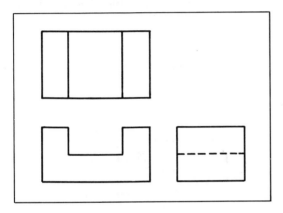

1. Start with the front view of the object. Place it nearer one end of the paper.

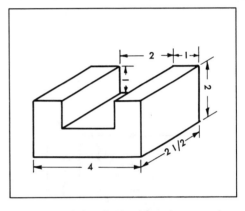

3. Mark off the depth along these angular lines. Add lines parallel to the front to complete the picture.

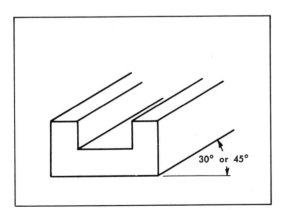

2. Extend the points in this view back, to the right or left. Use 30° or 45°. These lines are parallel.

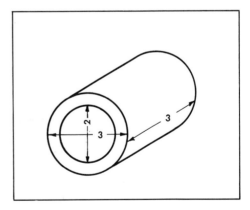

4. Put circles and curves in the front view to show their true shape.
Note the dimensioning.

you do you will find real fun and satisfaction in working with materials.

The Product Designer. The professional who designs products to be manufactured in industry is the product designer. Sometimes he is known as an industrial designer. He often works in ceramics, woods, jewelry, graphic arts, textiles, and other fields. The craftsman-designer is usually interested in creating only one or a few of a kind, which he himself makes. A product designer is a combination of professionals. He is an engineer for his understanding of materials, principles of engineer-

ing, and production. He is an inventor and creator for his ability to come up with new ideas. He is an artist for his ability to design beautifully.

How a Product Designer Works. The product designer works in this manner, although not always in this order:

1. He *defines* the problem. He must know exactly what the problem is which he is to solve. This clarifies the function or purpose of the product. Knowing this he can proceed.

2. He *determines* the limitations under which he must work. For example, he must know who

Fig. 2–17. Dimensioning according to United States of America Standards Institute practice.

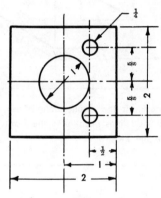

The common method. Vertical dimensions read from the bottom up.

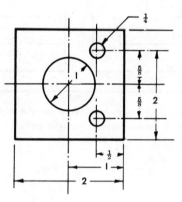

The new method. It is not necessary to turn the drawing to letter or to read it.

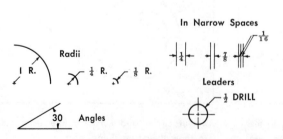

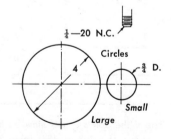

Fig. 2–18. Details of a camper's frypan and griddle. They go with the grill shown in Fig. 4–56. The details are shown to indicate how cross sectioning can be used in drawings. They can be made much larger. The frypan should have a thicker bottom and sides if it is to be larger. This will simplify the casting process. Draft in a pattern is a taper which permits the pattern to be removed easily from the sand.

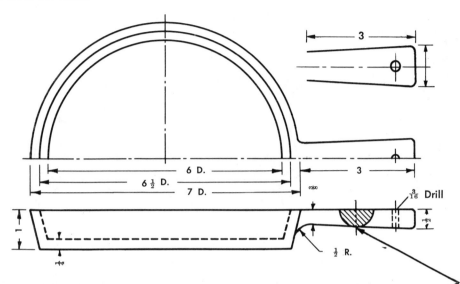

Patterns for castings should be made slightly larger to allow for the shrinkage of the metal and for the machining. Add $\frac{1}{8}''$ to the diameter.

1/8" to the diameter.

This "revolved" section shows the shape through the handle. It is crosshatched here, as sections should be. Why is it necessary to draw both halves?

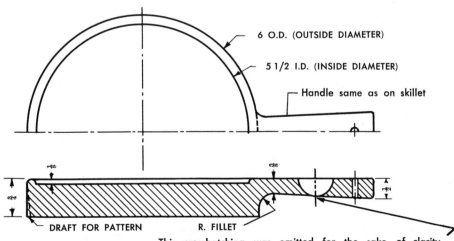

6 O.D. (OUTSIDE DIAMETER)

5 1/2 I.D. (INSIDE DIAMETER)

Handle same as on skillet

DRAFT FOR PATTERN   R. FILLET

This crosshatching was omitted for the sake of clarity. A draftsman uses his own judgment in such cases.

will use the product—child, youth, or adult. He must know the price range in which it will be marketed.

3. He *researches* the field. He studies competitive products and investigates different suitable materials and structures.

4. He *develops* several possible proposals in sketches, engineering drawings, and in models.

These are presented to a jury of designers, engineers, and management personnel for reactions. This may be a brainstorming session to assist in the search for a better idea. The jury accepts, rejects, offers alternatives and suggestions for improvement.

5. He *refines* the selected proposal. Drawings, including pictorial renderings in color, and models

Fig. 2–19. An alphabet for drawings.

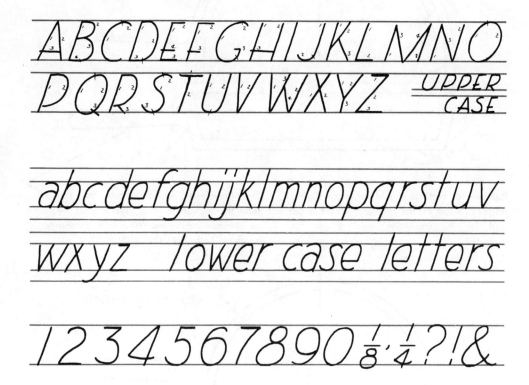

VERTICAL LETTERING IS DONE IN THE SAME
MANNER. COUNT THE STROKES IN RHYTHM AS
YOU LETTER.

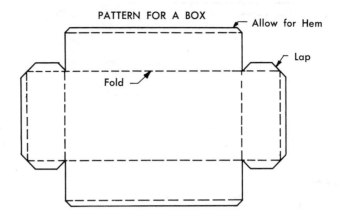

PATTERN FOR A BOX

Allow for Hem

Lap

Fold

PATTERN FOR A CONE

Lap

RADIUS

Step off
Circumference

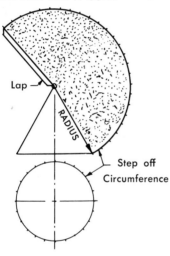

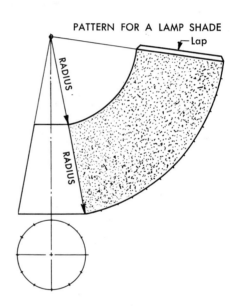

PATTERN FOR A LAMP SHADE

Lap

RADIUS

RADIUS

Fig. 2–20. Developing patterns.

are made. The models are *scaled* so that they have accurate proportion. They may be full size, smaller, or larger than actual. *Prototypes,* the first working models, may be made and field tested (tested in actual use). Further refinement may then be necessary.

6. He *submits* the proposed product to the client. If acceptable, it is ready for the next step. If not, the necessary changes are made.

7. He *provides* the necessary working drawings so that the molds, dies, jigs, or fixtures can be made.

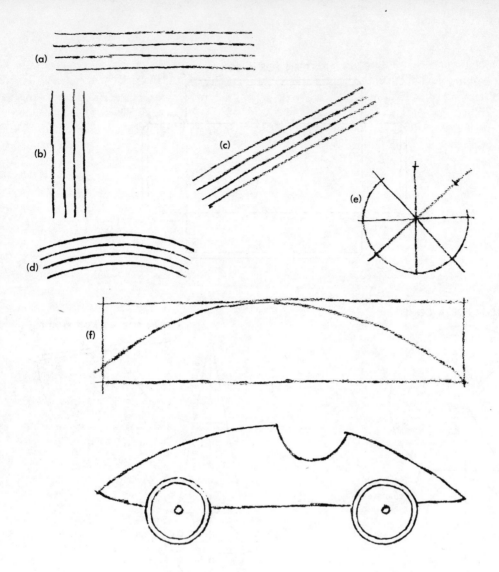

Fig. 2–21. *Freehand sketching.* Freehand sketching uses no drawing instruments. Only a soft pencil, 2B or 4B, and eraser are needed. Sketching is especially useful in expressing ideas quickly. Sketches can be as useful as instrument drawings. Sketching is not the same as drawing. Lines and scale are not as accurate. But the idea can be illustrated just as accurately. Follow these suggestions. (a) Pencil strokes are short, fast, relaxed, overlapping. Try horizontal lines first. (b) For vertical lines turn the paper so you sketch horizontally. (c) Do the same for angular lines. (d) For curves let your arm swing, as a radius. (e) For circles lay out several diameters. Mark off the estimated radius on each. (f) To sketch an object, first lay out the box to contain it. Then block out the main parts. Add details. Clean lightly with an eraser. Retrace the lines.

8. He *watches* the consumer acceptance of the product to determine the effectiveness of his ideas. From time to time improvements are called for.

Redesign is a large part of the work of the product designer. This involves taking an existing product and improving it. Common examples are kitchen appliances, bicycles, and automobiles. When a product is given only a face-lifting to improve its appearance from one model year to the next, it is *restyled*.

DESIGN THEORIES

To design a project that is functional and useful is relatively simple. To design one that is beautiful is difficult. But to create one that is both functional and beautiful is the real challenge. How can you know when it is beautiful or when it is ugly? Perhaps there is more to this than simply liking or disliking the object? You can like it because you made it. But does that mean it is really beautiful? Underlying these decisions are two points of view or philosophies of design. One says that the form of the object should be drawn from, should originate in, the function. This means that function determines the form, or *form follows function*. Examples of design based on this view are airplanes, automobiles, and Apollo rockets. Any beauty these have originates with the function they serve in traveling through air and space. They were designed to be functional. But they have a beauty all their own.

The opposite view is that *function follows form*. This means that an object is designed purely to have a beautiful form. The use made of it will depend on that form. A house could be designed and built in a setting with beauty the key factor. The family buying it would then adapt its living

Fig. 2–22. Typical layout for drawing paper—has border and title block.

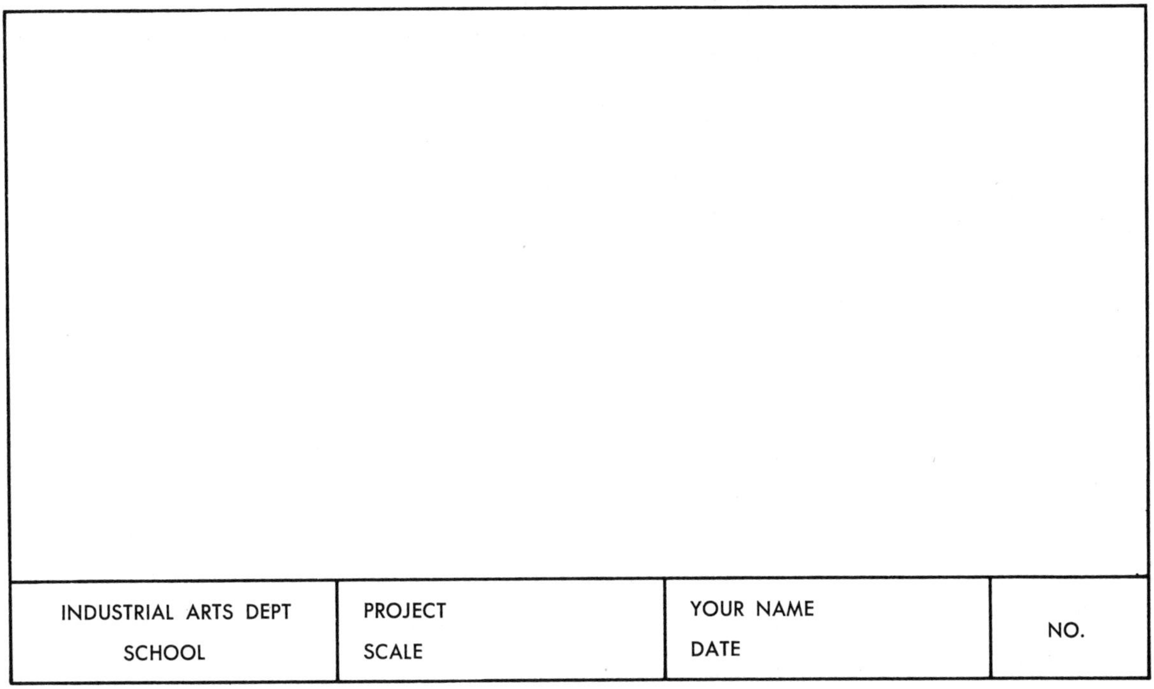

INDUSTRIAL ARTS DEPT	PROJECT	YOUR NAME	NO.
SCHOOL	SCALE	DATE	

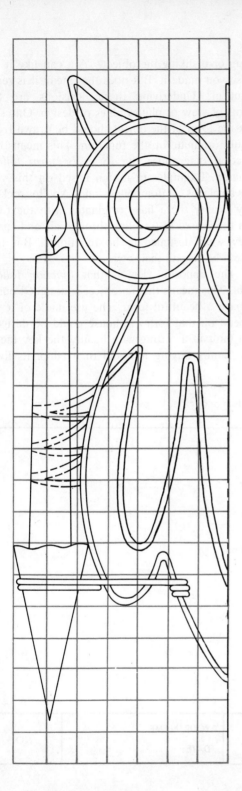

to fit the house. A ceramic pot can be designed to be beautiful. Its function can be simply to be beautiful. Such a piece would be called a work of art.

We have these two opposing views with which to work. One is not necessarily right and the other wrong. Each has its place and you as a designer will choose which is to be your guide for the particular project design. Designing is generally easier when you begin with the function.

DESIGNING FOR BEAUTY

Beauty is said to be in the eye of the beholder. This suggests that you may see beauty in an object when someone else doesn't. It suggests, too, that the sense of beauty is a personal feeling. Beauty in an object may not be obvious to everyone. Sometimes it may be necessary for the beholder to study and to think about the object before the beauty becomes apparent. We suggest that you make a special effort to look for beauty in things and places as you go about your daily tasks. See if you can find it in Nature as well as in man-made things. Nature is gifted in creating the beautiful. What about man? Look at our cities, factories, waters, harbors, homes, buildings, junkyards, schools. If we as a nation were really concerned about beautifying this planet what would we do? If you were interested in beautifying your home,

Fig. 2–23. One-half an owl. Your problem is to draw the other half. First, decide on the overall size. Then divide the length and width into the same number of squares as used here. Black iron wire, which can be bent with fingers and pliers, is suggested for the figure. The eyes could be small copper concave discs with enamel fused on for color. Weld or braze the joints. To a designer the owl does have to look real. But it still suggests an owl.

Fig. 2–24. Enlarging and reducing drawings. Draw squares over the original. Draw another set of squares enlarged or reduced as you wish. Locate points of intersection on the original. Transfer these to the other squares. Connect the points.

neighborhood, or industrial arts laboratory, what would you recommend? The action begins with you. Beauty may be considered the highest achievement of man.

YARDSTICKS FOR DESIGN

Good design in an object has several elements or components. Recognizing these will increase your sensitivity to beauty as well as to function. They will serve as guidelines to designing and as yardsticks for judging design. In actual practice you must often compromise in their use because of the concern for the function of the object. As the designer, you have the right and authority to express your own thinking in the design.

However, you may have to change a design if you wish it to please another person. Note that most of the following elements are *feelings*. They are guidelines for your reactions and your sensitivities.

1. *Emphasis* in a design means that it has a center of interest, that one part is dominant. All other parts are subordinated to it but they support it. If these parts compete for attention, the design becomes a hodgepodge. Emphasis can be achieved by form, size, color, texture, location.

Figure 2–25. An architect's design and drawings of a small club house. The perspective (picture) drawing shows the general idea.

Technical Drawing and Design

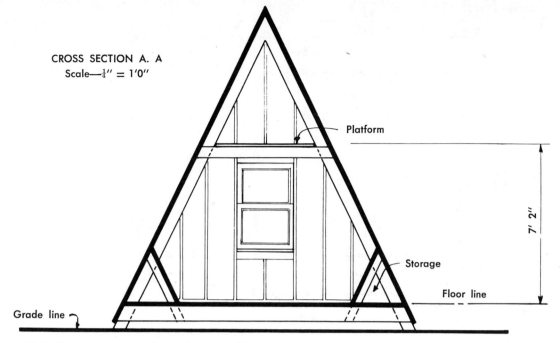

CROSS SECTION A. A
Scale—$\frac{1}{4}'' = 1'0''$

Platform

7' 2"

Storage

Floor line

Grade line

Fig. 2–25(a). The cross section shows the typical interior construction.

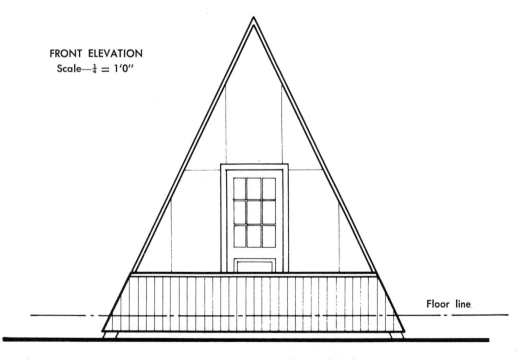

FRONT ELEVATION
Scale—$\frac{1}{4} = 1'0''$

Floor line.

Fig. 2–25(b). The front elevation is the front view as in orthographic drawing.

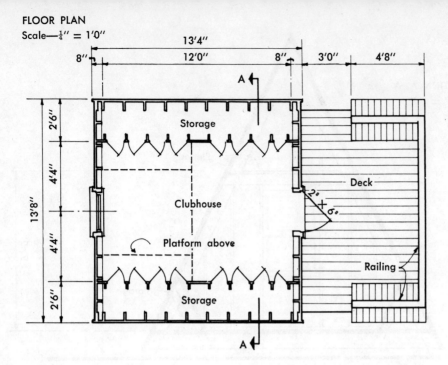

Fig. 2–25(c). The floor plan is a top view of the inside arrangement.

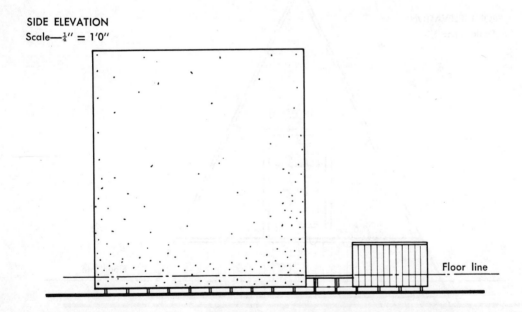

Fig. 2–25(d). The side elevation is a side view as in orthographic drawing.

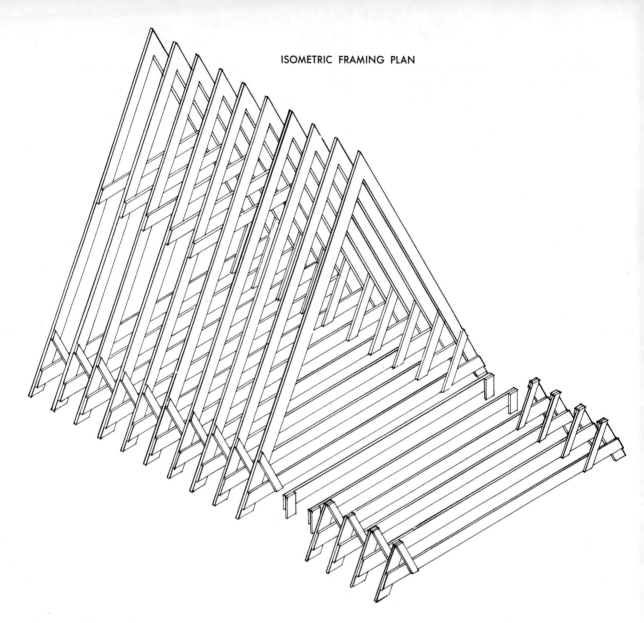

ISOMETRIC FRAMING PLAN

Fig. 2–25(e). The framing plan drawn in isometric shows the structure.

2. *Proportion* is the relationship of dimensions. It is expressed as a ratio. For example, if the width of a table top is 20 inches and the length, 30 inches, the proportion is 2:3 or 3:2. Odd ratios such as 2:3 and 5:7 are usually preferable to even ratios such as 2:4. Rectangles are generally more interesting than squares, and ovals more so than circles. Proper proportion often depends on the function of the object.

3. *Balance* is the condition of equality within a design. When you sense that it has good balance, it is neither top-heavy, bottom-heavy, nor lopsided.

55
Technical Drawing and Design

A design arranged so that a centerline divides it equally has *symmetrical* balance, or *symmetry*. When a design gives the feeling of balance without being symmetrical it has *informal* or *asymmetrical* balance. For example, a small bit of a bright color can balance a large area of a dull color.

4. *Unity* means a oneness in design. It is present when one's eye moves easily through the design without having to jump from part to part. All of the parts belong in it. If one is left out, the whole design is weakened.

5. *Rhythm* is a feeling of pleasing motion within the design. It is obtained by repetition of lines, curves, forms, colors, and textures. A static design, one without this movement, is boring to look at. There should be just enough *repetition* to provide a gentle movement. How is rhythm in music obtained?

6. *Variety.* Variety is the "spice of life" in a design. Too much repetition of the same kind is tiresome and boring. A bit of salt or pepper can perk up the flavor in food. Too much is not good. Variety is secured by contrast in form, size, color, texture, materials.

7. *Harmony* is a feeling of togetherness. All of the parts get along well together. Colors behave well with other colors, textures with the forms, and so on. It makes a happy design effect.

8. *Texture* is a condition of the surface of a material. We say that a plain sheet of metal has no texture but that a piece of burlap has a coarse texture. Texture can be felt with the fingers. Some materials have natural textures. Oak and teak woods have more than do apple and cherry. Textures can be applied, as to a piece of clay or a wood carving. When used they should be appropriate to the design. If the piece is better with the texture, leave it; if not, remove it.

9. *Simplicity.* A beginner in design often overdesigns, overdoes it. He puts too much stuff into the design. Too much color, texture, detail, fanciness, material. It may be too complicated. Simplicity in design is the mark of the expert. After you have done your design on paper, in a model, or a prototype, begin its simplification. See what you can leave out or change to improve it.

10. *Personality.* Design doesn't actually have a personality. Rather it reflects the personality of the designer. Your designs are you. They show your personality. The more you design the more you are likely to develop a style that people will recognize as yours.

When a designer has cared enough to do his very best he has put much of himself into the design. Imagination, cleverness, individuality, and quality are evident. An item designed by a person should look as though a human did it. A machine-

Fig. 2–26. In the early stages of the automotive design process, sketches and renderings are useful for working out basic themes and design directions. Dozens of preliminary sketches may be executed by the members of each studio design team as the basic shape of a new model line evolves. The real test comes, however, when an idea emerges in solid, three-dimensional form—the full-scale clay model.

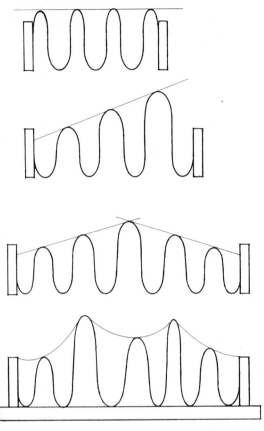

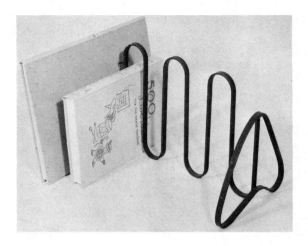

Fig. 2–27. A contemporary book rack employing the principle that "form follows function." The triangular ends of the book rack are simple in form, yet fully functional. Note that the rack is off center, yet balance is adequate. In the front view of the book rack, rhythm, repetition, and balance are easily seen.

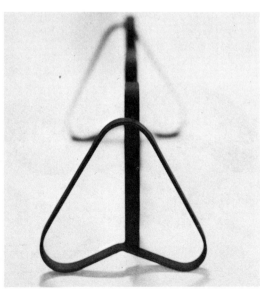

made product must be different. When you design a piece to be one-of-a-kind you have full freedom to express your personality. A piece designed for mass production must be adaptable to the machines that produce it. We don't assume that machine-made products are the yardstick for judging the man-made. Why?

COLOR SELECTION

Good color selection is necessary to good design. Selecting the colors can be a major problem in a design. Poor choice as in paints, finishes, glazes, and enamels can spoil an otherwise good design. But using color is fun, so enjoy yourself.

Color Qualities. Colors have qualities that can help you in their selection. Yellows, oranges, and reds are *warm* colors. They suggest warmth, fire, cheerfulness, action, courage. They shout and can easily become too noisy when used to excess.

Blues and greens are the *cool* colors. They lend such feelings as quiet, peace, coldness, formality. Colors have seasonal qualities, too. Reds, browns, and deep dull green suggest the fall season. Light yellows, bright greens, and lavenders suggest spring.

These suggestions can help you select color combinations.

1. Choose one color to be dominant. If only one is needed, it should be appropriate to the object and contribute to its function. The color for a kitchen stool probably would not be the same as for a living room coffee table. Colors for a football game poster may be loud and active. Those for a PTA meeting should be more formal. Don't

Fig. 2–28. (top left) The "now" furniture. This emphasizes simplicity and function. It uses Space Age materials such as chrome, acrylic plastics, glass, and exotic woods. **Fig. 2–29.** (bottom left) Colonial design in furniture is known as traditional. It is patterned after furniture made in America in the days of the colonies. **Fig. 2–30.** (bottom right) Some unusual shapes in wood. The two half bowls fit together on their straight sides. It is good design experience to adapt odd-shaped blocks to interesting uses.

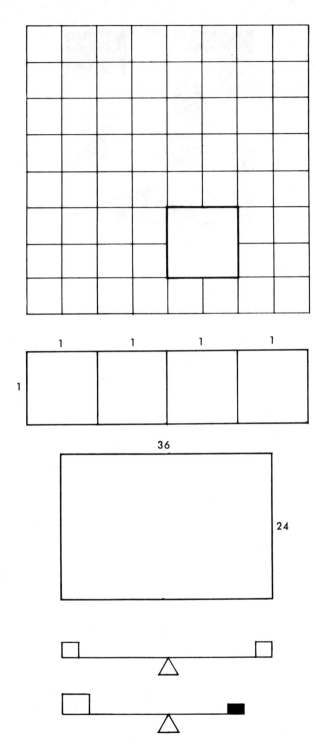

Fig. 2–31. (top left) *Emphasis*. A checkerboard is a mechanical repetition of a unit, a square. There is no design emphasis. But introduce a large square and it immediately "takes over." It becomes the center of interest where there was none. **Fig. 2–32.** (middle left) *Proportion*. The proportion or ratio is 1/4 or 4/1. A tabletop 24 x 36 has a proportion of 2/3. **Fig. 2–33.** (bottom left) *Balance*. A "50-50" situation gives formal or symmetrical balance. Informal balance is an optical relationship. The eye senses a feeling of balance. In this case a small block of bold color seems to balance a large block of weak color. **Fig. 2–34.** (top right) *Unity*. The six items in the first illustration are all different and all are in competition with one another. There is no unity or harmony. In the second arrangement all six become parts of a whole design. They have unity. There are many other possible arrangements to get unity. Cut out some shapes and try it.

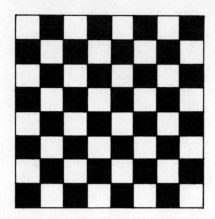

Fig. 2–35. *Rhythm, Variety, Harmony.* There is a rhythm, variety, and harmony in a checkerboard. But there is also a stalemate. There are just as many red squares as black, just as many horizontal lines as vertical. This is a good example of formal balance. Could it be any more formal? Now divide the space into unequal squares and let them overlap. Do you get a different feeling? There is rhythm because there is repetition. There is variety in size, shape, location. There is harmony and unity. All add up to interest.

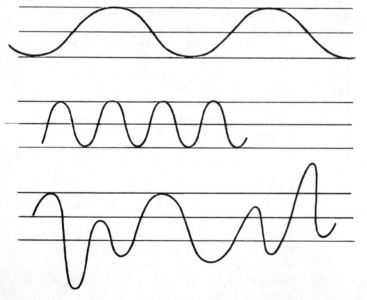

(a) Rhythm suggests movement. Like the rising and falling of waves. This one gives a slow, lazy feeling.

(b) This pattern is more active.

(c) But this pattern combines different frequencies or rhythms into one which has repetition plus enough variety to make it different from the others. Which do you prefer? Why?

Fig. 2–36. Free forms are useful in design. But not all free forms are good.

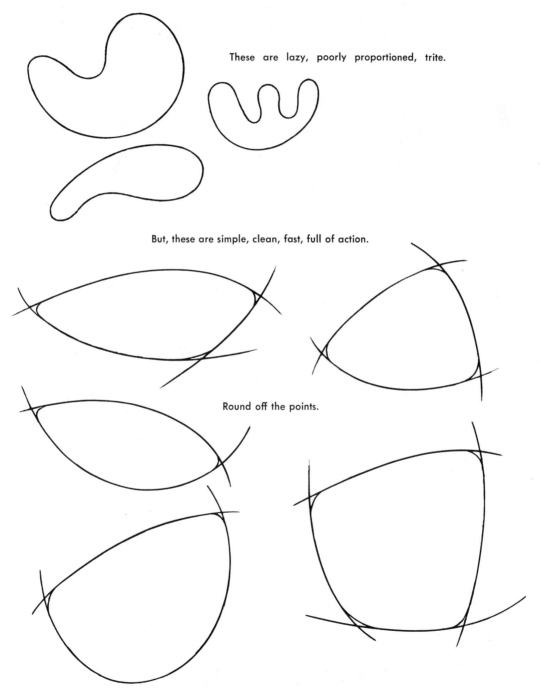

These are lazy, poorly proportioned, trite.

But, these are simple, clean, fast, full of action.

Round off the points.

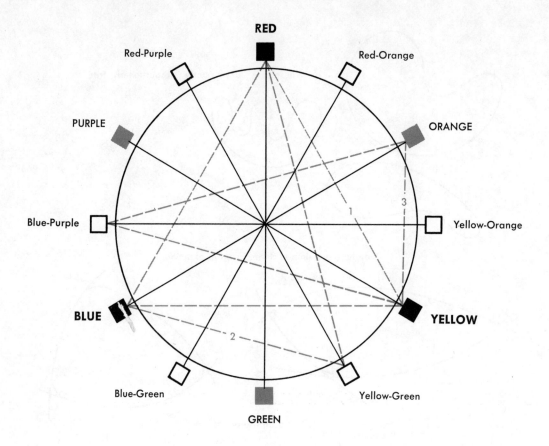

Fig. 2–37. *Color wheel.* Short dashline (1) shows a triad (three colors) equidistant on the wheel. Dashline (2) as a combination may be preferable. Triangle (3) is a split-complementary scheme.

overwork your favorite color. It may be a poor choice in many places. *Dominance* is gained with the amount used as well as with the intensity of the color.

2. To use several colors, make one dominant. Select the others to help it be more beautiful and attractive. The color wheel can be helpful here. Lay out a large one, 12 to 18 inches in diameter. Be careful to select the right *hues,* or colors.

Two-color combinations can be a pair of neighbor hues on the wheel. They get along well because they are closely related. For example, yellow and yellow-green, blue and blue-purple. These pairs are called *analogous* colors.

Two opposite hues on the wheel can be used together. Keep one dominant and use the other sparingly. This combination is called *complementary.*

Three-color combinations can include analogous hues on the wheel, as blue-green, blue, and blue-purple. Three hues equidistant around the wheel are called a *triad*. You can try a *split-complementary* combination, too. Select the dominant color and then add the one on either side of its complement.

Neutrals. Black, white, gray, tan, brown, ivory, and several other colors are called neutrals because they can be used in combinations without conflict.

Try out several color selections on scrap material to make certain they are effective. Try several combinations so that you can compare them.

A CHECK LIST FOR DESIGN QUALITY

With this scale you can rate products made by industry. You can also rate industrial arts projects. Let each of the ten measures have a possible ten points (see page 23). All of the measures may not fit every product or project. You may wish to put more weight on some of them than on others. Add other questions to each measure.

1. Function

How well does it serve its purpose? Is it easily operated? Is the purpose easily apparent?

2. Durability

Is it as durable as it should be for its function and cost? Will it last as long as desired? Is the finish appropriate?

3. Economy

Is it economical in use of materials and labor? Is it as good as it can be for the money? Could less expensive materials have been used?

4. Materials

Is the material suitable for the function? Is the material readily identifiable? Can the material be recycled? When discarded, will it cause a problem of waste disposal?

5. Construction

Is the workmanship of good quality? Is the structure appropriate to the materials and function? Are the methods of fastening appropriate?

6. Maintenance

Can it be easily and economically maintained? Can it be by do-it-yourself? Can it be easily cleaned?

7. Simplicity

Could it be more simply made? Is the appearance cluttered up or is it clean? Is it too simple, too plain to be interesting?

8. Safety

Is it structurally sound? Are the operator and passengers adequately protected? Is it safe in the presence of others? Is pollution (water, air, soil, noise) eliminated or minimized?

9. Beauty

Do the shape, proportion, size relate well to the function? Is it a beauty that will last? Is the beauty fresh and imaginative? Are the color and texture appropriate?

10. Personality

Does it appear that the designer really cared? Does it stimulate and hold your interest? Will it enhance its surroundings? Is it a good expression of a good idea?

Total Score   **Convert to Percentage**

SIGNS AND POSTERS

Making effective posters or more permanent signs requires a knowledge of many designing factors we have learned as well as a skill in lettering.

Signs and posters are silent salesmen. They have something to sell or to tell. The first duty of a poster is to draw attention to itself and then to get people to read its message.

Attention is attracted by the location of the poster or sign and by its color, layout, and letter style. A poster may be perfectly made, but if it is not placed where it can be easily seen, it cannot do its telling.

Choose colors that are most appropriate for the particular sign or poster. Two or three colors, including that of the background, are sufficient. The more additional colors you use, the more difficult it becomes to handle them well. The layout, or arrangement, of the words and illustrations is important. The words must be so arranged that they can be easily read. Use as few as possible, too. Use illustrations to draw attention and to help explain the message, and be sure that they are related to it. Simple letter styles are most effective. Use one style only in your first signs and posters. To be good at lettering, you will need to study and practice just as you must to play a musical instrument or drop-kick a football well. Start out with the alphabet shown in Fig. 2–19. Rule guide lines for various sizes of letters, then take each letter of the alphabet and form it many times with a soft pencil. Make the strokes rapidly instead of trying to draw them. Check your work often with the originals and with your teacher.

DRAWING AND OCCUPATIONS

The expert in technical drawing is a draftsman. He usually specializes in a certain type of drawing. This may be architectural, electrical, or engineering, for example. Engineers today usually have draftsmen make their drawings. They submit sketches of their ideas and the draftsman makes the finished drawings. Draftsmen are employed in all types of industries. Every industry represented in this book uses draftsmen. A draftsman usually gets his training in college, technical institute, or in a vocational trade school. He can get it by himself with a home study course.

Skill in design is always in demand. Industries and businesses are constantly in search of new and better ideas. This is what designing is all about—better ideas. The engineer and architect, as well as the product designer, need this talent. So does the commercial artist. A professional product designer usually has a college degree. It may be in engineering, art, or in product design.

The commercial artist draws, sketches, letters, and paints. His work is usually in the field of advertising, in ad design and layout, signs and posters, billboards, brochures, and such. He may get his training in college, technical institute, trade school, home study courses, or he can get it by himself with the aid of books.

Here are some of the common occupational fields for which drawing and designing are necessary.

Technical Drafting	Product Design	Commercial Art
Structural steel	Furniture	Newspaper ads
Architecture	Toys	Magazine ads
Electricity-electronics	Appliances	Cartooning
Marine	Sports equipment	Animation, TV
Aeronautical	Jewelry	Animation, movies
Mechanical	Automobiles	Signs
Civil engineering	Clothing	Posters
Patents	Pottery, dinnerware	Billboards

DRAWING AND RECREATION

Do you like to draw, to sketch, letter, paint, doodle, or dream up ideas with a pencil? If you like to make things in your home workshop you probably draw and plan them. Any of these interests needs lots of practice if you will develop skill.

Have you noticed how often people say they can't draw? They can't letter, or design something? They seem to envy the person who can?

If you like to draw, perhaps you can take more courses in industrial arts or art. There are excellent books on all types of drawing and painting. See if you can find a professional in your neighborhood. He may be glad to criticize your drawings and help you. Ask your teacher for his suggestions on how you may add to your skills.

Save every drawing you make. Mount each in a folder or scrap book. They help you to measure your progress. Start collecting a file of pictures and drawings. Make clippings from magazines of drawings, designs, paintings, and lettering which you especially like. File them by type for reference when you are looking for ideas.

Carry a small sketch pad and a pencil with you. When you see a good idea, make a sketch of it. If you have a camera, take a photo of it. Later on try to make a sketch from the photo. Perhaps you are the one who should be promoting shows and exhibits of drawings in your school?

DRAWING AND ENVIRONMENT

Drawing is a tool, not an industry. As a tool, it can be a powerful one for developing and showing ideas. It is useful in environmental planning. For example, there may be an unsightly neighborhood or area in your community. You could organize your class to study this. To begin you might take a complete set of photos of the situation. You could use some as color slides and some as black and white prints. Use these to analyze the situation. When you get to the point of making recommendations for improvement, drawing and sketches become important. From these, 3-dimensional scale models can be made. Drawings and models help to sell your ideas for the improvements. Seeing them, civic clubs might advance funds for the materials needed in making the actual improvements. Your class could be involved in this, too. Then, a good series of photos of the completed project should be taken. Altogether the photos, drawings, and models would tell a great story.

TO EXPLORE AND DISCOVER

1. Why decimals are used rather than fractions in some dimensioning.
2. What perspective drawing is.
3. The names of some important industrial designers and the products they design.
4. Why you should learn to design things rather than copy them.
5. What aircraft "lofting" means.
6. How to read a set of house plans.
7. How the full-size patterns for some airplane parts are made photographically.
8. What are hieroglyphics? Who used them? Why are they important?
9. Get some examples of American Indian sign language. What symbols were used? What were the translations?

10. If in the U.S. we change to the metric system of measurement, what problems will it cause? What problems will it solve? Is the change a good idea?
11. Make a full-size drawing and template (a cardboard full-size profile pattern) for a milk pitcher to be made of clay.
12. Make a stretch-out pattern full size for a sheet metal small-parts box.
13. Make working drawings for any of the projects illustrated in any of the chapters in this book.
14. Design a greeting card for linoleum block or stencil printing.
15. Design a poster announcing an industrial arts exhibit, ball game, or other activity.
16. Design several free-form copper dishes. Make the patterns, too.
17. Design and draw a cabin you would like to have and make a model of it.
18. Draw a floor plan for a dream workshop you would like to have at home. Make an equipment list. What is the cost?
19. Make either a cipher or a monogram of your initials in a circle, diamond, triangle. Paint them with tempera colors.
20. Make a 3-view drawing of a boat you would like to build. Better read up on boat design. Make a scale model of it. Check it out with a boat expert.

FOR GROUP ACTION

1. Visit a building under construction and look over the prints from which it is being built.
2. Visit the engineering department of a local industry, an architect, or the highway engineering department to watch engineers and designers at work.
3. Collect pictures of old and new autos and discuss the changes in engineering and body styling.
4. Design and build a scale model of the city of the future. What problems will you eliminate?
5. Design and build scale models of a space station and a space ship for interplanetary travel.

FOR RESEARCH AND DEVELOPMENT

1. A coffee table top is but a plane suspended in space. How many ways are there of suspending the top other than with four legs?
2. How are subsea and space vehicles different because of the media in which they travel?
3. Could automobiles be so made that it would be impossible for two of them to collide?
4. You are shipwrecked on an uninhabited island in the South Seas, with only an axe and a shovel for tools. You have built a cabin on the side of a hill which

has a spring at its foot. You want running water in the cabin. How could you get it without carrying it up from the spring?

5. Design and construct a working model of an air car.

6. What do you think the safety specifications will be for the car of 1980? What may it look like?

7. Using only one sheet of typing paper plus glue, make a built-up structure which will carry a load of at least 500 pounds.

8. Design a barbecue wagon for home use which "has everything" but which is easily portable and compact for storing.

9. Start with scale drawings of any car (such as are found in motor magazines) and restyle the body for 1980.

10. Make an environmental survey of your neighborhood, community, or area of your city. Include a 3-dimensional scale model of the present situation. Determine possibilities for redesign and renewal. Consider provisions for such as residential areas, mass transportation system, sidewalks, bicycle paths, recreational facilities, parks and playgrounds, schools, churches, shopping centers, industries. Set up specifications for zoning, building, and land use to insure the most desirable environment. Construct a scale model of your proposal. Make a trial presentation to your class and to the school faculty. Then make this to the PTA and civic groups. Record the reactions and evaluate your proposal.

FOR MORE IDEAS AND INFORMATION

Books

1. Anderson, Arthur D. *A Designer's Notebook.* Bloomington, Ill.: McKnight & McKnight Publishing Co., 1966.

2. Brown, Walter. *Drafting,* rev. ed. S. Holland, Ill.: Goodheart-Willcox Co., Inc., 1968.

3. Coover, Shriver L. *Drawing and Blue Print Reading,* 3rd ed. New York: McGraw-Hill Book Co., 1966.

4. Downer, Marion. *The Story of Design.* New York: Lothrop, Lee and Shepherd Co., 1963.

5. Dunning, W. J., and L. P. Robin. *Home Planning and Architectural Drawing.* New York: John Wiley and Sons, Inc., 1966.

6. Lindbeck, John. *Design Textbook.* Bloomington, Ill.: McKnight & McKnight Publishing Co., 1962.

7. Walker, John R., and E. J. Plevyak. *Industrial Arts Drafting.* S. Holland, Ill.: Goodheart-Willcox Co., Inc., 1964.

Booklets

1. *Catalog of American Standards.* Lists available standards for drawing, engineer-

Technical Drawing and Design

ing, materials, etc. United States of America Standards Institute, 10 E. 40th St., New York, N. Y. 10016

2. *Die Leistung* (The Accomplishment). The story of the J. S. Staedtler Company and of pencilmaking. J. S. Staedtler, Inc., Montville, N. J. 07045

Charts

1. *Lettering Exercises.* (Set of 6.)
2. *Single Stroke Gothic Alphabets.*
   Hunt Manufacturing Co., 1405 Locust St., Philadelphia, Pa. 19102
3. *Interchem—The Color Tree.* Story of color.
   Director of Advertising, Graphics and Colorant Systems, Interchemical Corp., 67 W. 44th St., New York, N. Y. 10036

Film

1. *Lewis Mumford on the City. Part II. The City—Cars or People.* 28 min., black and white. *Part VI. The City and the Future.* 28 min., black and white. Sterling Educational Films, Inc., 241 E. 34th St., New York, N. Y. 10016

GLOSSARY—DRAWING

Analogous colors colors adjacent on the color wheel; related colors.

Asymmetry not symmetrical.

Balance in design, a state of equilibrium, harmonious proportions.

Compass a tool for drawing circles.

Crosshatching in drawing, angular parallel lines to indicate a cross section.

Hue color.

Isometric drawing a form of pictorial drawing constructed on three equidistant axes.

Oblique drawing a type of pictorial drawing made from the front orthographic view.

Pattern in drawing, a full-size outline; in foundry, the full-sized model used in making a mold.

Proportion in design, the relation or ratio of two dimensions.

Rhythm in design, the sense of pleasing movement as the eye follows easily from part to part.

Scale in drawing, a measuring instrument; the ratio of one inch, or other unit on a drawing to a unit of measure on the object.

Schematic diagram in drawing, a plan showing the layout of a system or circuit.

Sections in drawing, an interior view.

Symmetry in design, the similarity of form on either side of the centerline.

Tangent touching at one point but not intersecting.

Texture the state or the condition of the surface of a material.

Tracing an exact copy of a drawing made on translucent paper, film, or cloth.

Triad a group of three colors equidistant on the color wheel.

Unity an arrangement of parts to produce a harmonious whole, a oneness in a design, a completeness.

Woods **3** Technology

For centuries, all over the world, trees were used chiefly for lumber and fuel. Today they are the raw material used in many huge industries. Trees provide material for lumber, veneer, wallboard, insulation, paper, chemicals, plastics, and foods. These are used in more products than we could list on this entire page. Today the lumberjack and the carpenter are but a few of the men who earn their living with trees and lumber. Scientists, chemists, engineers, and designers have developed new materials and products from trees. They have created new jobs that did not exist a few years ago.

About Trees. Lumber-producing trees are classed as *hardwoods* and *softwoods*. Hardwood trees have broad leaves that are shed in the fall. They are called *deciduous*. Most softwoods come from trees that have needles.

A broad outline of the woods industries follows. The kinds, qualities, and sources of many woods used for commercial products are listed. Different kinds of woods possess various qualities in varying degrees, of course. You will find some of the common products made of wood as well as the processes used.

THE WOODS INDUSTRIES

	Common Commercial Woods		Common Manufacturing Processes Used with Woods
Kinds	**Qualities**	**Sources**	**Manual**
Hardwoods	*Weight*	*Hardwoods*	Boring
Ash	Balsa is light in weight.	Generally found in the	Carving
Basswood	Hickory is heavy.	north central and mid-	Decorating
Birch		western states.	Fabricating
Cherry	*Flexibility*		Filing
Elm		*Softwoods*	Fitting
Hickory	Spruce is flexible.		Forming
Maple	Fir is stiff.	Yellow pine grows in the	Jointing
Oak		Southeast.	Planing
Walnut	*Resiliency*	Red cedar comes from the	Sanding
		middle states.	Sawing
Softwoods	Balsa is very absorbent of	Western cedar, fir, spruce,	Scraping
	shock.	and redwood are from the	
Balsa	The dense woods are less so.	Northwest.	**Mechanical**
Cedar		White pines are common to	
Fir	*Color*	the northern states.	Boring
Hemlock			Forming
Redwood	Redwood has an intense	*Foreign woods*	Jointing
Spruce	color.		Laminating
White pine	Basswood is almost white.	Most of these come from	Planing
Yellow pine		Central America, Africa,	Sanding
	Grain	India, and Southeast Asia.	Sawing
Foreign woods			Shaping
	Fir has a bold, harsh grain		Texturing
Ebony	pattern.		Turning
Granadilla	Walnut has a pleasing pat-		Veneering
Mahogany	tern.		
Rosewood			**Chemical**
Satinwood	*Permanence*		
Teak	Teak is almost impervious		Bleaching
	to moisture.		Distilling
	Basswood decays quickly in		Dyeing
	moisture.		Finishing
			Fireproofing
			Gluing
			Plasticizing
			Preserving
			Thermal
			Burning
			Drying
			Steaming
			Electrical-Electronic
			Bonding

Typical Products of American Forests and of American Technology

Saw Logs	Veneers	Wood Chemistry	Miscellaneous
Construction lumber	*Package veneer*	*Pulp wood for paper*	*Sawdust*
Houses	Boxes	Newsprint	Insulation block
Public buildings	Crates	Books	Insulation firebrick
Private buildings	Baskets	Printing	
Concrete forms		Writing	*Sap and gum*
	Construction	Wrapping	Maple sugar
Industrial lumber	Plywood	Container board	Rosin
Containers	Prefab houses	Wallboard	Turpentine
Furniture	Furniture		
Musical instruments	Concrete forms	*Paper conversion*	*Edibles*
Toys		Abrasives	Fruits
Sports equipment	*Fancy veneers*	Laminates	Nuts
Boats	Furniture	Towels	
Machinery	Wall paneling	Wallpaper	*Bark*
			Dyes
	Boat plywood	*Byproducts*	Tannins
	Sloops	Chemicals	Flavorings
	Canoes	Adhesives	
	Racing shells	Oils	*Roots*
		Fertilizers	Oils
	Compregnated plywood	Dyes	Pipes
	Airplane propellors	Plastics	
	Bearings	Explosives	*Hardboards*
	Gears		Paneling
	Table tops	*Waste wood*	Toys
	Furniture	Wax	Furniture
		Oil	
		Tar	*Bolts*
		Creosote	Ball bats
			Tooth picks
			Barrel stoves
			Timbers
			Ties
			Poles
			Posts

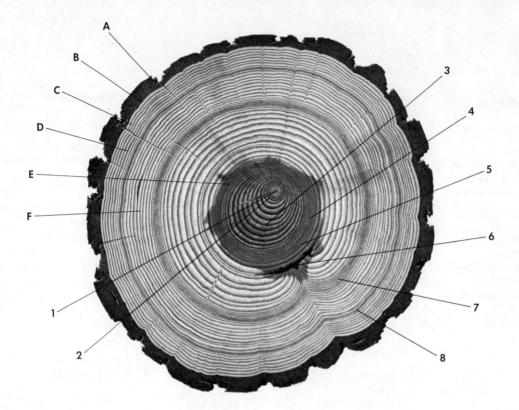

Fig. 3–1. A cross-section of a tree trunk showing the story of its growth. The parts:

A. Outer Bark—tree's protective armor.

B. Inner Bark—carries food to branches, buds, trunk, and roots.

C. Cambium Layer—each season new wood forms on the inside of this layer and new bark on the outside.

D. Sapwood—the outer rings of wood that conduct sap from roots to leaves.

E. Heartwood—the strong, mature core, once sapwood.

F. Medullary Rays—channels bearing food and moisture across the stem, making beautiful patterns when wood is sawed. Each spring new, light wood is added. This grows fast and consists of large cells. In summer the wood growth is darker and it has smaller cells.

The rings tell the age. For a fascinating story, follow the numbers in the next column against those on the chart.

1. The loblolly pine tree begins its life.

2. Evenly-spaced rings at the center show that the tree grew for five years normally with good sun and rain.

3. In its sixth year, something heavy pushed against one side of the tree, perhaps a large rock or fallen tree trunk. The rings became wider on the opposite side of this, as the tree built "reaction wood" to help support itself.

4. Narrow rings show that the forest grew up around the tree. This growth took water away from the tree's roots and sun from its branches.

5. Later the surrounding trees were cut and good growth conditions let the rings become wider.

6. Fire! Our tree was scarred on one side but in subsequent years new rings healed the scar.

7. Again a period of narrow rings, possibly caused by a long dry spell.

8. After several normal years, the rings narrow once more, which may have been due to bad effects of insects.

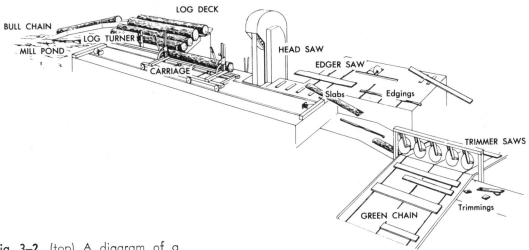

BULL CHAIN LOG TURNER LOG DECK HEAD SAW EDGER SAW CARRIAGE MILL POND Slabs Edgings TRIMMER SAWS GREEN CHAIN Trimmings

Fig. 3–2. (top) A diagram of a sawmill. Logs enter from the pond where they have been soaked. A carriage grips a log and moves it into the band saw. Bark slabs are carried away by a conveyor. The boards are then moved through machines with multiple saws that rip them to width and cut them to length. **Fig. 3–3.** (right) This machine debarks the logs. The bark is used in other products.

HOW A TREE GROWS

A tree grows by adding new growth just underneath the bark and on the tips of roots and branches. The age of a tree can be told quite accurately by counting the *annual rings* in the stump. The area from one ring to the next is a year's growth. (See Fig. 3–1.)

About Lumber. Most of the wood in trees cut commercially is made into lumber and pulp. Boards that are sawed directly from the log are called *rough lumber*. After seasoning, or air drying, the rough lumber is usually *kiln-dried* (a kiln is a large oven) to make it hold its shape and size better. After being surfaced to thickness, width, and length, it is ready for use.

Lumber is usually sold by the board foot. This is a unit measuring 1 inch by 12 inches by 12 inches. Some building lumber is sold by the lineal foot. To find board feet, multiply the length in feet by the width in feet by the thickness in inches.

Lumber is graded by different systems. Softwoods for building purposes are classed as *select* for finish and trim, or *common*. Hardwoods are graded as firsts, seconds, and common. Lumber is available as rough or surfaced. S2S means "surfaced 2 sides."

Plywood is built up of layers of veneer, or thin wood sheets with the grain of one sheet at right angles to the next. They are glued and pressed into panels. The veneer is generally cut from the log by peeling, just as paper is unwound from a roll. A long blade peels it off as the log rotates. Plywood is sold in sheets of standard sizes such as 4 by 12, 4 by 10, or 4 by 8 ft. and smaller. Panels may be faced with cabinet wood veneers such as walnut, oak, and mahogany.

Pressed woods are new forms of sheet wood made from wood chips that have been processed and pressed into sheets. They are more resistant to warping than is plywood of the same thickness and are less costly. Many interesting patterns and textures are available.

Defects in Lumber. Common defects in lumber are easily recognized. There is warping or twisting. Cupping means that one side of the board is hollow and the other high. Checking means the small cracks on edges, ends, or faces.

Conserving Forests and Woods. Less than fifty years ago it appeared that the forests would soon be used up. This would have been true had not a tremendous program of tree planting been started. Now our new wood growth approximately equals the amount that is cut or destroyed each year. Tree farming has become a going business.

Forest fires are more effectively controlled today, too. Yet in a recent year more than 7 million acres of forest were burned. Each year such fires cause needless, inexcusable waste of trees, and loss of life and property. Nine out of ten

Fig. 3–4. Giant lathes peel thin sheets of wood, called "veneer," in a long, continuous sheet. Cut to prescribed sizes and bonded together, the veneer produces strong, light plywood.

forest fires are caused by the carelessness of people.

Industry uses more of the tree today than it did formerly. Limbs, twigs, and even bark, leaves, and sawdust are converted into other materials and products.

DESIGN AND WOODS

Pretwentieth century styles are still used in wood products. We sometimes assume that projects we make must be modeled after those styles to be good. This is not a true assumption. Using modern materials, processes, and ideas, you should so design your projects that they serve today's purposes. Such a procedure does not show a lack of appreciation for the work of those cabinetmakers. They designed for their times. Let us design for ours. See if you can "dream up" some unusual ideas for wood projects—some that are different and better than any you have seen. Building period furniture is not designing. It is copying. Try to bring out the deepdown beauty of the wood. Remember, no design is so good that no one can improve on it.

Fig. 3–6. Fires destroy trees on thousands of acres of land each year. Most fires, caused by carelessness, can be averted if persons would realize their responsibility and exercise caution while in or near woods.

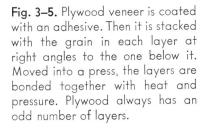

Fig. 3–5. Plywood veneer is coated with an adhesive. Then it is stacked with the grain in each layer at right angles to the one below it. Moved into a press, the layers are bonded together with heat and pressure. Plywood always has an odd number of layers.

Fig. 3–7. Laminated wood arches in sculptured forms are used not only for their beauty and warmth of feeling but for structural strength as well. See Fig. 3–41. This is St. Paul's Episcopal Church, Shreveport, La.

WHEN YOU WORK WITH WOOD

Wood is a friendly material. The trees from which the wood came were friendly to all. They sheltered the birds, fed the squirrels, and shaded the cattle from the hot sun. They slowed and calmed destructive winds and water and supplied wood for millions of homes. But wood can be very obstinate and "ornery" when you don't treat it right. It refuses to be smoothed with dull tools. It goes to pieces when nailed too close to the edge. It shrinks when cold and dry and swells when warm and damp. When the wrong wood is used, your project is less useful and beautiful than it could be.

The best way to get acquainted with wood is to work with it. Give it a chance to show how well it can serve you. Some wood is soft and some is hard; some is dark and some is light. Some is splintery and some is smooth; some is strong and some is weak. Some stand the weather and some cannot. Your teacher will show you various kinds of woods and will help you to select the best one for each project you make. You will need to get well acquainted with the tools described on the following pages. The more skillful you become with them, the more you will enjoy woodworking. Remember, *the best piece of wood is not always in the middle of the board.*

WOODWORKING TOOLS

Many different tools have been designed for working with wood. Each of them is intended to perform a certain process. The oldest was probably a sharp stone or shell for cutting and scraping. Hand tools are rarely invented today; the greatest developments now are in machine processes. An example of the latter is electronic gluing. In this, synthetic glues are cured electrically to form perfect bonds. The tools and machines described here perform common, much used processes.

MEASURING TOOLS

Measuring tools are designed to help you work accurately to dimension.

The Rule.   A steel rule is used for measuring and as a straight edge for marking. If you use a wood rule, set it on edge when measuring.

The Try Square.   This tool is used for measuring, for checking right angles (checking for square), and as a straight edge. A six-inch blade is best for your purposes. The steel square is larger.

The T Bevel.   This is an adjustable device for transferring angles. The blade and handle are set at the desired angle and locked. The T bevel is especially handy for checking chamfers and bevels.

The Calipers. These are tools for measuring inside and outside diameters. When setting, the caliper is laid on a rule to get the dimension.

The Dividers. This tool is similar to a compass used for laying out circles and arcs, except it does not hold a pencil. It is useful for finding centers, for stepping off distances, and for dividing lines.

HAND TOOLS AND PROCESSES

Handsaws

Handsaws are used for rough cutting. The crosscut saw cuts across the grain, and the ripsaw cuts with it.

Sizes of Handsaws. Blade lengths vary from 16 to 26 inches. The coarseness of the teeth is given in *points per inch.* A general purpose crosscut saw is a 22 inch, 10 point saw, and a ripsaw, a 24 or 26 inch, 5 point saw.

Care of Handsaws. To cut straight, a saw must be straight and sharp. Use a handsaw only on wood. Hang it up when you are not using it. The saw handle breaks easily when the saw is dropped.

How to Crosscut
1. Square a line across the board.
2. Clamp the board flatwise in the vise.
3. The idea is to cut just outside the line. Place the saw teeth accordingly. Place your thumb as a guide, about two inches above the teeth. Start the cut with slow, short strokes.

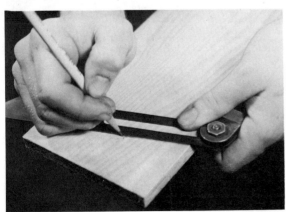

Fig. 3–8. (top left) Using the try square to square a line across a board. **Fig. 3–9.** (bottom left) Marking an angle with the sliding T bevel. **Fig. 3–10.** (bottom right) Measuring wood turning with calipers.

CROSSCUT SAW TEETH

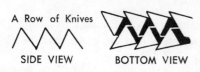

A Row of Knives

SIDE VIEW BOTTOM VIEW

RIPSAW TEETH

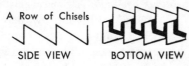

A Row of Chisels

SIDE VIEW BOTTOM VIEW

Fig. 3–11. A close-range view of saw teeth.

4. Gradually lengthen the strokes until most of the teeth are used. Move your thumb away. A sharp saw requires little pressure. If the saw sticks, you are pressing too hard.

5. Ease up with short strokes as you finish the cut. Hold the piece being cut off to avoid splitting.

How to Rip

1. Mark the line.

2. Clamp the board vertically in the vise and saw as for crosscutting.

Fig. 3–12. Start a handsaw with a pull stroke. This is a crosscut saw for sawing across the grain.

Fig. 3–13. A ripsaw is used for sawing a board in the direction of the grain.

1. A dull saw chews more fingers because it skids in starting.
2. Ease up as you finish the cut to avoid skinning your knuckles.
3. Hang the saw on the tool panel so that the teeth don't stick out ready to "bite" someone.

The Backsaw

The backsaw is used for making accurate cuts, as in joints, and for cutting to length. Blade lengths are usually from 10 to 16 inches. The coarseness of teeth for these lengths is usually 13 points per inch.

Care of the Saw. Never use the backsaw for rough cutting. Keep it hung up when not in use to prevent dulling of the teeth and bending of the blade or the back.

How to Cut a Board to Length
1. Square a line across the board.

2. Clamp the board flatwise in a vise.
3. Clamp a straight piece of wood over the mark for a guide.
4. Keep the blade snug against the guide as you draw it slowly back and forth over the face of the board.
5. Little pressure is needed on a sharp saw. Lift up slightly on the last few strokes so that the saw doesn't break through.

How to Cut a Dado
1. Using a square, mark out the dado (a cut or groove across the grain into which another piece of wood is fitted to form a dado joint).
2. With a guide, as above, cut to the depth on each side.
3. Clean out the center, using a wood chisel as near the width of the dado as possible.

How to Cut a Rabbet
1. Mark out the rabbet with a square and make the end cut.
2. Make the face cut using a guide.

Fig. 3–14. Use a backsaw to cut a board to accurate length. A guide block helps to keep the blade in the proper position.

Fig. 3–15. Cutting a dado with a backsaw and a guide block. The dado is marked out with the aid of a try square.

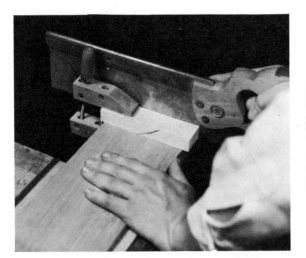

The Miter Box

The miter box is a device in which wood can be sawed accurately to desired angles. The saw is swung to the desired angle. With the wood held firmly against the fence, the cut is made. Be sure to figure the angle correctly before you begin to saw.

The Coping Saw

The coping saw is used for cutting curves in wood not thicker than an inch. It is not intended for cutting straight lines. It cuts best with a vertical, pull stroke as in a jigsaw rather than as a hand-saw.

Blades. Two types of blades are used: pin end and loop end. Round blades with spiral cutting edges are available. These cut in any direction without turning the blade. Some blades cut plastics and soft metals.

Fig. 3–16. Cutting the end grain for a rabbet. Note that two hands are required.

Fig. 3–17. Some cuts and joints in wood.

BEVEL—a slanting edge

CHAMFER—a cut-away corner

TAPER—a gradual decrease in thickness

BUTT JOINT—the pieces "butt" against each other

DADO—a rectangular groove cut <u>across</u> the grain

RABBET—a rectangular groove cut <u>along</u> the edge

PLOW CUT—a dado <u>with</u> the grain

DADO AND BUTT

RABBET AND BUTT JOINT

MORTISE AND TENON JOINT

Tenon   Mortise

Note the use of oblique drawings and symbols.

Dowel Pin

DOWEL JOINT

Inserting a Blade. Unscrew the handle three or four turns and press the saw frame against the bench until the blade can be slipped into the grooves. The teeth should point toward the handle. Then tighten the handle.

Cutting Curves. Clamp the board flat on the edge of the bench. Holding the saw vertically, start with short, quick strokes. Press forward only enough to keep the teeth cutting. If the blade sticks, you are pressing too hard. Keep the blade cutting as you turn the frame to go around the curves. For fine detail sawing, use a saddle or V block, to support the material. The blade should cut close to the V, as is illustrated in Fig. 3–19.

Cutting a Hole. Bore a ¼-inch hole near the line in the waste stock. Stick the blade through the hole and insert it in the frame.

Fig. 3–18. (top) Making an angular cut in a miter box. This cuts angles from 45 to 90 degrees. **Fig. 3–19.** (bottom) A V-block can be used to support the work when you use the coping saw. Hold the piece firmly so that it cannot jump up and down.

Fig. 3–20. To cut out a large hole, first bore a small one in which to insert the coping saw blade.

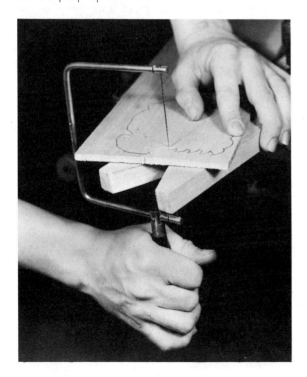

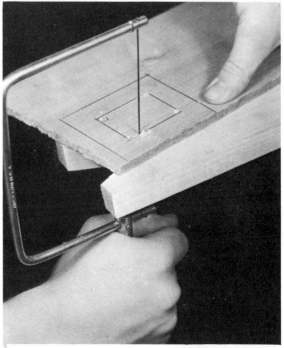

Hand Planes

Hand planes are designed to make wood surfaces smooth and flat. There are many types of planes, some for general use and others for special purposes. The following are those you are likely to use:

Block Plane. This is a small plane, about 6 inches long, which can be used in one hand. It is handy for model building, chamfers, and other light work.

Smooth Plane. This is the next larger plane, usually 8 to 9 inches long. It is your best all-round plane. Use both hands on it.

Jack Plane. This is the general purpose plane for a carpenter or cabinetmaker. Lengths vary from 11 to 15 inches.

Fig. 3–21. Four common hand planes. Use the one that best fits you and the job.

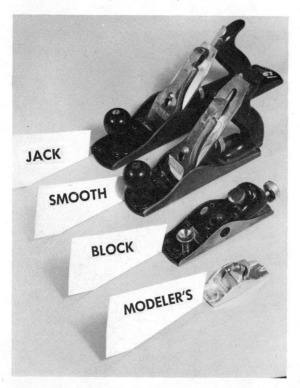

JACK

SMOOTH

BLOCK

MODELER'S

Some Tips on Using Planes. A plane is useless unless the blade is sharp and properly adjusted for the job. Your teacher will show you how a sharp, well-adjusted plane cuts. Note the shavings and that the plane shaves smoothly. It does not chew or tear the wood. Your teacher will show you how to keep the plane iron sharp and adjusted. When your plane is not in use, lay it on its side. Can you figure out why? Planes with cast-iron beds can be dropped only once. Why?

How to Plane the Face of a Board

1. Clamp the board on top of the bench or in a vise. Make a few trial cuts as you adjust the blade for thin shavings.
2. Now plane over the entire surface, in the direction of the grain. Check frequently for flatness with a try square. Should the grain roughen, plane from the opposite direction.
3. Hold the plane at a slight angle to the grain, but push with the grain. Lift the plane on the return stroke.

How to Plane an Edge

1. Mark the line to which you will plane and clamp the board in a vise, close to the line.
2. Hold the plane firmly and squarely on the edge and at a slight angle to it. Push it slowly from one end to the other.
3. Check for trueness with a straight edge. Then with a try square check the edge for squareness with the face.

How to Plane an End

1. The blade must be very sharp and set for a thinner cut than when planing an edge. Square a line across the board as near the end as possible. Clamp it in a vise, close to the mark.
2. Hold the plane firmly and squarely on the end and at a slight angle to it. Push it slowly and steadily. If you plane clear across, the wood will split at the far corner, so plane from each edge toward the middle.

How to Plane a Board Square

Using the suggested procedures for planing and checking for true with the try square, follow these steps:

1. Select the best side of the board for the No. 1 face. Test it for true. Plane off any high spots.

2. Plane an edge straight and square with this face.

3. Plane one end straight and square with the face as well as the edge.

4. Mark and saw off the board about $\frac{1}{16}$ inch longer than desired. Plane this end to the mark, making it square with the face and the edge.

5. Mark off the width from the true edge. Plane to this mark. Keep the edge square with the face and with each end. If there is too much wood to plane, rip off the excess with a saw.

6. Plane the No. 2 face true and the board to the desired thickness.

Chisels and Gouges

The Wood Chisel. The wood chisel is used for fitting wood joints and for shaping. It causes more injuries than any other shop tool because it is so often misused.

Types of Chisels. *Butt* chisels are short and husky; *socket* chisels are long and slender. Size is given as the width of the blade. Sizes range from $\frac{1}{4}$ to 2 inches.

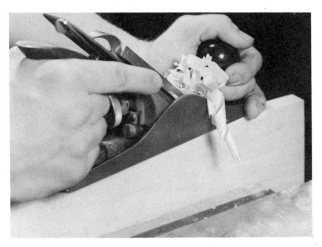

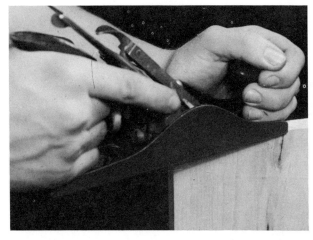

Fig. 3–22. (top left) Planing a face. Hold the plane at a slight angle to the grain of the wood. Push it with the grain. **Fig. 3–23.** (bottom left) Planing an edge. Hold the plane at a slight angle. Push along the full length of the edge. **Fig. 3–24.** (bottom right) Planing the end grain. Set the blade for a thin cut.

How NOT to Use a Chisel

1. A chisel is not for prying open cans of paint.
2. It is not for prying open boxes.
3. It is not a screwdriver.
4. It should not be carried in your pocket. Can you figure out why?

How to Use a Wood Chisel

1. Use only a sharp chisel. Ask your teacher how to sharpen it.
2. Use as wide a chisel as the job will permit and make sure that the handle is on securely.
3. Be sure to clamp the work tightly in a vise or on the bench so that you can use *both* hands on the chisel. Use a mallet on the chisel only when you must have the extra force.
4. When making *paring* or shaping cuts use one hand on the handle and the other on the blade so that you can control the cutting.
5. Never let one hand or finger get out in front of the chisel while cutting. Always cut away from yourself. When you lay the chisel down, place it so that the edge is protected.

Fig. 3–25. Some wood chisels and a set of carving tools.

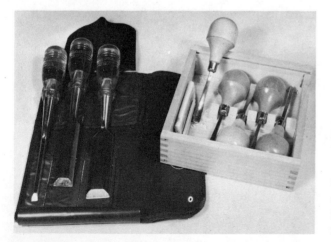

Gouges. A gouge is a chisel with a curved cutting edge. It is used as a chisel for hollowing and shaping.

Carving Chisels. These are small chisels and gouges with cutting edges in assorted shapes to suit various cuts. The chisels illustrated are used by hand only, not with a mallet. They will cut wood, linoleum, and other soft materials.

Knives

Knives are used mostly for whittling and carving. Those with replaceable blades are easiest to use and to keep sharp.

✚ SAFETY SENSE

Keep these pointers in mind and you won't cut yourself with a knife:

1. Use only a sharp knife. Dull ones cause most injuries.
2. Always cut away from yourself.
3. Never carry an open knife in your pocket, even for a moment.
4. Use a knife for cutting, not for prying or in place of a screwdriver.

Wood Files

Wood files are used for smoothing curves, usually on the edges of boards after they have been sawed out. The two common types are the *wood rasp* with coarse teeth for rough cutting, and the *wood file* (sometimes called the *cabinet file*). The latter has finer teeth for smoother cutting. The most used shape is the half round. It has one side round and the other flat. Wood files are also round or square. Common lengths are 8, 10, and 12 inches. *Surform files* cut much faster than common cabinet files. Use them if there is much wood to be removed. Do not use any file without a handle.

Fig. 3–26. (top left) A wood chisel is used to clean out the waste for a dado. The bevel edge should be down. Use both hands on the tool. **Fig. 3–27.** (top right) Hollowing out a tray with a gouge and a mallet. Do this before cutting the tray to shape.

Fig. 3–28. (bottom left) Using a carving tool to cut away the background of a carving. This tool is a gouge. The outlining was done with a V-shaped tool, a veining tool. **Fig. 3–29.** (bottom right) When cutting with a knife, always cut away from yourself.

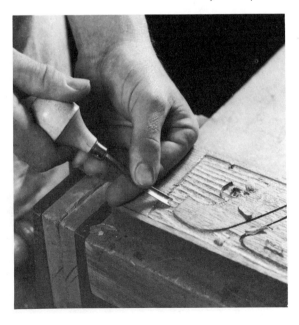

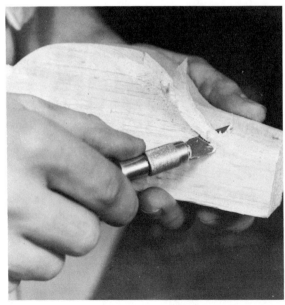

How to Shape an Edge with a File

1. Clamp the board in a vise with the edge to be filed as horizontal as possible. Clamp the edge close to the vise.

2. With one hand holding the point and the other the handle, push the file diagonally across the edge. Lift on the return stroke and repeat. A file cuts better this way than when rubbed back and forth.

3. Use the curved side for hollow curves and the flat side for flat edges and convex surfaces. Work down to the mark so that there will be a minimum of sanding to do. Clean the teeth frequently with a *file card*.

4. Watch the edge as you file. The wood should be quite smooth and cleancut. If the file roughens the edge, change the direction of the filing. File from the opposite side or at a different angle to the edge on which you are working.

Fig. 3–30. (left) Surform tools for removing wood rapidly. Fig. 3–31. (right) A wood file removes saw marks. Sandpaper removes file marks.

The Brace and Bit

The brace and bit together are used for boring holes in wood. The brace is used with other tools, as countersinks and screwdriver bits.

The Brace. Braces are made with or without ratchets (for working in close quarters). The ratchet is the device that permits the handle to be turned with short strokes in either direction. It is set by the knurled knob on the handle. The jaws can grip both round and square shanks. Brace size is given as the swing. This is the diameter of the circle swung by the handle as the brace is turned. A 12-inch swing is a common size.

Auger Bits. Auger bits are woodboring tools. The most common sizes bore holes from ¼ to 1¼ inches in diameter. The number on the shank is the diameter of the hole in sixteenths. For example, a No. 5 bores a $\frac{5}{16}$-inch hole.

Care of the Brace and Bits. A drop of oil is occasionally needed in the head of the brace. The bits need sharpening periodically. Ask your teacher to demonstrate this. If a bit gets bent, lay

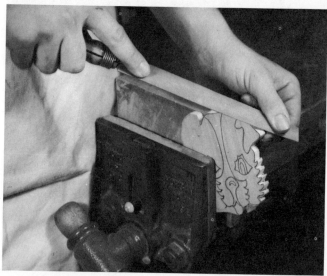

 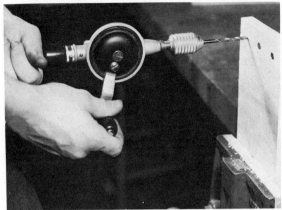

Fig. 3–32. (left) Boring a hole with an auger bit and brace. Fig. 3–33. (right) The hand drill makes small holes.

it on a scrap board and roll it over slowly until the high side comes up. Rap this with a soft mallet.

How to Bore Holes

1. Unscrew the chuck on the brace a few turns, insert the tang, and tighten.

2. Press the point of the lead screw in at the center of the desired hole. Press against the head of the brace with one hand. With the other, turn the handle clockwise. To cut fastest, press firmly but turn slowly.

3. When the screw point comes through the back side, stop and reverse the direction to remove the bit. Finish the boring from the opposite side. This procedure prevents splintering.

The Hand Drill

The hand drill is used for drilling small holes, usually not larger than ¼ inch, in wood, metal, plastics, or other material. It turns a straight shank twist drill bit. (See page 94.) The drill illustrated is typical; some have enclosed gears.

How to Use a Hand Drill

1. To insert a drill bit in the drill, grasp the chuck with one hand. With the other, turn the crank backward (counterclockwise) to open the jaws. Insert the bit, hold the chuck, and turn the crank in the opposite direction until it draws up snug.

2. Prick a center mark at the desired point and insert the point of the bit. Hold the drill handle with one hand; turn the crank clockwise. Use only enough pressure to keep the bit cutting. Ease up on the pressure as the bit breaks through the back of the material.

The Automatic Drill. This tool is used for drilling small holes in soft materials. Pumping the handle up and down makes the bit revolve. Its special bits are kept in the handle.

✚ SAFETY SENSE

Since you can get a finger pinched in the gears when tightening or loosening the chuck, watch where you hold the drill.

Fig. 3–34. To control the electric hand drill, use both hands.

Fractional Size	Wire Gauge	Letter
1/16	52*	
3/32	42*	
1/8	30*	
5/32	22*	
3/16	12*	
1/4		E
5/16		O*
3/8		V*
7/16		
1/2		

How to Use Twist Drills

1. Use only sharp drill bits. A dull bit rubs and gets hot. If forced it will break or burn. Ask your instructor to show you how to tell when a drill bit is cutting as it should.

2. Press only hard enough to keep the drill bit cutting. It should not bend under the pressure.

3. Oil is used as a coolant on the bit when drilling steel, and water is used on cast iron. No coolant is needed for wood, plastics, brass, copper, or aluminum.

Hammers

A hammer is probably the first woodworking tool you ever used. It is the handiest to have around. Several types are used:

The Claw Hammer. This is the hammer for driving and pulling nails. These hammers are sized according to the weight of the head. A 7-oz hammer is for small nails and brads. A 16-oz hammer is for general use. A claw hammer should not be used on cold chisels or any hard material that can mar the face. This makes nail driving difficult.

How to Drive a Nail

1. Grasp the hammer as shown in Fig. 3–35. Tap the nail lightly to start it, so that it stands by itself.

2. Drive the nail down flush with the surface with accurate blows. At first it takes a lot of courage to strike the nail hard. With practice you should be able to drive a 6d nail with three or four blows. Why not have a school nail-driving contest?

Drill Bits

Drill bits are used in hand drills, drill presses, and portable power drills. They drill holes in woods, metals, plastic, and the like. There are several types, the most common being the straight shank *twist drills*. Sizes are designated by fraction, number, or letter. Common sizes range from 1/32 inch to 1/2 inch by 64ths. They number from No. 1 to No. 80 in the wire gauge, and from *A* to *Z* by letter. The size is stamped on the shank, but after the drill has been used, the marks are difficult to read. In this case find the hole that fits in a drill holder to get the size. *Carbon steel twist drills* are used for soft materials such as wood and plastic. *High-speed twist drills* are used for metals.

Comparative Sizes of Drill Bits. These are but a few of the more than a hundred sizes up to 1/2 inch (* indicates nearest to fractional size):

Fig. 3–35. To drive a nail, use all of the hammer handle. Note that the nails are slanted. They hold better this way.

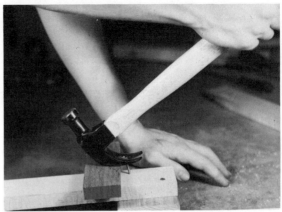

Fig. 3–36. To pull a large nail with a claw hammer, use a block under the head to increase leverage and to protect the surface of the wood.

Pulling a Nail. The claws of the hammer are intended for pulling nails. To protect the surface of the board, lay a block under the head. This procedure gives added leverage, too, and makes pulling easier.

Upholsterer's Hammer. This is a tack hammer with one end magnetized for holding and starting tacks. Once started, the tack is driven with the other end.

Soft-face Hammers and Mallets. Soft-face hammers have replaceable plastic faces. In woodwork they are used for the same purposes as are mallets. For example, they drive wood chisels and tap joints together. Mallets have soft heads, usually of wood, rubber, or rawhide.

The Nail Set. This is a slender punch for setting the heads of brads and finish nails below the surface. The holes may be hidden with filler. Place the set squarely on the head and, with the hammer, drive it down $1/32$ to $1/16$ inch.

Nails

Although the carpenter depends on nails for fastening lumber, the cabinetmaker uses glue and, now and then, small nails. The ordinary types of carpenters' nails are *common, casing,* and *finishing.* Sizes are given as 6d, 8d, and so on. The "d" is called "penny." Originally in England the price of nails was so many pennies per 100; for example, 8d per 100. The larger the nail, the more it costs.

The table below shows the approximate length in inches for all types of ordinary carpenters' nails. Box and common nails come in sizes larger than given in the table: the 20d is 4½ inches, 40d is 5 inches, and 60d is 6 inches long.

COMMON SIZES OF CARPENTERS' NAILS

Size	Approximate Length (in.)
2d	1
4	1½
6	2
8	2½
10	3
12	3½

Wire Nails and Brads. Wire nails are similar in appearance to common nails. Wire brads are similar to finishing nails. Wire nails and brads

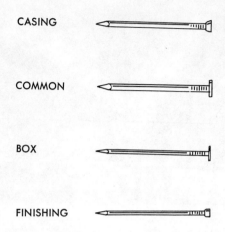

CASING

COMMON

BOX

FINISHING

Fig. 3–37. Kinds of nails. Wire brads resemble finish nails. Wire nails resemble common nails.

are made in a greater assortment of lengths and diameters, however. Sizes range usually from ¼ inch No. 20 to 3-inch No. 10. The larger the "No.", the smaller the diameter. A No. 15 brad is approximately $\frac{1}{16}$ inch in diameter and is about twice as thick as a No. 20 brad.

To Make Nails Hold Better. Slanting the nails gives them more holding power. (See Fig. 3–35.) Stagger them so that they don't fall into the same grain line and split the wood. Wire brads, casing nails, and finishing nails are usually set. (See above.)

Screwdrivers

The original use for a screwdriver was to drive wood screws. It is so handy for prying open cans, boxes, and such that manufacturers usually make some of them rugged enough to do this. There are two distinct types: those for *slotted* screws and those for *recessed* head screws (Phillips head). Sizes of the former are given by the length of the

blade. A 4-inch and a 6-inch screwdriver will do most of your work. For recessed screws, pick the screwdriver to fit.

How to Use a Screwdriver
1. Pick the screwdriver that fits the screw snugly.
2. Insert the bit in the slot and hold it there with one hand while you turn the screwdriver with the other hand. Keep the screwdriver in line with the screw.

Fig. 3–38. Installing a wood screw. Use a screwdriver that fits the screw head snugly both ways.

Wood Screws

Wood screws hold better in wood than do nails and they can be easily removed to permit taking a project apart. There are a great many kinds of wood screws. They vary by head, material, finish, slot, and thread as shown in this table.

Head .... flat, round, oval
Material .... steel, aluminum, brass
Finish .... bright, blue, cadmium, chromium
Slot .... straight, recessed (Phillips)
Thread .... regular, drive type

Those that you will use most often are the flat-head bright (F.H.B.) and the round-head blue (R.H.Bl.) with either slotted or recessed heads.

Sizes. Both lengths and diameters vary. The most common lengths are from ½ to 3 inches with

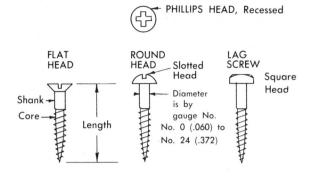

Fig. 3–40a. Common types of wood screws. Diameters for flat-head and round-head screws are by gauge number. Phillips heads are recessed. Lag screws have square heads for a wrench. Diameters for lag screws are in fractions of inches.

Fig. 3–40b. Ways of installing wood screws. A washer is used under the head of a lag screw. Counterboring is done when a wooden plug is used to conceal the screw.

Fig. 3–39. Countersinking a hole for a flat-head wood screw.

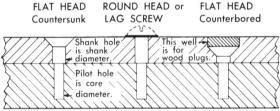

diameters from 4 to 12. A screw size is indicated as, for example, 1 inch No. 6, 1½ inches No. 8. The number is the diameter. The larger the number, the larger the diameter.

How to Install Wood Screws

1. Select the screw and the screwdriver.
2. Bore the shank hole in the outer piece.
3. Bore the pilot hole. These should be smaller in soft woods than in hard woods.

4. Countersink the shank hole, if needed. (See method below.)

5. Drive the screw.

DRILL BIT SIZES FOR WOOD SCREWS

Screw diameter	4	5	6	7	8	9	10	11	12
Shank drill	⅛	⅛	⁵⁄₃₂	⁵⁄₃₂	³⁄₁₆	³⁄₁₆	³⁄₁₆	⁷⁄₃₂	⁷⁄₃₂
Pilot drill	¹⁄₁₆	³⁄₃₂	³⁄₃₂	⅛	⅛	⅛	⅛	⁵⁄₃₂	⁵⁄₃₂

The Countersink (Brace Type). This tool, used in an auger brace, cuts a tapered hole to fit the head of a flathead wood screw.

How to Countersink a Hole

1. Insert the point in the screw hole and turn the brace clockwise.

2. After two or three turns try the screw. The head should lie flush with or just below the surface.

Screwdriver Bits. These bits are used in an auger brace for driving screws easier and faster. Blades are available for both types of screws.

ADHESIVES

Adhesives are glues and cements. They can make stronger joints in wood than can nails or screws. They can be used when nails or screws would be inappropriate. A good glue joint is actually stronger than the wood itself.

Several kinds of adhesives are available: animal, fish, casein, and synthetic. Animal and fish glues can be had in liquid or in flakes for melting. They are not waterproof. Casein glue is made from milk and is purchased in powder form to be mixed with water.

The synthetic adhesives are available in powder form to be mixed with water. They also may be had in liquid form. These adhesives are made water soluble, water resistant, or waterproof. Be sure to follow the manufacturer's instructions.

Contact cements are useful in adhering veneers and plastic laminates to table tops. A coat is applied to each surface and allowed to dry. Place a piece of wrapping paper over one surface and

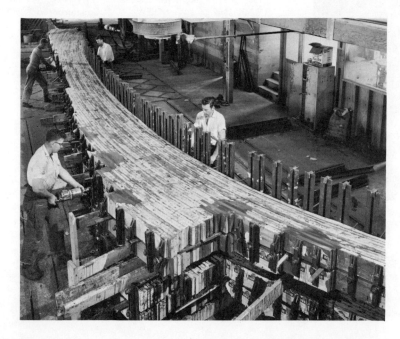

Fig. 3–41. Here four laminated arches are being glued and clamped into shape. This produces a very strong and beautiful arch for supporting the roof of a building.

Fig. 3—42. Parallel-jaw woodworkers' clamps are used to hold pieces while the glue sets.

Fig. 3—43. A table top held together with bar clamps while being glued. Pressure by clamping is necessary for strong joints. Newspapers catch any glue that drips.

position the pieces. Hold them in position while carefully withdrawing the paper. Once in contact the pieces cannot be moved. Roll or press the pieces into full contact.

Liquid plastic in tubes, like model airplane cement, is useful in the assembly of small pieces of wood. Because it dries so quickly it would be difficult to use for gluing boards together for a table top, for example.

Rubber cement, although not for gluing wood together, is a very handy adhesive. Use it for cementing paper, cloth, leather, paper patterns to wood and metal, and the like. When dry, it can be rubbed off the surface.

Some Tips on Gluing

1. When gluing boards edge to edge, alternate the grain to counteract warping.

2. Pressure is needed when gluing boards together. Use just enough to force a thin line of glue out of the joint. Adjust the clamps to fit before applying the glue. Do not remove the clamps until the glue is dry. Follow the manufacturer's recommendation.

3. Glue surfaces must be in full contact to get strong joints.

4. Apply glue quickly to both surfaces, using no more than necessary to cover. Excess glue oozes out. Wipe off the excess of water-mixed glues immediately with a damp cloth. Scrape off the excess of quick-drying cements.

CLAMPS AND CLAMPING

Clamps are used for holding and pressing when a vise is not suitable. Use the type and size best suited for the job. Protect the work from jaw marks by inserting pieces of wood between the project and the clamp.

"C" Clamps. These are general purpose clamps, shaped like a "C," for holding pieces of wood face to face. Sizes are given in inches, referring to the maximum opening.

Hand Screws. These are for general clamping. They are especially useful when the sides of the work being held are not parallel. Common sizes have openings ranging from 6 to 14 inches.

Bar Clamps. Bar clamps are for holding large work, especially for gluing boards edge to edge, as for table tops. Sizes are given in feet and refer to the maximum opening.

ABRASIVES

Sandpapering follows filing, planing, and such tool processes in order to make a smooth surface on wood. Tool marks are removed with coarse sandpaper. Finer grades of sandpaper are used to remove the scratches made by the coarser papers.

Sandpaper is not sand glued to paper. Rather, it is paper coated with tiny, sharp pieces of hard, crushed rock or other abrasive. Three kinds of sandpaper are common: *flint, garnet,* and *aluminum oxide.* Flint is least costly and is used mainly for home chores. Better abrasives would be ruined too quickly for such tasks as removing old paint from wood. This would gum up or clog the grit. Garnet cuts faster and lasts longer than flint. Aluminum oxide abrasive is made from bauxite in an electric furnace. It is much harder than flint or garnet rock and is superior to either.

In the comparison of grits given in the chart note that the larger the mesh number (50, 100, etc.) the finer the grit. These are the screen-mesh size through which the grit is sieved. The abrasives industry now classifies flint only by the designations of "extra fine" through "extra coarse." The industry uses mesh numbers rather than symbols (1/0, 2/0, etc.) but gives both designations for garnet. This is because garnet is used primarily as a wood-sanding abrasive. The furniture industry uses the traditional symbols.

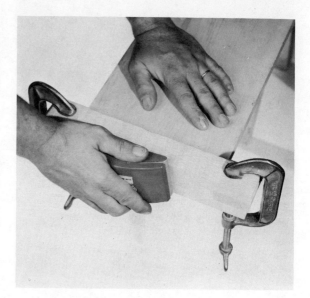

Fig. 3–44. Hand sanding with the abrasive on a sanding block.

Using Sandpaper. Cut the sheet into four approximate squares on a paper cutter (this keeps the cutter sharp). Use one piece at a time on a wood or rubber sanding block for smoothing flat surfaces. For inside curves, wrap the paper around a piece of dowel or other suitable shape. Always sand with the grain of the wood, using only enough pressure to keep the sandpaper cutting.

FINISHES

Finishes are applied to wood to protect and preserve it, to make it more useful, and for color and texture. A thorough sanding of the surface is usually necessary before a finish is applied. Consult with your teacher about the selection of a finish for your projects. Always read the instructions on the label of a can of finish before using it.

Shellac and Wax. This is an easy finish to apply. It is dustproof and leaves the wood in its natural color. Brush on a coat of white shellac thinned two parts of shellac to one of alcohol.

A COMPARISON OF GRITS

(adapted from Behr-Manning classification)

Flint	Garnet	Aluminum Oxide
Coarse	No. 50 (1)	No. 40
		50
		60
Medium	80 (0)	80
		100
		120
Fine	120 (3/0)	150
		180
		220
Very fine	220 (6/0)	240
		280

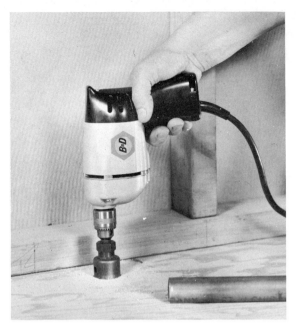 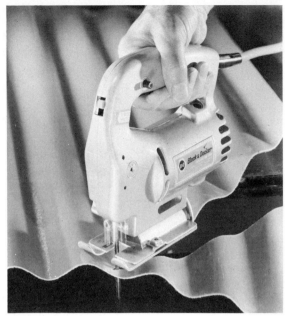

Fig. 3–45a. (left) The portable electric drill can be used to cut holes with a hole saw.

Fig. 3–45b. (right) The saber saw has many uses. Here it is cutting fiber glass. With a proper blade it can cut metal.

When it is dry, which will usually take about 20–25 minutes, rub lightly with 6-0 sandpaper until the surface feels smooth. Repeat this with two or more coats. Let each dry thoroughly.

Apply several coats of paste polishing wax. Let each dry and then rub it to a luster before applying the next. Four or five coats are sufficient. Put the wax inside a soft cloth when applying it. Clean shellac brushes in alcohol.

Oil and Wax. Oil finishes bring out the natural beauty of woods most effectively. They are easiest to apply, dry dust free, but need much rubbing.

With a rag, swab on a coating of a mixture of two parts warm linseed oil and one part of turpentine or mineral spirits. After this has soaked in for about an hour, wipe off the excess. Let it dry overnight, then rub it hard with a soft cloth until it shines.

Apply as many coats as you wish, one per day, but at least three or four. When thoroughly dry and rubbed, apply two coats of wax as for the shellac finish. Paste shoe wax is excellent.

Tempera. *Tempera paints* are watermixed opaque colors used for decorative purposes. They are easier to use than is enamel for painting designs. First seal the surface of the wood with a thin coat of shellac. Sand it smooth when dry. Stir the tempera and add water until it brushes smoothly. It should cover in one coat. Tempera that is too thick will peel.

When the design is dry, spray on a coat of clear lacquer or plastic to protect the tempera. Cut the lacquer with lacquer thinner until it sprays

Fig. 3–46. (left) Applying an oil stain.
Fig. 3–47. (right) Wiping the oil stain to even the color.

easily. A fly spray gun will do. Tempera brushes are cleaned with water.

Oil Stain. To change the color of wood, use an oil stain. Test the color on a piece of scrap. Diluting the stain with paint thinner lightens the color. Colors can be mixed. Apply with a brush and wipe with a cloth before it dries.

Enamels and Varnishes. *Enamels* are opaque, waterproof colors used when a slow-drying, weather-resistant finish is desired. Some are made of natural oils and gums and some are synthetic, so be sure to read the instructions on the label.

Enamels are commonly brushed or sprayed on the bare wood. After 24 hours of drying they are hard enough to sand lightly with fine sandpaper. Lay on a heavy second coat, brushing from the center of the surface out to each end. This coat should be just heavy enough that brush marks will level out, thus leaving a mirrorlike surface. When possible, let the enamel set for an hour in a horizontal position to avoid runs. On vertical sur-

faces paint with vertical strokes. Two coats are usually adequate.

Varnish is a clear enamel and is applied in the same manner. Spar varnish is waterproof. Brushes are cleaned in paint thinner followed by hot water and soap.

Plastic Varnishes. New plastic finishes frequently appear on the market. They are all different enough that it is important to follow the directions on the can. A recent type, the polyurethanes, can be applied by brush or spray. Special solvents are required.

Spray Cans. Quick-dry lacquers and enamels in "fizz" cans are useful for small jobs. For even coating, the nozzle should not be closer than 6 to 8 inches from the work. Spraying should be done indoors only if there is good ventilation. The fumes are inflammable. Clean the nozzle by spraying for a few seconds with the can upside down.

Lacquer. *Lacquer* is a synthetic liquid plastic. It is waterproof and weather resistant and dries in 15 to 20 minutes. Airplane dope is lacquer. For brushing, use a slow-drying lacquer thinner to cut the lacquer. For spraying, use a fast-drying thinner. Lacquers spray better than they brush.

On porous woods, like balsa, mix some whiting with the first coat to act as a filler. Brush this on. When dry, sand it smooth and add six to eight coats of the desired color. Allow about 15 to 20 minutes between them. No sanding is done between these coats because each coat dissolves the one beneath. This results in one thick coat. When the last coat is dry, apply a coat of thinner to help smooth the surface. Let this dry. Then rub well with fine rubbing and polishing compound until you have a soft, lustrous, automobile-type finish. Clean brushes and spray gun with lacquer thinner.

Lacquer-type or plastic finishes are common today. Be sure to follow the manufacturer's instructions. Use the solvents recommended.

PORTABLE POWER TOOLS

There was a time when there were only hand tools. Today there are several portable electric power tools that do the work of hand tools.

The Electric Hand Drill

This is a two-hand tool for drilling holes. It requires high-speed straight shank twist drills. Its drilling capacity is limited by the chuck. For example, a ¼ inch drill takes drill bits as large as ¼ inch. Follow these steps:

1. Insert the bit fully into the chuck. Tighten it with the special chuck wrench or key.

2. Hold the work securely, preferably in a vise or by a clamp. Use a scrap block of wood under the work whenever possible.

3. Make a small starting hole in the work. If wood or plastic, use an awl. If metal, use a center punch.

4. Place the bit in the starting hole and hold the drill in the proper position.

5. Touch the switch to operate the drill for an instant to get the bit started. Then proceed to drill.

6. Push on the drill only hard enough to keep it cutting. Ease up just before it cuts through. If the bit doesn't cut it probably needs sharpening.

✚ SAFETY SENSE

1. Do not overload the drill. Too much pressure can break the bit or stall the motor.
2. Be cautious about the many special attachments offered for this tool. Some are not only ineffective, but dangerous.
3. When using any electric power tool make sure it has a ground connection to the electric outlet.

The Saber Saw

The saber saw cuts curves, bevels, and rips and crosscuts. Blades for cutting different materials are available.

For general wood	7 or 8 teeth per inch
For plywood, hardboard, veneers, plastics	10 teeth per inch
For soft sheet metals, brass, copper, aluminum	14 teeth per inch
For mild steel sheet	24 teeth per inch
For solid steel rods and bars	32 teeth per inch

Follow these suggestions for cutting

1. Select and install the proper blade.

2. Place the front of the saw shoe plate firmly on the material. The blade should just touch the stock. Turn on the motor.

3. Push the saw slowly into the work to start the cut. Then push it only as fast as it can cut. Do not force it.

4. When the cut is complete turn off the motor. Let the blade stop before removing it from the cut.

5. A saber saw can start its own cut in wood. Boring a hole for the blade is not always necessary. Tip the saw forward so that it rests on the

sole plate. Let the tip of the blade touch the material. Hold it firmly in this position, turn on the motor. Let the blade cut its way through the stock. This is called a *plunge* cut. Bring the saw gradually back to a vertical position as the blade cuts through.

6. To bevel cut, set the shoe plate at the desired angle.

7. Remember to saw on the outside of the mark.

✚ SAFETY SENSE

1. Turn off the motor as soon as the cut is completed.
2. The blade should be stopped before it is lifted from the cut.
3. Keep the power cord behind the saw.

The Portable Power Sander

There are two types of these sanders. The abrasive on the *orbital* sander moves in a small circle. The *reciprocating* sander moves the abrasive back and forth in a line. The first is for rough sanding and the second, for finish sanding. For either, follow these steps:

1. Be sure the proper abrasive is on the machine.
2. Place the sander in position. Hold it firmly with both hands.
3. Flip the switch and guide the movement back and forth. Put no pressure on it. Its weight is sufficient to assure cutting.
4. Change grades of abrasive as necessary.

✚ SAFETY SENSE

1. Keep the power cord out of the way. The abrasive can cut through the insulation.
2. Don't "ride" the sander. Only guide it. If it doesn't cut, replace the abrasive.
3. Use a dust collector. If not available, use a respirator.

POWER TOOLS AND PROCESSES

The Jigsaw

The jigsaw is a power coping saw. It is also called a *scroll saw*. A short blade moves up and down, cutting on the down stroke. It saws wood, cardboard, plastics, and such soft metals as aluminum, copper, and brass. Jigsaws are available in bench and floor models in sizes ranging from 12 inches for model making to 36 inches or more for industrial use. A saw with a swing of 18 inches or 24 inches (it will cut stock this wide) is best for all-around use.

Limitations. It is important to know what *not* to cut on a jigsaw.

1. The jigsaw is not designed for cutting straight lines. Use a handsaw or circular saw.

2. It is not made to cut heavy stock, like 2 by 4's. Let one-inch hardwood be the maximum.

3. It cannot do every job that a coping saw can. You will still need to use a coping saw now and then.

4. It does not cut as fast as a band saw, so take your time. *The longer you can saw without breaking a blade, the more expert you are.*

Cutting Speeds. The saw cuts best when run at the proper speed for the material. If the speed is too slow, you will tend to force the cutting. If it is too fast, you will not keep up with it. The fastest speeds are used on woods and the slowest on metals. Your teacher will show you how to set the speed according to the material used.

Blades. Jigsaw blades are usually 5 inches in length and are available in various combinations of thickness, widths, and t.p.i. (teeth per inch). The blade should be selected for the kind of sawing to be done. The thinner the material being sawed, the greater the number of t.p.i. The sharper the curves, the narrower the blade. Sheet metal requires a blade with finer teeth than does wood. For your usual sawing in wood, a blade about 0.110-inches wide, 0.020-inches thick, and with 10 t.p.i. is recommended. This blade can be used

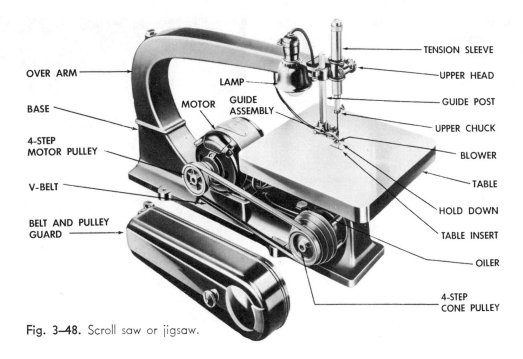

Fig. 3–48. Scroll saw or jigsaw.

OVER ARM

BASE

4-STEP
MOTOR PULLEY

V-BELT

BELT AND PULLEY
GUARD

MOTOR

LAMP

GUIDE
ASSEMBLY

TENSION SLEEVE

UPPER HEAD

GUIDE POST

UPPER CHUCK

BLOWER

TABLE

HOLD DOWN

TABLE INSERT

OILER

4-STEP
CONE PULLEY

on plastics. For sheet metals, use blades with 20 to 30 t.p.i.

How to Install a Blade

1. Remove the table insert and turn the motor by hand to get the vise at the bottom of the stroke.

2. Insert the blade, teeth pointing down, in the lower vise. Align the blade in the guide and tighten the vise. Then clamp the top end securely. Do not use a wrench or pliers unless the vises are designed for them. There should be not less than ½-inch of blade in a vise.

3. Turn the motor slowly by hand to see that the blade is in correctly. If it bends, the upper vise must be moved down the blade or the upper housing raised slightly to add tension. Replace the table insert. Have your teacher check the blade before you turn the motor on.

Order of Sawing. Sometimes you cannot saw completely around a pattern without stopping. This is because of extra sharp curves and abrupt changes in direction. These make several separate

Fig. 3–49. For very sharp curves with the jigsaw, make relief cuts first. These are the cuts made from the edge to the bottom of the curves.

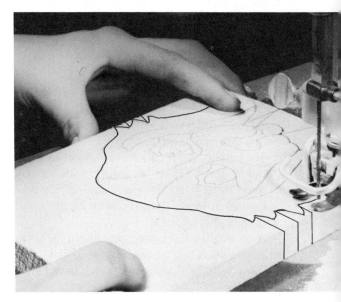

cuts necessary. Plan the order of cutting before you start sawing so that you won't have to do much backing out.

How to Saw

1. Check for the proper blade and speed.
2. Adjust the hold-down fingers to press the work lightly against the table.
3. Guiding the work with both hands, push it just fast enough to keep it cutting. When making a turn, keep the blade cutting or it will twist and break.
4. To cut out holes, drill a ¼-inch hole near the line in the waste stock and install the blade through it.

The Band Saw

The band saw is a versatile machine. It has an endless blade around two wheels like a flat belt around two pulleys. It cuts curves and straight lines in thin and thick materials such as woods, plastics, and metals. It cuts faster but not as finely as does a jigsaw. There are two common types, the wood cutting and the metal cutting. The size is measured in inches from the blade to the frame across the table. (See Fig. 3–51.)

A 14-inch band saw is recommended for home and school use. Contrast this with band saws used in lumber mills for ripping logs, which have blades a foot wide.

Fig. 3–50. To cut out a hole, the jigsaw blade is installed through a small drilled hole.

Limitations. The limitations are in the nature of overloading:

1. Wide blades cannot make as sharp turns as can narrow blades. A ¼-inch blade will cut curves of 1-inch radius and larger. Curves that are too sharp bind the blade.
2. Heavy work can overload the blade. The saw may be able to accommodate a piece of balsa wood 6 inches thick but not be able to cut a thickness of more than 3 inches of hard maple.

Blades. Blades are purchased to fit the saw and the total length is specified. For a 14-inch saw, blades from ⅛ to ½ inch wide are available for cutting wood, or metal, or plastic with the appropriate tooth style.

How to Use a Band Saw

Before using the band saw for the first time have your teacher "check you out." Each time you use

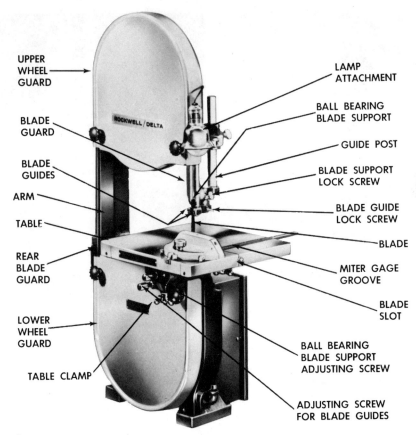

UPPER WHEEL GUARD

BLADE GUARD

BLADE GUIDES

ARM

TABLE

REAR BLADE GUARD

LOWER WHEEL GUARD

TABLE CLAMP

LAMP ATTACHMENT

BALL BEARING BLADE SUPPORT

GUIDE POST

BLADE SUPPORT LOCK SCREW

BLADE GUIDE LOCK SCREW

BLADE

MITER GAGE GROOVE

BLADE SLOT

BALL BEARING BLADE SUPPORT ADJUSTING SCREW

ADJUSTING SCREW FOR BLADE GUIDES

Fig. 3–51. Band saw.

it make sure that all adjustments have been made beforehand.

1. Set the top blade guide so that it just clears the work.

2. Turn on the switch and, holding the work with both hands, feed it slowly into the teeth.

3. Cut on the outside of the line, using only enough pressure to keep the blade cutting. The blade must be cutting as you follow a curve or it will bind and break.

4. Follow a plan for cutting as described for the jigsaw. You can pull the blade off the wheels backing out of tight places.

✚ SAFETY SENSE

A band saw can be hard on fingers. Watch where you hold it.

1. If the blade is dull, don't use it. It will burn and break.

2. Keep the light focused on the spot where the blade is cutting.

3. Keep your fingers to the side of the blade, never in front of it.

4. Make no adjustments on the machine while it is running.

5. When you are finished, shut off the motor

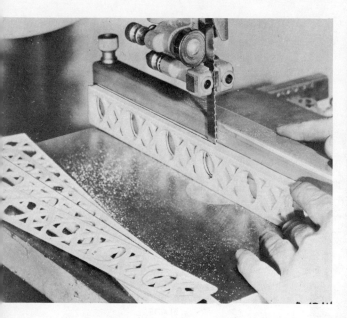

Fig. 3–52. The band saw slices wood. Note the use of the fence for accurate cutting.

Fig. 3–53. The band saw cuts curves in heavy stock.

and, after the saw stops, drop the blade guide down close to the table.

6. To watch someone operate the saw, stand behind him and to the right. Do not speak to him while he is sawing.

7. If the blade breaks or comes off, stop the motor.

The Drill Press

Originally the drill press was for drilling holes. With the many attachments now available, it is also used for boring, mortising, routing, shaping, sanding, grinding, and buffing. Drill presses are made in a wide range of sizes to suit a wide variety of jobs. The *single spindle type* (it drills one hole at a time) may be small enough to drill holes so tiny you can hardly see them or large enough to take locomotive parts. The *multiple spindle type* drills two or more holes at once and is an industrial production machine.

Limitations. There are limitations on the use of a drill press. These are determined by the drill chuck, the size and shape of the work that can be accommodated, and the drill bit itself. A different chuck is required to hold a No. 80 drill bit from that required for a ½-inch size. Highspeed bits should be used for drilling metals.

Drill Speeds. Different size drill bits require different speeds. Different materials being drilled may require different speeds. See your teacher's chart on drill speeds. He will show you how to change speeds on your drill press.

Lubricants. It is necessary to use lubricants to protect the drill bit when drilling certain materials. Cutting produces heat. If this heat is not carried away, the bit may get hot enough to turn blue. This means that it has lost its hardness and will no longer hold an edge. On hard steel, use kerosene; on mild steel, cutting oil. Cast iron, aluminum, copper, plastics, and wood should be drilled dry.

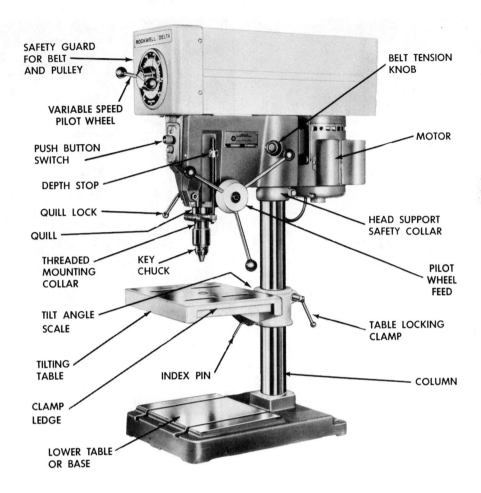

SAFETY GUARD
FOR BELT
AND PULLEY

VARIABLE SPEED
PILOT WHEEL

PUSH BUTTON
SWITCH

DEPTH STOP

QUILL LOCK

QUILL

THREADED
MOUNTING
COLLAR

KEY
CHUCK

TILT ANGLE
SCALE

TILTING
TABLE

INDEX PIN

CLAMP
LEDGE

LOWER TABLE
OR BASE

BELT TENSION
KNOB

MOTOR

HEAD SUPPORT
SAFETY COLLAR

PILOT
WHEEL
FEED

TABLE LOCKING
CLAMP

COLUMN

Fig. 3–54. Drill press. This is a bench or table model.

How to drill

These are general suggestions for drilling all materials:

1. Insert the drill bit and tighten the chuck with the key. Adjust the belt for the proper speed.

2. When possible, clamp the work to the drill press table. Use a V block to hold round stock. Always clamp metals. Put a piece of scrap wood under the work whenever possible to prevent drilling into the table.

3. Metals must be center punched for drilling.

4. Raise the table as near to the drill bit as possible and lock it in position with its center hole directly under the bit.

5. Turn on the motor and start the cut with only slight pressure until you are sure the location is correct.

6. If a lubricant is needed, keep it dripping slowly on the bit as it cuts. Use only enough pressure to keep the bit cutting. When drilling wood,

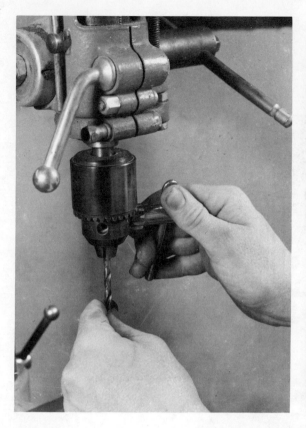

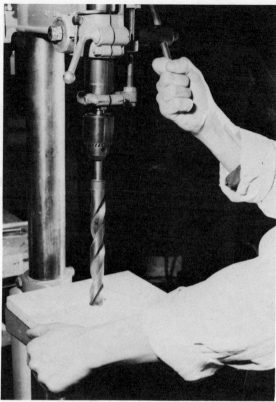

Fig. 3–55. Inserting a straight shank drill bit into the chuck. Always use the chuck key to tighten the chuck.

Fig. 3–56. Drilling a hole with the drill press. Small pieces of wood should be held with a clamp or in a drill press vise.

raise the bit from the hole occasionally to clear out the chips.

7. Ease up on the pressure as the bit cuts through the other side. Raise the bit and shut off the motor.

✚ SAFETY SENSE

The drill press is one of the safest of shop machines. Follow these suggestions to keep it that way.

1. The work must always be clamped or other- wise held securely, so it will not be jerked away from you.

2. The belts must be guarded to protect fingers and hair.

3. Use an eye shield to guard against flying chips of metal.

4. Be sure you get the necessary instruction be- fore you use any attachment on the drill press.

Drilling Troubles. The following will help you troubleshoot drilling problems:

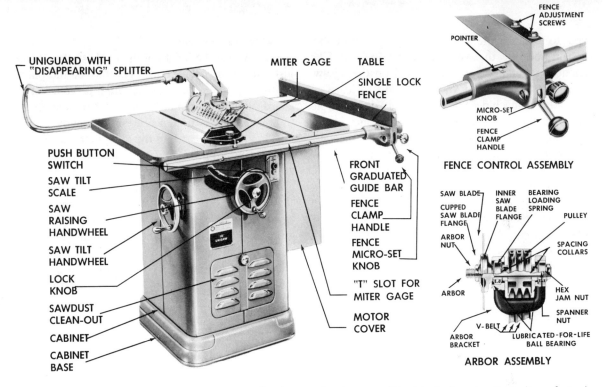

Fig. 3–57. A school-type circular saw. The blade is a toothed disc of steel.

Trouble	Remedy
Drill gets too hot.	Use a sharp drill.
	Use a lubricant (when drilling steel).
	Clean chips from the drill bit.
	Use a faster feed to cut faster.
Drill doesn't cut.	Sharpen the drill.
	Use a slower speed.
Drill squeals.	Use a sharp drill.
	Keep the drill cutting.
	Use a lubricant (on steel only).
Drill breaks.	Use less pressure on the bit.
	Hold the work steady.

The Circular Saw

The circular saw, also known as a bench or table saw, is a most useful machine for woodworking. It deserves considerable respect. Anyone who takes lightly the common sense rules for its use is likely to get hurt. If your teacher does not permit

Fig. 3–58. Adjust the height of the saw blade so that it will just cut through the material. Do this when making the saw setup before turning on the motor.

you to operate the circular saw, it is probably because he feels that you are not ready for it. The chances for injury are greater than on the jigsaw. When he decides that you are ready to learn to operate this machine, he will probably demonstrate crosscutting and ripping first. The following is general information to acquaint you with the saw.

About the Circular Saw. The circular saw has a round, flat blade with teeth cut into the rim. The common types of blades are the *rip,* the *crosscut,* and the *combination.* The blade turns at high speed and the work is fed into it only as fast as the blade will cut. A guard is kept over the blade. In some states this is required by law. A movable *fence* on the table is the guide for ripping. A *miter gauge,* which slides back and forth in a groove, is for crosscutting. The size of a circular saw is the diameter of the blade. An 8-inch or 10-inch saw is a favorite for home and school shops.

Fig. 3–59. When crosscutting, use the cut-off attachment to hold the stock at the proper place.

How to crosscut

1. Adjust the height of the blade so that it will just cut through the board. The farther the blade protrudes, the greater the invitation to injury. The guard should be in the proper position covering the blade. Move the fence away, or remove it.

2. Lay the board on the table with its straight edge against the miter gauge and align the mark at which it will be cut with the blade. A blade usually removes about ⅛ inch of wood.

3. Turn on the motor and hold the board against the gauge. Push the miter gauge along the groove, slowly feeding the board into the saw. Stand behind the miter gauge, not in line with the blade. Push the board past the saw until it comes out from under the guard. Then the guard drops

Fig. 3–60. Ripping a piece. The stock is held against the fence. Note the anti-kickback device. Its teeth keep the saw from pushing the work toward you. The guard is removed to show these details. The scrap piece is not held.

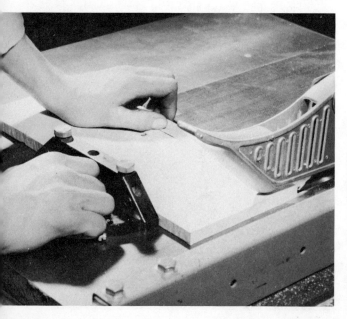

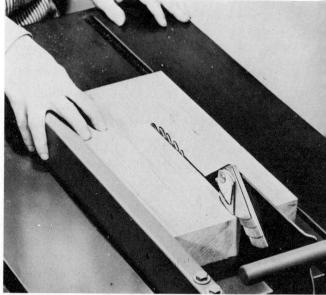

down on the table. Shut off the motor but *do not pick up the end sawed off until the blade has stopped.*

How to rip

1. Remove the miter gauge and hang it up. Each time this falls to the floor some of its accuracy is lost. Set the blade as before and check the position of the guard.

2. Move the fence into position, using the scale on the front guide bar. Then with a rule check the distance between the fence and the closest saw teeth. Lock the fence.

3. Start the motor and, holding the board against the fence, push it slowly and steadily into the blade. Hold only the piece between the fence and the blade. When this piece is 3 or 4 inches wide, use one hand to hold it. When narrower, use a push stick.

4. Push the board through until it is free of the guard. Turn off the motor, and if the pieces are

Fig. 3–61. Use a push stick when ripping narrow pieces.

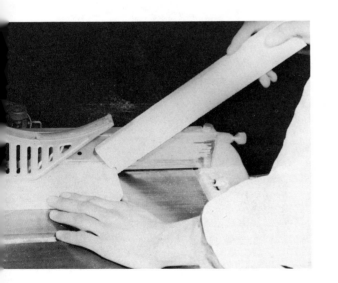

on the table do not pick them up until the blade has stopped.

✚ SAFETY SENSE

1. Always have your teacher check your saw setup before you turn on the motor, until he tells you that you can do it on your own. Machine adjustments are always made before the motor is started.

2. If the blade is dull, don't use it. Stand to one side of the blade when sawing.

3. Do not try to crosscut pieces that are too short to be held securely against the miter gauge. Do not rip pieces less than 6 inches long. Short, thick stock is easily jerked out of your hands.

4. Make sure that you have mastered ripping and crosscutting before trying other cuts, even though you have a saw at home.

5. No one other than the instructor should be standing near you as you saw. No one should talk to you, nor should you talk to anyone.

6. Know where the blade is and where your fingers are at all times.

The Jointer

The jointer trues edges, planes faces and tapers, and bevels. It cuts very smooth rabbets. (See Fig. 3–66.) The knives are sharpened like plane irons and revolve at very high speed. When the stock is held accurately, the cuts are true. This machine is especially good for jointing edges for gluing.

Limitations. The width of the cutter limits the width of the stock that can be surfaced. A six-inch jointer, for example, takes boards as wide as six inches. The depth of a single cut should be limited. You will probably get a true surface with a light cut. Let $\frac{1}{16}''$ be the maximum depth. If you have much wood to remove, it may be faster to use the circular saw.

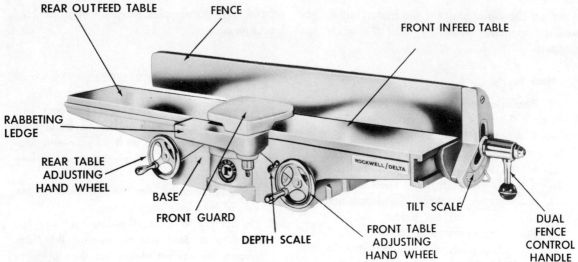

REAR OUTFEED TABLE   FENCE   FRONT INFEED TABLE

RABBETING LEDGE

REAR TABLE ADJUSTING HAND WHEEL

BASE

FRONT GUARD

DEPTH SCALE

FRONT TABLE ADJUSTING HAND WHEEL

TILT SCALE

DUAL FENCE CONTROL HANDLE

ROCKWELL/DELTA

Fig. 3–62. A jointer and its principal parts.

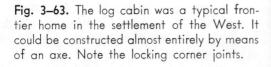

BEARING HOUSING

BALL BEARING

KNIFE

KNIFE LOCK BAR

KNIFE

HEX HEAD SET SCREW

BALL BEARING

SHAFT

How to joint an edge

1. Set the depth of cut for no more than $\frac{1}{16}''$.

2. Check the fence for squareness with the table.

3. See that the guard is in the proper position.

4. Stand squarely on both feet so that you are well balanced.

Fig. 3–63. The log cabin was a typical frontier home in the settlement of the West. It could be constructed almost entirely by means of an axe. Note the locking corner joints.

5. Hold the stock firmly on the table and against the fence. Push it slowly over the cutter and past the guard so that the guard can close.

6. One cut may true an edge. See if the entire surface was planed.

7. The end grain may be jointed. Be sure the stock is at least four or five times as wide as the throat opening. This splinters the edge, so do it before the edge is jointed.

How to plane a face
1. Set the table for a light cut, $\frac{1}{32}''$ or less.
2. Position the fence to the width of the stock.

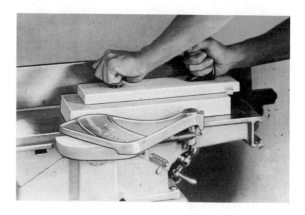

Fig. 3–65. Planing a face. Note the use of a push stick to keep the fingers away from the cutter.

Fig. 3–66. Rabbeting on the jointer. During the cutting, the stock becomes the cutter guard. *You must have permission from your instructor to make this cut.*

Fig. 3–64. Jointing an edge on the jointer. Notice that the young man has a good stance. He is not likely to get off balance as he pushes the board through. Note, too, his safety glasses.

3. The cut should be made with the grain.

4. With the push stick, slowly pass the stock across the cutter until the guard closes.

How to rabbet

1. Adjust the guard for the necessary clearance.

2. Move the fence toward the rabbeting ledge. Set it for the width of the rabbet. Set the depth of cut for the full depth of the rabbet.

3. Hold the stock firmly on the ledge and move it slowly over the cutter.

For beveling and chamfering the fence is tilted. Then the stock is held as for jointing.

✚ SAFETY SENSE

1. Do not use pieces shorter than 12 inches without the approval of your instructor.

2. Use a push stick whenever possible.

3. Take light cuts. They are easier, truer, and safer.

4. Stand so that you cannot lose your balance when reaching across the cutter.

5. Always keep track of your fingers. Keep them clear of the cutter unless you want them planed!

The Wood Lathe

The wood lathe is a machine on which wood is shaped into round and cylindrical forms, such as bowls and legs. This is done by means of a tool held and manipulated by the operator. Production lathes used in furniture industries are automatic. The operator pulls a lever that moves a cutter into the wood, thus making the entire cut at once. Sizes of wood lathes are given as *swing,* and as the *distance between centers.* On a lathe with a swing of 11 inches you can turn a piece as large as 11 inches in diameter. Shaping on the wood lathe is called *turning.*

Limitations. The wood lathe turns round shapes only, since the wood revolves against a tool. For large-diameter, heavy turning a heavy duty lathe is needed to control vibration.

Turning Tools. Wood-turning tools are shaped to make particular types of cuts. A dull tool is useless. It makes dust instead of shavings. Tools must be sharpened frequently if they are much used, so you should learn how. When a tool is burned, nicked, or blunted by honing, it must be ground. Between grindings the tools are honed to keep the edges sharp. A *slip stone* and an *oil stone* are used. The sharper you keep your tools, the better turning you can do. Your teacher can show you how to keep your tools sharp.

Lathe Speed. As a rule, the faster the wood turns, the smoother the cut. Certain speed limits must be observed, however. If the wood is not perfectly centered and balanced, too much speed may cause it to fly out of the lathe. The larger the diameter of the piece, the slower it should revolve. A long piece such as a baseball bat billet should revolve more slowly than a short piece of the same thickness.

APPROXIMATE TURNING SPEEDS FOR FACE PLATE WORK (R.P.M.)

Diameter of work	Roughing	Shaping	Sanding	Finishing
3"–4"	500–1000	1000–1500	1500–2000	2000
5"–6"	400–800	800–1200	1200–1500	1500
8"–10"	200–400	300–500	500–700	700

FOR SPINDLE TURNING

Diameter of work	Roughing	Shaping	Sanding	Finishing
1"–2"	500–1000	1000–2000	2000–2500	2500
3"–4"	400–800	800–1200	1200–1500	1500
5"–6"	300–500	500–800	800–1200	1200

If your lathe has but four speeds, do the roughing at the slowest, and then increase the speed gradually through the stages for shaping, sanding, and finishing.

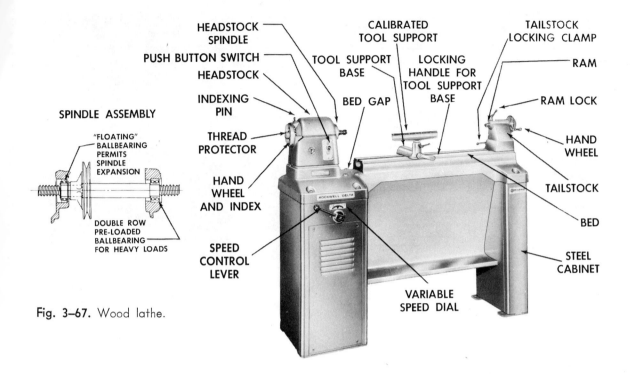

Fig. 3–67. Wood lathe.

Labels in figure:

- HEADSTOCK SPINDLE
- PUSH BUTTON SWITCH
- HEADSTOCK
- INDEXING PIN
- THREAD PROTECTOR
- HAND WHEEL AND INDEX
- SPEED CONTROL LEVER
- SPINDLE ASSEMBLY
- "FLOATING" BALLBEARING PERMITS SPINDLE EXPANSION
- DOUBLE ROW PRE-LOADED BALLBEARING FOR HEAVY LOADS
- TOOL SUPPORT BASE
- BED GAP
- CALIBRATED TOOL SUPPORT
- LOCKING HANDLE FOR TOOL SUPPORT BASE
- VARIABLE SPEED DIAL
- TAILSTOCK LOCKING CLAMP
- RAM
- RAM LOCK
- HAND WHEEL
- TAILSTOCK
- BED
- STEEL CABINET

✚ SAFETY SENSE

Wood turning is fun; don't spoil it by being careless.

1. Roll up your sleeves and remove your tie if there is any chance of its getting caught on the work. (This saves wear and tear on the chin.)

2. Always make the setup and any adjustments in it before starting the motor.

3. Stop the motor and check the work frequently to see that it has not come loose. If you hear any strange noises, stop the lathe. The work may be loose.

4. Try to hold the turning tools so that the chips strike your fingers instead of your face. Always wear a face shield or safety glasses.

5. If dust bothers you, use a respirator. Remember, a sharp tool makes little dust.

6. Always remove the tool rest before sanding.

7. When you leave the machine, turn off the motor.

Lathe Turning. There are two types of turning on the lathe: *face plate* and *spindle*. Face-plate work is best to start on, so let's try to make a bowl.

1. Get a piece of hardwood, like cherry or maple, from which a disc about 2 inches thick and 6 inches in diameter can be cut. Plane one side flat. This is the side next to the face plate. Cut out the disc on the band saw.

2. Make a full-size drawing of the shape you plan to turn. From this, make a template for the outside and for the inside.

3. Saw out a disc of scrap wood about 1 inch thick and ⅛ inch smaller in diameter than the foot of the bowl. Cut a disc of wrapping paper the same size. Now glue this block to the center of the bowl stock with the paper in between. Clamp them and let them dry.

4. Screw the small disc to the face plate. Do not screw into the bowl stock. Screw the face plate on the spindle.

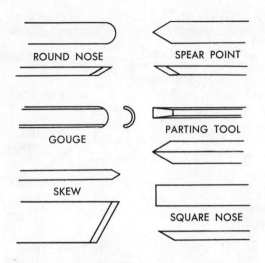

ROUND NOSE

SPEAR POINT

GOUGE

PARTING TOOL

SKEW

SQUARE NOSE

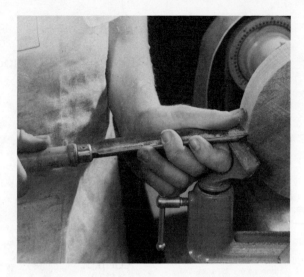

Fig. 3–68. (top) Set of wood-turning tools.
Fig. 3–69. (bottom) Fastening the face plate to the block.

Fig. 3–70. Rough turning the bowl to true it, with a gouge.

5. Set the proper lathe speed. Adjust the tool rest parallel to the edge of the block at the center line. The block should just clear when turned by hand. Screw the tail center against the block.

6. Turn on the motor and hold a gouge firmly in both hands, on the tool rest square with the stock. Slide it back and forth, using a forefinger as a guide. Make thin cuts until the block is round. Move the tool rest when necessary to keep it close to the work.

7. Increase the lathe speed one step now for the shaping. Use a round nose tool. Move it from side to side as you peel off the shavings. Check the shape with the template now and then, with the lathe stopped.

8. Slide the tool rest out of the way and rough-sand the outside. Use a piece of 1-0 garnet paper on a felt pad. Keep the sandpaper moving from side to side below the center line. Follow with successively finer paper until the surface is free from all scratches.

9. To hollow the bowl, stop the lathe. Set the tool rest across the front of the block on the center

Fig. 3–71. Hollowing the bowl with the round-nose tool. Keep the tool rest close to the work.

Fig. 3–72. Sanding the bowl as it spins. The tool rest has been removed.

line and about ¼ inch away. Use the round nose tool again. Slide it back and forth from the left edge to the center. If you go beyond the center the wood tries to pick up the tool and hand it to you. Use the inside template frequently.

10. Sand the inside as you did the outside.

Finishes. Use a finish that adds to the natural beauty of the wood. Try one of the following:

1. For a wax finish, which is the simplest of all, use a piece of soft cloth folded into a small pad. Put paste polishing-wax inside the cloth and apply it to the revolving bowl. Polish each coat before applying the next. The more coats, the better, until there are about a dozen.

A piece of beeswax may be held against the bowl. The heat of friction melts the wax. When thoroughly covered, press a pad against it to force it into the wood. Repeat if desired.

2. The French polish is an old-time cabinet-maker's finish. On a pad of cloth, lay about a teaspoonful of white shellac. On this place four or five drops of boiled linseed oil. Hold the pad against the spinning bowl. Cover the surface, using

moderate pressure. When the pad gets dry, add more shellac and oil. Repeat until you have a mirrorlike finish. Protect the finish with two coats of paste polishing wax.

3. For a waterproof finish use clear lacquer. Remove the face plate and apply 10 to 12 coats of lacquer. When it is dry, place it on the lathe again. Get a thick cloth pad and some lacquer polishing compound. Put a little of the compound on the pad and hold it lightly against the spinning bowl.

4. Salad bowls and trays may be soaked in warm cooking oil. They should be rubbed dry when saturated to prevent gumminess.

Spindle Turning. Spindle turning is done between centers. It enables you to make lamp bases, table legs, ball bats, and the like. Here's how to turn a ball bat:

1. Make a drawing of the bat with the necessary dimensions.

2. For the stock, use straight-grained ash. This can be purchased in billet form for bats.

3. With a backsaw make shallow cuts across

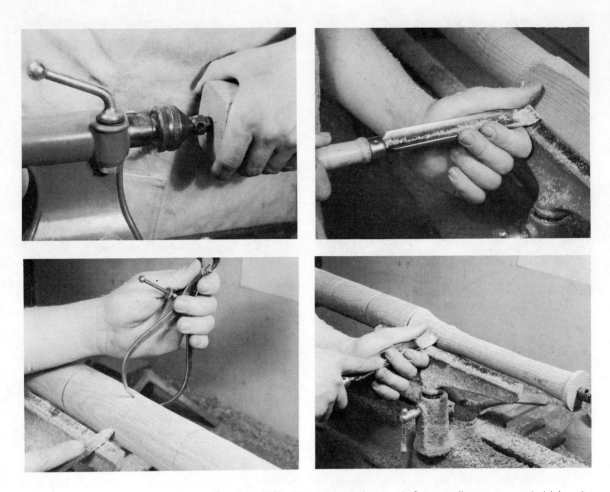

Fig. 3–73. (top left) One end of the stock for spindle turning is held by the tail center. **Fig. 3–74.** (top right) Rough turn the piece using a heavy gouge. **Fig. 3–75.** (bottom left) To turn to a particular diameter, use the outside caliper for measuring. Cut to the diameter at one point with the parting tool. **Fig. 3–76.** (bottom right) Shaping with a gouge.

corners on each end of the billet. This marks the location of the lathe centers. Tap the headstock center into one end with a mallet. Insert this in the headstock. Run the tail center into the other end and draw it up snug. Turn the stock a few times by hand to see that it is secure.

4. Set the tool rest to within ½ inch of the stock, on the center line and parallel to it. Use the slowest roughing speed. Have your teacher check the setup for accuracy.

5. Rough-turn a section at a time with the gouge, using it as you did for the bowl. When the

Fig. 3–77. (left) This machine turns out a bat in seconds. It is first turned round from a square billet and then shaped. **Fig. 3–78.** (right) Branding a finished ball bat. Care is taken to put the trademark in the proper place. This helps the batter in handling the bat correctly.

stock is round, mark out the stations on it according to your drawing. Set a pair of calipers to the diameter of the first station, plus $\frac{1}{16}$ inch. Cut into the diameter with a parting tool. Do this at each station. Now cut away the waste stock with a gouge. Use the round nose tool to shape the handle.

6. Sand the bat smooth and burn in your monogram for the trademark on a flat-grain side. Saw off the stubs and apply a hot oil finish.

7. When it is dry, try it for a home run like the big-leaguers.

TREES, WOOD, AND OCCUPATIONS

The nature of occupations within the field of trees and woods is changing just as in other fields. One reason is the changing uses of the materials. For example, hardwoods have been used chiefly in furniture for many years. It is now predicted that by 1980, 85 percent of the furniture will be made of plastic. But at the same time, the plastic itself will be made from wood. The nature of the occupations involved will of course change. The typical occupations in trees and woods can be seen in the following chart. Which of these require no training? Which require special training? Which require college?

TYPES OF OCCUPATIONS

Forestry-tree culture
Nurserymen
Foresters
Rangers
Tree farmers
Botanists
Landscapers
Pest Control
Park managers
Forest recreation
    leaders
Environment
    specialists

Lumber manufacture *
Loggers
Lumber jacks
Saw mill operators
Heavy equipment
    operators
Machine operators
Mechanics
Machinists
Welders

Wood Products Manufacture *

Produce designers	Contractors
Cabinet makers	Architects
Carpenters, finish	Draftsmen
Carpenters, rough	Machinists
Engineers	Mechanics
Wood finishers	Welders
Painters	Electricians

Chemicals
Chemists
Chemical engineers
Chemical manufacturing
Adhesives manufac-
    turing
Plastics manufacturing
Finishes, manufacturing
Preservatives,
    manufacturing
Machine operators
Plumbers
Mechanics
Welders
Machinists

Management
Business administration
Industrial engineering
Production planning
Material-control
Sales and marketing
Safety engineers
Inspection
Maintenance
Training

Professional, Private, Occupations
Custom furniture making
Jewelry design, making
Toy design, making
Wood carver
Industrial arts teacher
Trades and industries teacher
Boat making
Architectural model making

* See chart, Typical Products of American Industries, p. 73.

TREES AND ENVIRONMENT

Trees are one of Nature's most valuable contributions to man. They convert carbon dioxide into oxygen, which is essential to human life. Have you ever noticed how invigorating the air is in a deep woods? Trees conserve moisture in the earth by holding water among their roots. They provide shade and humidity in summer to keep temperatures cooler. A tree-shaded house is cooler in summer. With trees for a windbreak a house is warmer in winter. Trees provide a beauty with their forms, color, blossoms, fruit, and fragrance. They add value to real estate and provide an environment fit for humans to live in. Woods and forests are homes for wildlife. We hope that during your course in industrial arts you will be able to plant a tree. In later years its growth will be a sense of pride and satisfaction to you. The County Agriculture Office and State Forester's Office can provide instructions for tree selection and planting.

Trees have enemies. These are usually people, although Nature's bugs and blights also take their toll. As earlier mentioned, most forest fires are caused by people. Noxious gases and dust and foreign chemicals in water are fatal to some trees. Unless we can reduce this pollution it will be necessary to find pollution-resistant varieties. Recently a cherry tree growing in Turkey has been found to be pollution resistant. A chestnut tree from China was found to resist the blight that destroyed the American chestnut. The Dutch elm disease is expected to make the beautiful American elm extinct. American black walnut is fast disappearing due to overcutting and underplanting. Trees aid in our fight against pollution. The leaves serve as dust catchers. In some cities they must be washed down with detergents to remove the collected dust. Right now we need to ask ourselves if we really love trees enough to assure their survival.

TREES, WOOD, AND RECREATION

A great variety of hobby activities can be found in this area of trees and woods. Some require tools. Others need a home workshop. Some require no special equipment. Each of the following suggestions is a source of many other ideas. For example, wood turning can include several individual hobbies. One could specialize in bowls, trays, lamps, candleholders, salt cellars, music boxes, and such. Try opening up the other suggestions for ideas.

Trees

Growing seedlings from seeds
Planting seedlings and transplanting young trees. (Do you know the story of Johnny Appleseed?)
Tree identification

Woods

General woodworking
Wood carving
Wood turning
Toys
Tools, utensils
Furniture
Woods identification, collection

Furniture finishing, refinishing, repairing
Antique restoration
Sports equipment: skiis, tennis racquets
Bric-a-brac: corner shelves, containers, decorative tiles
Wood clocks
Lamps from old weatherbeaten wood
Book and album covers

TO EXPLORE AND DISCOVER

1. What a carpenter means by "toe nailing."
2. How a tree grows. What is the oldest tree in your neighborhood?
3. What the difference is between plywood and veneer.
4. Which wood is the heaviest of all? The lightest?
5. What a mortise and tenon joint is. Find examples.
6. How many different kinds of trees there are in the United States.
7. How fast forest fires travel. How they are fought—and how they can be prevented.
8. The functions of the United States Forest Service.
9. Why wood insulates well against heat and cold.
10. Why wood splits more easily with the grain than against it.
11. Visit an upholsterer's shop to watch him cover a piece of furniture.
12. Plan what tools and machines for woodworking you would want in your workshop. Which should you get first?
13. Make a list of words that are part of the language of the woodworker.
14. Find out some of the plastics that are made from trees.
15. Learn to identify several common woods.
16. Find out what products are made from sawdust.
17. Make a collection of various types and sizes of nails. Identify them and display them attractively.
18. Find out why the American chestnut tree became extinct.
19. Suggest what can be done to preserve the American elm from extinction.
20. Explain why trees are important to our environment.

FOR GROUP ACTION

1. Invite a forest ranger, furniture designer, architect, chemist, or other professional in the field of woods to talk to your class.
2. Visit a tree farm or state forest to see the cultivation of trees; then plant some seedlings on the school grounds, in the park, or at home.
3. Visit a museum to see tools and wood products made and used long ago.
4. Arrange an exhibit of wood products available today. Your furniture dealer may be glad to have you put it in his store window.
5. Visit a house under construction to see how lumber is used.
6. Prepare a display showing how many people are employed in the various industries directly dependent on forest materials.
7. Build several bird feeders to be placed outside the windows of the elementary grades classrooms, on the school grounds, or in the park.
8. Draw up a set of safety rules for each machine used in your woodworking laboratory.
9. Repair or build some furniture or toys for an underprivileged family in your community.

10. Set up a wood products manufacturing plant in your school shop to mass produce a quantity of items needed by the Red Cross. Contact your local Chapter.

FOR RESEARCH AND DEVELOPMENT

1. Compare chair designs from the earliest to the present. Use photos and drawings. Make the comparison to include styles and materials from as many countries as possible. This can be an excellent instructional aid to be left with your teacher.
2. Make a study of the architecture of houses in your community. Find the oldest house. Get such information as: date of construction, name of original owner, builder, name of present owner. Photograph the house and its details. Make drawings to show structures. Arrange the materials on a display board. Repeat this study for other houses.
3. What wooden tools and utensils were used in the early American home? Get photos and drawings. Which of these could you reproduce? Try it.
4. Make an occupational survey in your community. What industries depend on or use wood in their production? What occupations are involved. The local Chamber of Commerce may be of assistance in this. Find out the following: the opportunities, qualifications, wages and salaries, fringe benefits, hazards.
5. Find out how wood is laminated from veneer. Make a test piece ½ inch thick by 1 inch wide and 12 inches long. Make a solid piece of the same kind of wood and dimension. Compare the breaking strengths by hanging weights at their centers. When would laminated wood be preferred over solid? When the reverse?

FOR MORE IDEAS AND INFORMATION

Books

1. Aller, Doris. *Wood Carving Book*. Menlo Park, Calif.: Lane Books, 1966.
2. Capron, J. Hugh. *Wood Laminating*. Bloomington, Ill.: McKnight & McKnight Publishing Co., 1963.
3. Coleman, D. G. *Woodworking Factbook*. New York: Robert Speller and Sons, 1966.
4. Edlin, H. L. *What Wood Is That?* New York: The Viking Press, 1969.
5. Feirer, John L. *Industrial Arts Woodworking*. Peoria, Ill.: Charles A. Bennett Co., Inc., 1965.
6. Fryklund, Verne C., and Armand J. LaBerge. *General Shop Woodworking*. Bloomington, Ill.: McKnight & McKnight Publishing Co., 1965.
7. Groneman, Chris H. *General Woodworking*. New York: McGraw-Hill Book Co., 1965.
8. Olson, Delmar W. *Woods and Woodworking for Industrial Arts*. Englewood Cliffs, N. J.: Prentice-Hall, Inc., 1965.

9. *The Use of Hand Woodworking Tools.* New York: Delmar Publishers, Inc., 1962.
10. Wagner, Willis H. *Woodworking.* S. Holland, Ill.: Goodheart-Willcox Co., Inc., 1964.
11. Wolansky, W. D. *Woodworking Fundamentals.* New York: McGraw-Hill Book Co., 1964.

Booklets

1. *ABC's of Hand Tools.* General Motors Corp., Public Relations Staff, 3044 W. Grand Blvd., Detroit, Mich., 48202
2. *Fine Furniture Woods and Their Care.* The Fine Hardwoods Association. 666 Lakeshore Drive, Chicago, Ill. 60611
3. *Hints to the Handyman* (Form 6544-2A); and *Masonite Hardboards for Home, Business, and Industry* (Form 6326). Masonite Corp., 29 N. Wacker Dr., Chicago, Ill.: 60606
4. *Plywood Construction Guide; How to Finish Plywood; How to Work Plywood.* American Plywood Association, 1119 A St., Tacoma, Wash. 98401
5. *Redwood—Properties and Uses.* California Redwood Association, 617 Montgomery St., San Francisco, Calif. 94111
6. *Information About Our Forests and Products.* Western Wood Products Association, Yeon Bldg., Portland, Oregon 97204. A fine information packet available to teachers. It contains pamphlets, charts, film sources.
7. *Lumber and Wood Products Literature.* Bibliography. National Forest Products Association, 1619 Massachusetts Ave., N. W., Washington, D. C. 20036
8. *School Packet.* American Forest Institute, 1619 Massachusetts Ave., N. W., Washington, D. C. 20036. A collection of interesting, helpful booklets, maps, and charts, as well as a bibliography of other free materials.

Charts

1. *Forests and Trees of the United States.*
2. *Growth of a Tree.* American Forest Institute, 1835 K St., N. W., Washington, D. C., 20006
3. *How a Tree Grows.*
4. *Products of American Forests.*
5. *What We Get from Trees.* Division of Information, Office of Management Services, U. S. Dept. of Agriculture, Washington, D. C. 20250

Films

1. *ABC's of Hand Tools,* see under Booklets.
2. *The Forever Living Forests.* Color and sound, 27 min., 16 mm. California Redwood Association, 617 Montgomery St., San Francisco, Calif. 94111
3. *Free Educational Films from Industry.* (catalog) See *Wonderful World of Hard Woods* and others. Modern Talking Picture Service, 1212 Avenue of the Americas, New York, N.Y. 10036

GLOSSARY—WOODS

Abrasives hard cutting particles usually bonded to cloth or paper.

Adhesives bonding agents, as glue.

Aluminum oxide a manufactured abrasive.

Annual ring the annual growth in the diameter of a tree.

Auger bit boring tool for wood.

Back saw a handsaw with a stiff back.

Band saw a machine tool using a continuous band blade.

Bevel an angular cut other than a right angle.

Bit a boring or drilling tool.

Bore to make a circular hole in wood with an auger bit.

Brace hand tool for holding and turning auger bits.

Brad a small headed nail, usually in short lengths.

Butt the square end of a board.

Cambium layer the seasonal new growth of a tree just under the bark.

Chamfer the flat surface made when a square edge or corner is planed off.

Chisel a long slender hand tool for cutting wood with the aid of a mallet.

Circular saw a mechanical saw using a disc blade.

Countersink to make a conical hole for the head of a wood screw; the tool for making the cut.

Crosscut to saw across the grain of a board.

Dado a "u" shaped cut across the grain of a board into which a butt is fitted.

Deciduous a class of trees that shed their leaves annually.

Dowel a round wood or metal peg used to align parts.

Drill a tool for cutting circular holes.

Drill press a machine that turns drill bits.

Enamel varnish with pigment.

Faceplate a device on a wood lathe to which stock is fastened for turning.

File card a brush for cleaning file teeth.

Garnet natural rock used as an abrasive.

Gauge a hollow curved chisel.

Hardwood a class of woods usually from deciduous trees used for furniture.

Heartwood the mature core of wood at the center of a tree.

Jig saw a mechanical saw for cutting intricate shapes, also a scroll saw.

Jointer a machine tool for truing the edges of boards.

Kiln an oven for drying and heat treating wood.

Knot a defect in lumber caused by the growth of a branch.

Lacquer a cellulose nitrate finish for woods and metals.

Medullary ray the channels that carry food and moisture across the stem of a tree.

Miter to fit two pieces of wood to form a corner, each is cut at 45 degrees.

Mortise opening cut into a piece of wood into which a tenon fits.

Nail wire fastener with one head, driven into wood.

Plane to smooth a wood surface with a cutting tool as a hand plane or a machine planer.

Plywood a sandwich of layers of sheet woods bonded together.

Rabbet a step cut in the edge or end of a board.

Rasp a coarse file for rough shaping.

Rip to saw with the grain of the wood.

Sapwood outer rings of wood in a tree, which conduct sap from roots to leaves.

Shellac natural rosin dissolved in alcohol for a wood finish.

Softwood usually the lumber from needle-bearing trees.

T-bevel an adjustable tool for transferring angle measurements.

Tempera water mixed opaque colors.

Tenon the stub that fits into a mortise.

Try-square a measuring tool for checking squareness.

Turn to cut or shape stock on a lathe.

Varnish a finish of resinous substances and oils.

Veneer thin layer of sheet wood.

Warp disforming of lumber due to internal strains.

Wood lathe a machine tool that revolves the stock for shaping by cutting tools.

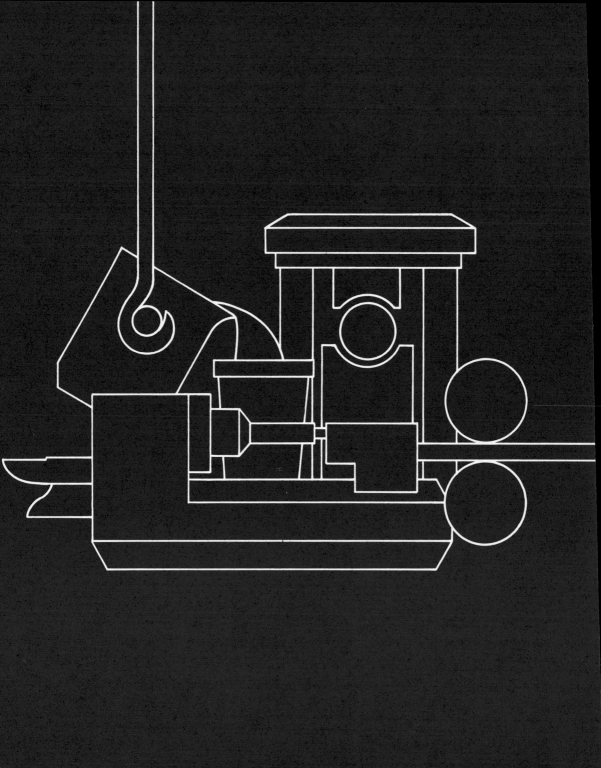

Metals Technology 4

When early man discovered that he could shape copper implements with heat he started the metals industries. Eventually he learned to increase the heat of his fire by forced air. He used goat-skin bellows and was then able to melt iron from its ore. Hand- and foot-operated bellows were used for centuries until James Watt made his steam engine. With engine-powered bellows, greater heats were reached. This made possible a better refining of ores. Today's steel plants use huge blowers to supply air. It is heated for blast furnaces to 3500°F or more.

The first ironmaking in America was probably near Jamestown, Virginia, at about 1640. But copper was being used in America before Columbus arrived. Pittsburgh became the center of steelmaking about 1800 because both its iron ore and coking coal were convenient. Today the major center of our steelmaking is in the Pittsburgh-Youngstown-Cleveland area. The supply of rich iron ore in Minnesota has been exhausted. A low-grade ore called *taconite* is now being mined there. It is ground and pressed into pellets. Modern steelmaking methods can use this ore satisfactorily. Increasing amounts of ore are being imported as our steel demands are increasing.

(from *Machinery's Handbook*, courtesy The Industrial Press)

Metal	Melting Point Degrees F.	Pounds per Cubic Foot
Aluminum	1220	168.5
Brass (80% cop., 20% zinc) ..	1823	536.6
Bronze (90% cop., 10% tin) ..	1841	547.9
Copper	1981	554.7
Gold	1945	1204.3
Iron, cast	1990–2300	438.7–482.4
Lead	621	707.7
Magnesium	1204	108.6
Silver	1761	650.2–657.1
Steel	2500	486.7

Fig. 4–1. The Decade 70 House of New Concepts in Houston, Texas. It has copper roofs, bronze floors, and a pollution-free all-electric environmental control system. Copper, the first metal refined by man, is today even more important as a material. It is alloyed with many other metals from steel to gold.

Metals are refined from mineral ores dug from the earth, although some free metals are frequently found in the form of nuggets. Metals have luster and the ability to conduct heat and electricity. They can be fused together. Some are magnetic (iron and steel); some are very heavy (lead and mercury); and some are very light (aluminum and magnesium). That quality of a metal enabling it to be drawn out into a wire is called *ductility*. *Malleability* is the characteristic that allows it to be hammered or rolled into sheets.

If someone offered you a gold brick measuring 6 inches by 6 inches by 12 inches if you could carry it away, could you lift it? How much would it weigh?

Alloys. An alloy is the metal that results when two or more other metals are melted together. It has different characteristics from the metals of which it is made. Such qualities as hardness, toughness, and resistance to heat and corrosion are controlled by alloying. Steel is an alloy of iron and carbon. It is further alloyed with chromium, nickel, manganese, tungsten, or other metals to produce desired qualities. Most of the metals used in products today are alloys. Here are some common ones:

Babbit—tin alloyed with copper and antimony
Brass—copper, with zinc
Bronze—copper, with tin
Pewter—tin with lead, bismuth, antimony, copper
Soft solder—tin and lead, usually 50/50
Stainless steel—Steel with chromium, nickel, molybdenum
Sterling silver—silver with copper

Pure gold is 24 K. The "K" stands for "carat," the unit measurement of fineness in gold. If your ring is 14 K. gold, it is 14 parts gold and 10 of other metal, usually copper.

Metallurgy. The science of metals is called *metallurgy*. The scientist in metals is the metallurgist. He knows the chemistry and physics of metals and develops new and improved alloys in his laboratory. He can change the characteristics of metals. For example, he can make steel stronger but lighter in weight. He has created dozens of

Fig. 4–2. Iron ore, being near the surface of the earth, is dug from open pit mines. Here it is being carried out by a conveyor from a mine near Calumet, Minnesota.

SHEET THICKNESSES
(from *Machinery's Handbook,* courtesy
The Industrial Press)

Steel			Aluminum and Copper		
	Gauge* Manu- facturers'				Oz/Sq ft
Inches	Standard	Inches	Gauge**		(copper)
0.0120	30	0.005	36		4
.0220	30	.010	30		
.0149	28	.0126	28		
.0179	26	.0159	26		
.0239	24	.0201	24		16
.0299	22	.0253	22		
.0359	20	.0320	20		24
.0478	18	.0403	18		
.0598	16	.0508	16		
.0747	14	.0641	14		48

* Manufacturers' Standard Gauge.
** Browne & Sharpe Gauge.

HOW IRON IS MADE

Iron is made in a *blast furnace*. Alternate layers of coke (partially burned coal), iron ore, and limestone are dumped into the furnace. The coke burns in a blast of hot air, reaching a temperature of about 3000°F. This melts the iron from the ore.

different stainless steels and tool steels to suit different purposes.

Common Metal Shapes and Sizes. The following list shows the usual ways in which measurements are given for metals.

Sheet—Thickness is given by gauge number, in thousandths of an inch or in ounces per square foot.
Plate—This is in sheet form but usually ³⁄₁₆ inch or more thick.
Wire—Diameters of steel and copper wire are given by gauge number; of aluminum, in thousandths of an inch.
Rod—Diameters are in fractions of inches.
Band—Width and thickness is given in inches.
Pipe and Tube—Either inside diameter (I.D.) or outside diameter (O.D.) is given.
Angle—Thickness of stock and width of sides is given.

Thickness of aircraft aluminum is given in thousandths of an inch.

Fig. 4–3. Common metal forms.

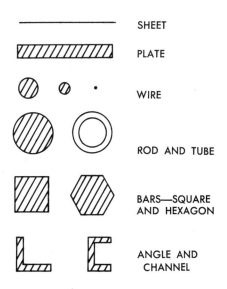

SHEET

PLATE

WIRE

ROD AND TUBE

BARS—SQUARE
AND HEXAGON

ANGLE AND
CHANNEL

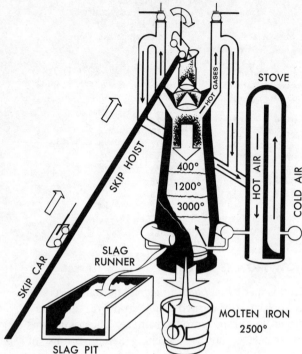

HOW STEEL IS MADE

Steel is iron at its strongest. The difference between steel and cast iron is in the amount of carbon they contain. Low-carbon steels are soft and easily formed. They contain only about 0.25 percent of carbon. High-carbon steels are hard and stiff and contain as much as 1.7 percent carbon. Cast iron has more than 1.7 percent carbon and is brittle and porous.

Bessemer Steel. Low-grade steels are made in the Bessemer converter. Molten iron from the blast furnace is poured into the converter. Air is forced through it to burn out impurities and carbon.

Open-hearth Steel. Most steel is made in open-hearth furnaces. The hearth is a huge dish lined with firebrick. Scrap iron and limestone are

Fig. 4–4. A simplified diagram of a blast furnace. The skip car dumps raw materials into the top of the furnace. The molten iron that runs out at the bottom may be poured into molds as pig iron or it can be further processed into steel. Every four to five hours a "cast" of iron is tapped, ranging from 150 to 350 tons.

The limestone melts and collects impurities that float on the surface of the molten iron. Then the iron is drawn off and is either cast into *pigs,* as *pig iron,* or used immediately for making steel.

Pig iron is further heated and refined to make *cast iron.* This is cast into many useful products such as machine parts, fire hydrants, boilers, and automobile engine blocks. Parts of cast iron must be large and heavy because it is brittle and breaks easily.

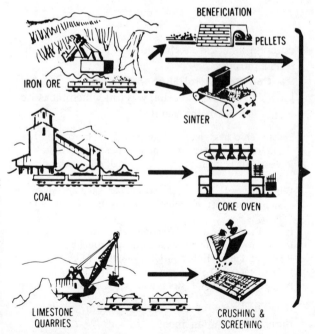

loaded into it. Oil or gas flames are passed over it to melt the iron. Then molten pig iron from the blast furnace is added and the mixture is heated to about 3000°F. This burns out carbon and impurities. The furnace is then tapped and the steel is cast into huge chunks called *ingots*.

Basic Oxygen Process. The open-hearth process almost replaced the Bessemer process because it produced better steel at lower cost. At present the basic oxygen process is replacing the open hearth. It uses the Bessemer principle but

Fig. 4–5. A blast furnace photographed at dusk. The two streams of light are moving cars of molten iron being transported from the blast furnace to a nearby open-hearth furnace.

Fig. 4–6. A flow chart of steelmaking, from ore to finished forms.

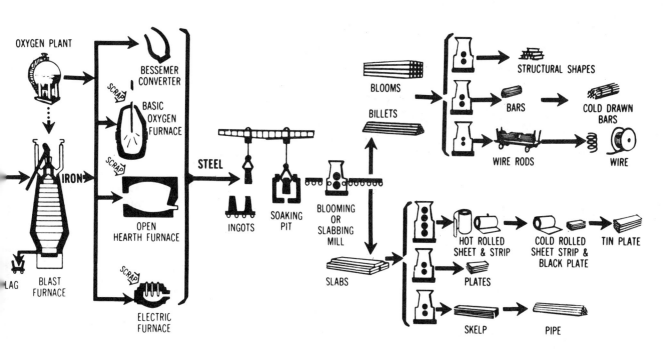

forces oxygen instead of air through the molten metal. This creates higher temperatures and produces a *heat* of steel in less than an hour. The former process required six to eight hours.

Vacuum Degassing. Steel quality is further improved by removing unwanted gases. The molten steel is poured into a ladle with a sealed lid. While it is being agitated, vacuum pumps draw off the gases.

Continuous Casting. The newest in steel mills use continuous casting. The molten steel is poured into bottomless molds. Here it congeals into a continuous length and is fed into the rolling mill. Steelmaking is now becoming a continuous, computer-controlled process.

Electric Furnace Steel. The highest quality steel is made in electric furnaces. Here the melting and refining can be most accurately controlled. The heat is produced by the current that arcs between the furnace's electrodes.

From Ingots to Usable Steel. Steel sheet, plate, rod, bars, strips, angles, and other such forms are rolled from ingots in rolling mills. The hot ingot is run through a series of huge wringerlike rolls. They squeeze it down into the desired thickness and shape. Some steel is dipped in molten zinc to give it a protective coating called *galvanize.* You have seen this on pails, tubs, and garbage cans. When coated with tin, the steel sheet is called *tinplate,* the stock from which tin cans are made. There are enough tin cans made each year in the United States to supply each person with about 2000 of them. So we have a litter problem.

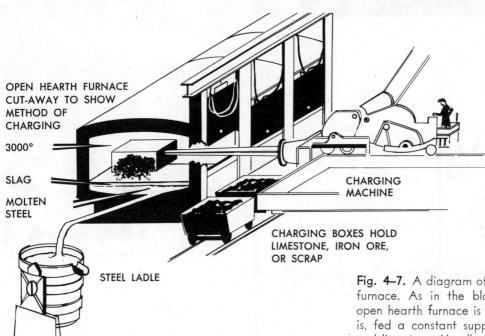

OPEN HEARTH FURNACE
CUT-AWAY TO SHOW
METHOD OF
CHARGING

3000°

SLAG

MOLTEN
STEEL

STEEL LADLE

CHARGING
MACHINE

CHARGING BOXES HOLD
LIMESTONE, IRON ORE,
OR SCRAP

Fig. 4–7. A diagram of an open hearth furnace. As in the blast furnace, the open hearth furnace is "charged"; that is, fed a constant supply of ore, coke, and limestone. Usually from 100 to 300 tons of molten iron are produced at one time, but a "heat" can run up to 600 tons. Flames sweep across the metal ore, the slag rises to the top and spills over into the slag pot. Refining takes eight to ten hours.

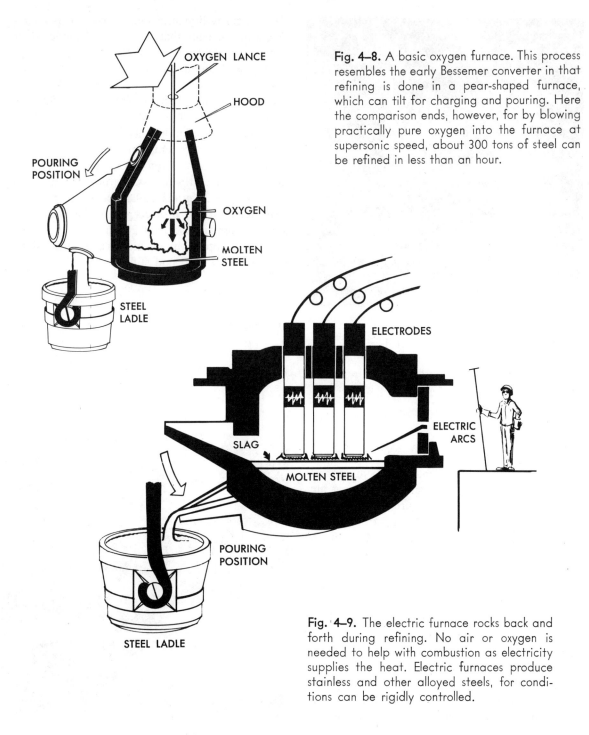

OXYGEN LANCE

HOOD

POURING
POSITION

OXYGEN

MOLTEN
STEEL

STEEL
LADLE

Fig. 4–8. A basic oxygen furnace. This process resembles the early Bessemer converter in that refining is done in a pear-shaped furnace, which can tilt for charging and pouring. Here the comparison ends, however, for by blowing practically pure oxygen into the furnace at supersonic speed, about 300 tons of steel can be refined in less than an hour.

ELECTRODES

SLAG

ELECTRIC
ARCS

MOLTEN STEEL

POURING
POSITION

STEEL LADLE

Fig. 4–9. The electric furnace rocks back and forth during refining. No air or oxygen is needed to help with combustion as electricity supplies the heat. Electric furnaces produce stainless and other alloyed steels, for conditions can be rigidly controlled.

Fig. 4–10. Molten iron from a blast furnace is being poured from this giant ladle into an open hearth furnace. The ladle and contents weigh nearly 200 tons. How many pounds is that?

ALUMINUM

Aluminum is an abundant material, being present in ordinary clay. This is actually an oxide of aluminum. It is not found as a pure metal and it is not yet practical to refine the metal from clay. It is obtained instead from *bauxite*. Most of the bauxite which is used in the United States comes from Surinam, South America, and Jamaica. Of that mined in the United States, 95 percent comes from Arkansas.

Because of its strength with lightness, aluminum is used extensively in aircraft. Because it conducts heat well, it is used in cooking utensils. Because it is a good conductor of electricity and is light in weight, it is used for cross-country electric power transmission lines. It weathers well, so it is used as a building material. Aluminum is *duc-*

tile, it is easily drawn out into a wire. It is *malleable,* which means that it is easily hammered or formed into shape.

In 1852 aluminum was valued at $545 per pound because it was difficult to refine. With the discovery of an electrical process in 1886 by Charles Martin Hall, a student at Oberlin College, the refining was simplified. Today, basic aluminum costs only about 25 cents per pound.

Pure aluminum is too soft for many uses. It is alloyed to give it stiffness, hardness, and great strength. The 2S grade (99 percent pure) is recommended for the projects you make because it is easily formed. 17S and 24S are aircraft alloys containing copper, manganese, and magnesium. They are very hard and stiff.

Fig. 4–11. A white hot strip of steel rushes out of the strip mill at a speed of 2300 feet per minute. It started as an ingot and has been rolled into the desired dimensions. The man in the foreground controls the strip mill from his console.

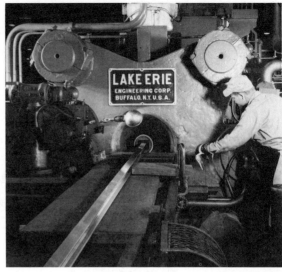

Fig. 4–12. (top) Molten aluminum is being poured from a crucible to form 700-pound ingots. From these are made sheet, plate, rod, tube, and other products. **Fig. 4–13.** (bottom). A rolling mill for producing aluminum foil. It is wound into a roll and is then annealed. The rolling process makes the foil stiff and hard. This effect is called "work hardening."

Fig. 4–14. (top) An aluminum extrusion emerges from a hydraulic press. Since aluminum is one of the most workable metals, extrusions of it can be made in an unlimited variety of shapes for all sorts of uses. Here it is being forced through a die, much as toothpaste is squeezed from a tube. **Fig. 4–15.** (bottom) These bars are pure copper, refined from the ore by means of electric furnaces. From such as these all copper products are made.

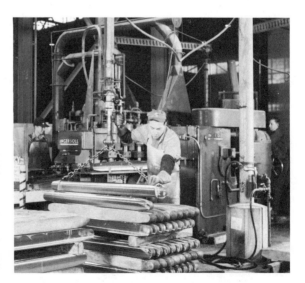

COPPER

Copper is easily identified by its rich, reddish-brown color. It, too, is very ductile and malleable, but it is also very heavy. (See page 126.) It tarnishes (oxidizes) quickly. You can polish copper until it gleams, only to have it dull in a day or so. The polish can be preserved by means of clear lacquer. Copper conducts heat well. This makes it useful for such things as pots and pans, automobile radiators, and electric refrigerators. Almost one-half of the copper produced in the United States goes into electrical uses. It is an excellent conductor of electricity. It is used for alloying with steel, aluminum, gold, silver, and other metals. We get most of our ore from the western states, chiefly from Arizona, Montana, Utah, and Nevada. The largest known deposits are in Chile.

Copper Core Coins. Copper core coins, first minted in 1965, were designed to conserve silver. Look at the edge of a dime, quarter, or half dollar to see the sandwich effect of copper between silver. The coin stock is laid up in layers that total approximately 5 inches at the start. They are called *clads* and are fed into a machine where a controlled continuous explosion bonds them permanently. Next they are hot rolled and then precision rolled to the required coin thicknesses, for example .041 inch for the dime. After annealing, the stock is sent to mints. Here the coin blanks are punched out. These are fed into stamping machines, which imprint the designs. (See page 169 and Fig. 4–84.)

INDUSTRIAL PROCESSES FOR FORMING METALS

The following processes are actually groups of related processes. Each includes several variations. The latter are increasing with manufacturing research and development.

Casting. Molten metal is poured or forced into cavities of the desired shapes in sand, plaster,

Fig. 4–16. (top) Pouring a vacuum-degassed ingot. This was the largest ever poured from an electric furnace and was made into one of the largest electric generator rotors ever made. **Fig. 4–17.** (bottom) Under a 7,500-ton hydraulic press, the hot ingot weighing 638,400 pounds is being forged into rough shape as the first step in shaping it into the generator rotor.

Fig. 4-18. (top) The giant rotor assumes the rough outlines of its ultimate shape under the pressure of the forging press. **Fig. 4-19.** (bottom) The final machining of the rotor forging on a huge lathe. After final testing, it will go to the customer for finish machining and assembly.

or other metals. When cool, it is hard and has the shape of the cavity. This is the process used in foundries. Metal molds are called *dies* and the casting process is *die casting*.

Rolling.   Metals can be formed by squeezing between heavy rollers. Sheet and plate are rolled to thickness. Bars and railroad rails are rolled to shape.

Extruding.   Molten metal is forced through dies that give it shape, like toothpaste from a tube.

Drawing.   Metal is pulled through dies to change its shape or size. Different diameters of wire are obtained by drawing.

Forging.   This is the hammering process, as described on pages 152–153.

Machining.   These are the finishing processes by which a formed part is cut to the final shape and size. It is a machine process and sometimes it is done automatically.

Pressing, Stamping.   Metal sheets, hot or cold, can be shaped by pressing into steel dies.

The lower die is made to the shape of the desired part. The metal is laid over it and is pressed to that shape by the upper die. Sheet metal parts for toys, automobiles, airplanes, and many other products are formed by pressing. (See Fig. 4–20 below.)

Sintering.   Powdered metal is also pressed into solid shapes in dies. Then it is *sintered,* or heat treated, to give it strength.

Spinning.   Spinning is the process for shaping circular sheets of metal into bowl-like forms. The disc is pressed into shape over a form as both revolve in a lathe. Some of your mother's pans have probably been spun. Light reflectors, airplane propeller spinners, and such concave forms of light metal are spun. (See Fig. 4–21 below.)

Dies and Molds.   Dies and molds are devices for forming materials. They produce identical parts. In the metals products industries, dies are usually blocks of steel in two parts with a cavity in each. Material placed in the cavity is pressed

Fig. 4–20. The upper part of the die in this automatic press forces the metal sheet against the lower half. Here it is stamping out roof panels for a new car.

Fig. 4–21. In the spinning process shown here, a circular aluminum blank or shell is pressed against a chuck rotating at high speed on a lathe. Pressure applied manually or mechanically with forming tools shapes the blank to fit the chuck.

into shape as the two parts are brought together. For example, after the shape of an automobile hood is *stamped out* in a flat steel sheet, it is formed in dies. There are dies for casting, also. Molten metal is injected into the cavity to solidify and take the shape of the cavity.

A mold is commonly used for the forming of sheet materials laid over it. Sheet plastic, fiber glass, and clay can be shaped this way. The terms "dies" and "molds" have somewhat different meanings in different industries. Hollow ware of clay is formed in plaster molds by the process of *slip casting.* (See page 284.) A similar process using liquid metals or plastics is called *die casting.*

HAND TOOLS FOR METALWORKING

Some common tools and processes are used with common metals. Get acquainted with these tools; they are the ABC's of metalworking.

Measuring Tools

Layout and Measuring. The first step in metalworking is usually the laying out of the work. Metal to be removed must be measured and marked off before cutting. Holes to be drilled must be located. Measuring and layout is done with *rules, scribers, squares, dividers,* and *calipers.* You can work to very close dimensions with metals, so measure accurately.

Rules. The twelve-inch steel bench rule is a convenient, accurate measuring tool. Remember, "Measure twice, cut once."

Scriber. The scriber is a thin, pencil-like tool of steel. Use it to mark fine lines when scratching the metal is not objectionable. A pencil is used on soft metals.

Square. With a square, lines can be marked at right angles to an edge. The square is also used to check flat surfaces and edges that are to be squared to other edges. If you drop a square, or use it as a hammer, it may no longer be square.

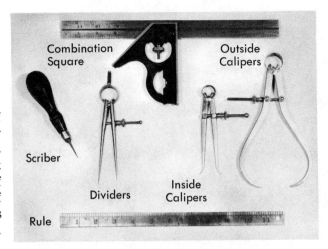

Fig. 4–22. Layout and measuring tools used by the metalworker.

Dividers. To *scribe* arcs and circles and to measure and step off distances, use the dividers. Set the two points at the desired distance on a rule. To scribe a circular curve, the center should be pricked slightly with a sharp center punch.

Calipers. The outside caliper is used to measure thickness. The inside caliper measures openings. To use a caliper, adjust it to slide snugly over the stock or into the opening. Then hold one of its legs on the end of the rule and read the dimension at the other.

Combination Square. This instrument is used to lay out lines and angles. It includes a *protractor,* which can be set at any angle, and a center head, which is used to locate centers on the ends of round stock.

Forming Tools

The Hacksaw. The hacksaw saws metal as the handsaw saws wood. Blades are made in different lengths and with fine, medium, or coarse teeth. Fine teeth are for sawing thin material and coarse teeth for thick. The blade should be placed

Files. Files are hardened pieces of steel with sharp ridges or teeth that scrape and sheer off the metal. They are made in a variety of shapes and sizes and for many different materials. Files are usually *single cut* or *double cut*. The first has all of its teeth slanting in the same direction. The other has two sets of teeth crossing each other. The single cut file is for fine, smooth cutting. The double cut is for fast rough work. Choose a file to suit the job, but never use it without a handle. Files are very brittle and often break when dropped. They should not be placed so that they can rub together. When the teeth get shiny, they are dull and the file should be replaced. If your file squeaks on the metal should you oil it?

Use both hands on the file. Insert the work in a vise with the mark close to the jaws. The file cuts on the push stroke and should be lifted on the return. Clean the teeth frequently with a *file cleaner.* When small pieces get stuck between the teeth, pry them out with the pick supplied with the file cleaner. A clogged file will scratch the surface of the work.

Fig. 4–23. (top) In using the hacksaw, use both hands.
Fig. 4–24. (bottom) Filing also requires the use of both hands.

Fig. 4–25. Various weights of hammers. Note the paperclip for comparison of sizes.

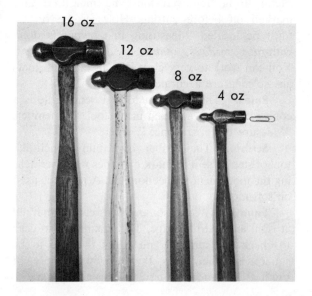

16 oz

12 oz

8 oz

4 oz

in the frame with the teeth pointing away from the handle and should be kept taut when sawing. The work should be held in a vise with the saw line close to the jaws to prevent chatter. With one hand holding the fore end and the other the handle, use long, slow strokes with just enough pressure on the blade to keep it cutting instead of rubbing. Ease up on the pressure as the blade cuts through the opposite side.

Fig. 4–26. Metal can be easily sheared off in a vise with a cold chisel. Hold it so that if the hammer misses, it won't hit your hand.

Draw-filing produces a smooth finish. Hold the file in both hands at right angles to the stroke. Then pull it toward you over the work.

Hammers. The *ball peen hammer* is the metalworker's hammer. There are many types of hammers for special purposes. (Some are described on pages 143 and 147.) Hammers are usually sized by the weight of the head as 12 ounce, 24 ounce, and so on. Try to select the hammer that best fits the job to be done and that you can use most easily and safely.

Cold Chisels. A cold chisel is a hardened piece of tool steel with the cutting end sharpened as a wedge. The other end is blunt to receive hammer blows. It is made to cut cold steel or other metals. Use as heavy a chisel as possible and hold it so that if the hammer misses the target, it won't hit your hand. Light metal such as band iron can be cut in a vise by shearing it off with a chisel. To cut round rod, notch it on opposite sides, then bend it back and forth until it breaks. A chisel's cutting edge should usually be ground to a 60-degree included angle.

Drilling. Small holes can be drilled with a hand drill. (See page 89.) However, a portable electric drill, or drill press, is much easier to control. (See page 89.) Straight-shank, high-speed twist drills are recommended. (See page 90.)

To drill a hole, first mark the location with a center punch to keep the drill from wandering. Select the correct drill and tighten it in the chuck. The work should be clamped to a bench or in a vise. Pressure must be applied to force the turning drill into the metal. Too much pressure bends and breaks small drills. Always ease up on the

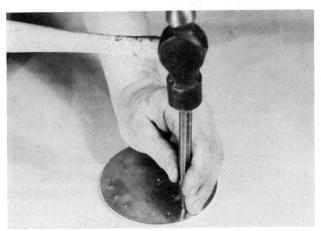

Fig. 4–27. (top) Before drilling a hole in metal, the mark should be centerpunched. Fig. 4–28. (bottom) A portable electric drill makes drilling easy.

pressure as the drill comes through, to keep it from grabbing. (See page 168 for information about lubricants.)

Forming Metal. Metal hoops, curves, and scrolls are formed most easily in a *bending jig.* There are several types. Adjust the jig to fit the metal band. Insert the band and bend it slightly. Move it into the jig a little and bend again. Repeat this process until the curve is complete. (See page 153 for information regarding the forming of hot metals.)

Punches. Punches are of the chisel family. There are many types to serve many purposes. A machinist's hand punch is for general use. Use it for removing rivets, bolts, and the like from holes and for punching holes in sheet metal. Lay the sheet on the end grain of a block of wood or on a block of lead, and with a ball peen hammer drive the punch through the sheet just far enough to make a hole.

Riveting. Riveting is the process of fastening pieces of metal together with rivets. These are made in several different heads. The common

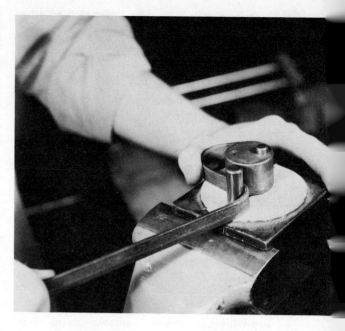

Fig. 4–29. Bend band iron in a jig to get smooth, even curves.

Fig. 4–30. (left) A bending jig made of dowel pegs set into 3/4 in. plywood. In this is formed the book rack described in Chapter 2. **Fig. 4–31.** (right) The jig used in forming the ends for the book rack in Fig. 4–30.

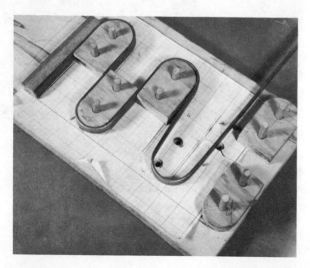

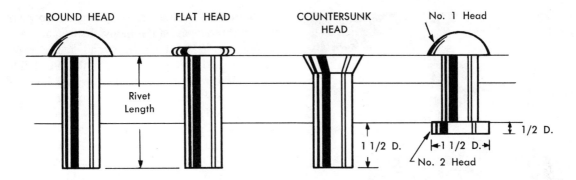

ROUND HEAD   FLAT HEAD   COUNTERSUNK HEAD   No. 1 Head

Rivet Length

1 1/2 D.   1 1/2 D.   1/2 D.

No. 2 Head

Fig. 4–32. Rivet types.

ones are round, flat, and countersunk or flush. Ordinarily the rivet is of the same metal as that being riveted.

A rivet should fit snugly into the hole. Drill the hole the same size as the rivet shank. The shank should be just long enough to extend beyond the work, a distance equal to one and one-half to two times the rivet diameter. A rivet block is used to hold the head and to absorb the hammer blows in forming the No. 2 head. Do this with the face of a ball peen hammer. The No. 2 head should be about twice the diameter of the rivet and half as thick as the rivet shank. To remove a rivet, center-punch the No. 1 head. Drill through it with a drill bit just under the diameter of the hole. Snap the head off with a *pin punch* (a slender punch).

Threads. The spiral cut on a bolt and in a nut is a thread. When the two threads match, the nut screws onto the bolt. Threads can be cut with *taps* and *dies*. The die cuts the thread on a bolt. The tap cuts the thread in the nut. The common thread types are National Coarse (N.C.) and National Fine (N.F.). A ¼-inch bolt with the N.C. thread has 20 t.p.i. (threads per inch), and one with the N.F. has 28 t.p.i.

To tap a hole it must first be drilled to the proper tap size. This is always smaller than the bolt diameter. Consult your teacher's drill chart to find the size. Tighten the tap in a tap wrench. Insert the end in the hole and turn the tap clockwise as you gently press down. If cutting oil is needed, use it frequently. (See page 168.) Turn the tap backward about a quarter turn for each revolution to clear out chips. Taps are very hard and brittle. The small ones break easily and plug the hole very effectively.

To thread a bolt or a rod, select the **correct** die and lock it into the *die stock* (the holder). Grind a slight chamfer on the end of the rod to make it easier to get the die started. Press the die into the stock and turn it slowly clockwise. Use a lubricant if needed. Reverse the direction a half turn occasionally to break up chips. The size indicated on a die is the diameter of the rod that it will thread. The size indicated on the tap is the diameter of the bolt.

ART METAL PRODUCTS AND TOOLS

Art metal products include bowls, trays, table service, jewelry, and the like. They are usually made from aluminum, brass, copper, silver, or pewter. Most of the common, hand metalworking tools are used, plus a number of special ones described here.

Jeweler's Saw. The jeweler's saw is similar to a coping saw except for its adjustable frame.

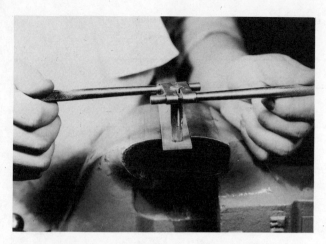

Fig. 4–33. (top) Tapping a hole in an aluminum lamp bracket. This tap cuts threads for the pipe nipple to which the socket screws.
Fig. 4–34. (bottom) Cutting threads on a rod with a threading die.

It is used to cut fine details in the design. Blades come in many widths, from horsehair fine to those $\frac{1}{16}$ inch wide. They have various numbers of teeth per inch.

Needle Files. Regular files are used for coarse cutting. Needle files are used for fine detail. Since they are slender and delicate, they must be used carefully.

Hammers. Hammers for art metal work are made in three types and in many shapes and sizes. *Raising hammers* usually have ball-shaped ends. They are used to tap and stretch the metal into rough, hollow forms from the inside. *Forming hammers* have rounded oval faces and are used to form round edges and for raising bowl shapes from the outside. *Planishing hammers* have flat faces and are round or square in shape. They are used to smooth and finish formed surfaces. Art metal hammers must not be used for anything else. If the faces get nicked, they will mar the surface of soft metals.

Stakes. Stakes are heavy pieces of steel or cast iron in different shapes, which are used as anvils. Metals are formed over them. Their surfaces must not be nicked.

Form Blocks. These are pieces of hard wood on which metal is roughly formed. Blocks for raising hollow pieces have shallow cups into which the metal is hammered. Blocks can be cut to shape.

Chasing Tools. These are small chisel-like tools for cutting lines and designs in metal. A soft-face hammer or mallet is used to tap them.

Daps and Dapping Dies. Daps are punch-like tools with a ball end for raising beads or dome shapes. The metal is placed on the dapping die and is tapped into the hole with a dap and a mallet.

BASIC TOOL PROCESSES

There are a few art metal processes that you will often use.

Sawing with the Jeweler's Saw

The measure of your skill with this tool is how close to your pattern you can cut and how much sawing you can do with one blade. Here is how:

1. Draw the design on paper and cement it to the metal with rubber cement.

2. Select the saw blade. The thinner the metal, the finer the teeth. For 18-gauge metal use a blade with 15 to 20 t.p.i.

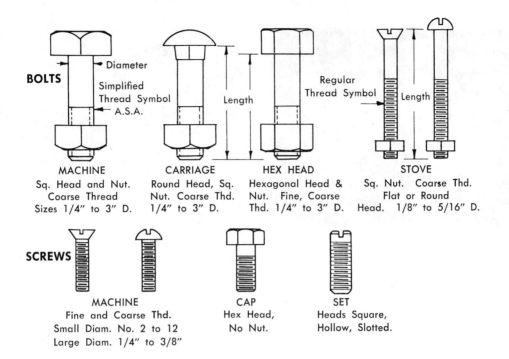

BOLTS

Diameter

Simplified
Thread Symbol
A.S.A.

Length

Regular
Thread Symbol

Length

MACHINE
Sq. Head and Nut.
Coarse Thread
Sizes 1/4" to 3" D.

CARRIAGE
Round Head, Sq.
Nut. Coarse Thd.
1/4" to 3" D.

HEX HEAD
Hexagonal Head &
Nut. Fine, Coarse
Thd. 1/4" to 3" D.

STOVE
Sq. Nut. Coarse Thd.
Flat or Round
Head. 1/8" to 5/16" D.

SCREWS

MACHINE
Fine and Coarse Thd.
Small Diam. No. 2 to 12
Large Diam. 1/4" to 3/8"

CAP
Hex Head,
No Nut.

SET
Heads Square,
Hollow, Slotted.

Fig. 4–35. (top) Some common fasteners. Note: these are common bolts and screws. There are many special kinds, such as aircraft bolts and miniature ones as in watches. **Fig. 4–36.** (lower left) A set of art metal hammers. From left to right: planishing, raising, and forming. **Fig. 4–37.** (lower right) Types of stakes over which art metals may be formed.

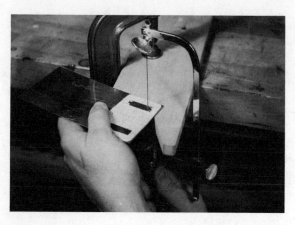

Fig. 4–38. Sawing with the jeweler's saw. Note the slender blade.

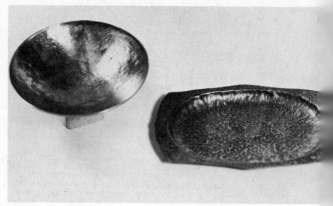

Fig. 4–39. Two dishes tapped from sheet copper. The round one was formed by tapping into a shallow hole in the end grain of a block of wood. The tray was formed into an oval mold cut in a piece of wood.

3. Clamp the blade in the upper vise of the frame with the teeth pointing toward the handle. Fasten the lower end and then draw the blade up taut.

4. Hold the metal flat on a V-block and saw with a slow, easy up-and-down stroke. Put only enough forward pressure on the blade to keep it cutting.

Piercing. Piercing is the sawing out of openings within the metal. Drill a hole in the waste. Insert the saw blade, clamp it in the frame as before, and then saw.

Filing. File parallel to the edge of sheet metal instead of across it when you are using a regular file. Needle files can be used in any direction. Select the shape that best fits the section being filed. Be sure to remove any burrs and sharp edges.

Forming. Shallow bowls and trays can be most easily made by forming them into openings cut to the desired shape in wood blocks.

1. Trace the outline of a tray on the block and cut out the inside on the jigsaw. Round the top inside edge a bit with a wood file. Nail this to another block for a base.

2. Cut out the metal about ½ inch larger than the opening on each side. Drive several wire brads into the form around the opening to hold the metal in place. Anneal the metal.

3. With a forming hammer tap lightly around the edge, stretching the metal down into the form. Tap around and around, gradually working down to the bottom. When the metal hardens and resists stretching, *anneal it.*

4. When the tray is fully shaped, remove it and trim the edges. You should then planish it on a flat stake.

Raising. Raising is a means for shaping sheet metal on a stake or form block. This is more difficult than the forming process just described, but it has greater possibilities.

1. With a pencil compass, mark several concentric circles ¾ inch apart around a 4-inch or 5-inch disc of 18-gauge copper or aluminum.

2. Hold the disc with the outer circle over the hollow in the raising block. With a raising hammer, tap around the outer circle. Turn the disc as you go, to space the dents evenly. Strike each blow with equal force but not so hard that the metal is stretched out of shape.

Fig. 4–40. A forming mold can be easily made by cutting a desired shape in a block of scrap wood and nailing it to another piece to make a die into which the metal can be beat down.

Fig. 4–41. Tapping a copper dish into the mold. 18-gauge sheet is just right for this work.

3. Follow around the circles tapping toward the bottom. If the bowl needs more shaping after you have gone over it once, repeat the process. Remember to anneal the metal when it begins to resist forming.

Annealing.　Annealing is a softening process. If you keep your metal soft, you can do much shaping with it. When it gets hard from hammering, it tears easily. To anneal copper and brass, heat them to a dull red in a flame large enough to cover the metal, then quench in water.

To anneal aluminum, first wipe some lubricating oil on it, then heat it until the oil burns off. A mark of blue carpenter's chalk turns white at the annealing temperature of aluminum. Quench in water.

Planishing.　Planishing smooths the surface and makes the metal stiff and hard. It leaves small, shallow, overlapping facets that are a mark of good craftsmanship. To planish, select a stake that fits the curve of your bowl as closely as possible. Place the bowl over the stake and with the planishing hammer tap over the surface. Work around the bowl and gradually move down to the bottom. Try to make the marks overlap evenly. Polish them with fine steel wool to check your progress.

Chasing.　Line designs are easily applied by chasing.

1. Draw the design in pencil on the metal and select the chasing tool that fits best.

2. Lay the metal on a board and tap the tool lightly with a mallet as you follow the design, moving toward yourself. Go over the lines several times until they are deep enough.

Etching.　Since some acids attack metals, they can be used to etch designs. Prepared etching solutions, called *mordants,* are safer to use than are raw acids and do a better job.

1. Clean the metal with fine steel wool and scouring powder. Try to keep your fingers off the clean surface.

2. Trace your design on the metal with a pencil and then paint out all areas that are not to be eaten away, with *asphaltum* paint.

Fig. 4-42. Raising a dish on a form block—a hollow cut in the end grain of a heavy block of wood.

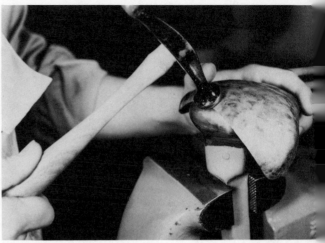

Fig. 4-43. Planishing the dish on a stake. This removes the large hammer marks and smooths the surface.

Fig. 4-44. Sweat soldering is generally used when the solder should be invisible in a seam. Tin the facing sides of metal to be joined and hold them together with old pliers over a flame until solder is melted. Remove from heat but maintain pressure until the molten solder hardens.

3. After you have read the directions on the bottle, place the object in a glass tray and cover it with the mordant.

4. After 10 or 15 minutes remove the metal, rinse it in water. Dissolve the asphaltum with kerosene or turpentine.

Soldering. Soldering is the process of joining pieces of metal by means of an easily fusible alloy, called *solder*. When melted this has an affinity for the metal. Ordinary soft solder is half lead and half tin. A flame is preferred for soldering art metal projects because they conduct the heat away so rapidly. (See page 150 for soldering with an iron.) Follow these suggestions:

1. Fit the pieces of metal closely together, then apply flux to the surfaces to be joined.

2. Heat each piece until it melts the solder when it is touched to the fluxed surface. Keep it just hot enough that the solder will flow over the surface. Use only enough solder to give a thin coating; this is known as *tinning*.

3. Place two tinned surfaces together and reheat until the solder flows. If you must solder a joint close to the one already soldered, paint a coat of clay slip on the first joint to stop melting.

Finishing. See page 171 for finishing suggestions.

SHEET METAL TOOLS

Sheet is one of the most commonly used forms of metal. You have seen the aluminum wrapper on breakfast cereal and candy bars. This is as thin as one usually finds sheet metal. It is called *foil*. Contrast this with armor plate and boiler plate. The first is a heavy steel that can shed bullets and shells. Boiler plate is usually one-half inch or more thick and when fabricated into a boiler can retain steam at hundreds of pounds pressure. The processes used in forming your sheet metal products are somewhat different from those for art metal products. Many of the tools described in other parts of this chapter are used for sheet metal working and others are especially designed.

Hammers. Riveting hammers are for *setting* rivets, that is, forming the second head (see page 143). Wooden *mallets* and soft-faced hammers are used in forming sheet metal because they do not stretch it as much as do metal hammers. An auto body repairman uses a *bumping hammer* to smooth out dents.

Stakes. Sheet metal can be formed by hammering or bending over T-shaped anvils of steel called stakes.

Punches. Holes in sheet metal are usually punched instead of drilled. This process is faster and more economical. You can do it by hand with a machinist's solid punch of the correct diameter, as described on page 142.

Groover. Seams are locked together with a groover. Use one that fits the seam and tap it with a mallet as you slide it along the seam. Lay the seam over an anvil while tapping.

Fig. 4-45. Hammers used in sheet metal work. From top: riveting hammer, raising hammer, setting hammer, hickory mallet. Mallets are made also of rubber, plastic, or are covered with rawhide; these do not damage thin metals.

Fig. 4-46. (top) Stakes for sheet metal work. A bench plate holds the stakes rigidly. (bottom) Holes of various sizes accommodate all sorts of stakes.

Tin Snips. These are heavy, scissor-type shears for cutting sheet metal. To cut, lay the sheet on the bench. Press down on the top handle and let the bench push against the bottom one.

Sheet Metal Machines. Sheet metals are processed by hand-driven and automatic machines. The *squaring shear* cuts off sheet to a straight line. Edges are turned and hems formed on *bar folders* and *brakes*. Sheet is formed into round cylinders on a *slip roll*. Turned and wired edges are made on *rotary machines*. Sheet is pressed into form on *presses* (see page 138). It is cut to shape or punched on *punch presses*.

SHEET METAL JOINING PROCESSES

An auto body is made up of several pieces of formed sheet fastened together by welding. This is the commonest process for joining sheet metals.

Fig. 4-48. Slip rolls put curves in sheet metal. The grooves at the end of the rolls will curve wire.

Fig. 4-47. Hand snips. From top: Compound leverage shears increase hand power; some models are made for cutting right or cutting left, and can make straight or irregular cuts. Extra heavy straight snips for smooth long cuts. Double cutting shears are used to cut holes or strips and for cutting cylinders to length.

Fig. 4-49. Tube, pipe, and tanks are made from metal sheet and plate by roll forming. Here large cylinders are formed from flat plate by rolling. The joints are later welded.

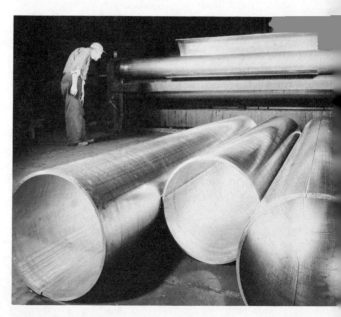

The joint may be welded continuously along its length or only in spots, called *spot welding*. Soft and hard soldering are used in place of welding in some products when little heat can be applied. Riveting is less common now than before the perfection of welding. (See page 172.) Hems and seams are the joints in such products as tin cans, pails, boxes, and heating ducts.

Hems and Seams. This may sound more like dressmaking and tailoring than sheet metal work. There is a similarity, for a seam is a fold, whether in cloth or in metal. You may not have thought of the medieval armorer as a tailor, but he was, and a metalworker, too. A hem in cloth is sewed in place. In metal, the ends are folded, or hemmed, then hooked together and locked, making a seam. To stiffen the joint even more and to make it watertight, solder is flowed into the folds.

Soldering. Galvanized and tin-plated steel can be soldered with a soldering iron (soldering

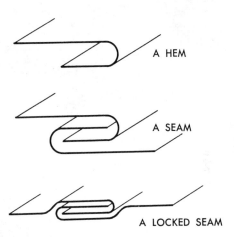

Fig. 4–51. Some common folds for sheet metal.

copper). These metals do not conduct heat away as rapidly as do copper and brass. For the latter, a flame is preferred if the joint is large. Follow these steps for soldering:

1. Heat the soldering copper until it will melt solder.

2. Wipe the tip clean with a damp cloth.

3. Dip the hot tip into flux and rub on a small amount of solder.

4. If the tip does not take the solder all over, rub it on a block of wood until it does. This is *tinning*. A well-tinned copper holds solder better and transfers more heat to the metal than does a dirty one.

And Now to Solder a Joint

1. The surfaces must be clean, but do not steel-wool the galvanizing or the tin plating. They are too thin. A good flux will usually do the cleaning.

2. Apply flux to the joint.

3. Hold the tip of the copper against the joint until the metal is hot enough to melt the solder, then add as much as is needed.

Fig. 4–50. The bar folder is used for making bends in sheet metal.

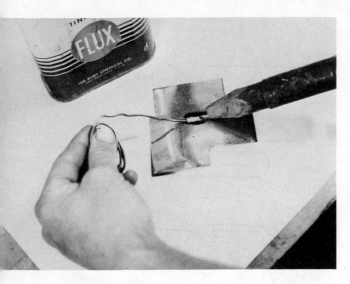

Fig. 4–52. Soldering with a soldering copper.

FORGING

Forging is the process of shaping hot metal by hammering. It makes metal stronger and tougher. The village blacksmith shaped horseshoes, wagon-wheel rims, and plowshares on his anvil. As he struck the hot steel with heavy blows, the sparks flew in all directions and glanced from his leather apron. In industry today, hand forging has almost disappeared. It has been replaced by machines, which, for example, can forge an automobile engine crankshaft with a few blows. You will enjoy forging. There is something fascinating about forming red-hot steel. It is soft and plastic.

The Forge. The forge is a furnace designed to heat metals quickly and at particular places. The old blacksmith used wood or coal for fuel and a bellows for forced air draft. Modern forges are generally gas fired or electric. Do not operate your forge until your teacher shows you how.

The Anvil. The anvil is a block of iron or steel with a flat face on top and a tapered horn on one end. (See Fig. 4–54.) On it most of the forming of hot metal is done in hand forging. The *face*

is for hammering metal flat and straight; the *horn* is used for making curves. The square hole in the top is the *hardy hole.* Tools for cutting and shaping are held in it. The round hole, the *pritchel hole,* is for punching holes and shaping rod. Striking the face of the anvil with a hammer will make nicks in it that will show up as scars on the metal being forged.

Tongs. Use tongs to hold the metal while it is hot. These are long-handled pliers with jaws shaped to hold various sizes of metal. Do not keep the tongs in the fire while the metal is being heated.

Hammers. An engineer's hammer with a short handle and heavy head is ideal for general forging. The short handle makes it easier to control your aim. Use as heavy a hammer as you can safely swing.

Basic Forging Processes

When steel is forged, it should be heated to a point between a bright cherry red and an orange. Too much heat burns the metal; too little causes it to crack under the hammering.

Fig. 4–53. Heating a piece of steel in a forge.

Fig. 4–54. Steel rod heated red hot becomes plastic. It is easily shaped with a hammer and anvil or bent in a vise. The early blacksmith made tools, horseshoes, and wagon parts by forging hot steel on his anvil.

Drawing Out. The first shaping in the making of tools such as chisels and punches is *drawing out*. Hold the hot metal on the anvil horn and strike it several times with the face of the hammer, rotating it after each blow and working down toward the hot end. This procedure lengthens and tapers the stock. After the rough shaping, hammer the metal on the face of the anvil to smooth and straighten it.

Bending and Forming. Angles can be bent over the edge of the anvil. For some bends a heavy vise is preferred. When making a bend a few inches from the end of the material, first heat it. Then quickly quench the end, leaving the section to be bent hot. To twist bars, clamp one end in a vise and turn the other with a heavy wrench.

Cutting and Shearing. Hot metal is cut on the anvil. Put a *hardy,* the blacksmith's cold chisel, in the square hole and lay the metal across the cutting edge. Strike it several blows with the hammer on both sides until it is nearly cut through. Then bend it back and forth until it breaks. Light bars can be sheared off in a heavy vise. Clamp the metal with the cut-off mark even with the jaws. With a cold chisel and hammer, shear it against the stationary jaw.

HEAT TREATING

Without heat treating, many tools and machines would be useless. It is the process by which soft steel can be made hard and hard steel, soft. In industry, special heat-treating furnaces heat metal with temperatures controlled automatically. You can do yours in the forge. Heat treating includes hardening, tempering, and annealing.

Hardening. Hardening is the first step in heat treatment. You have learned (page 130) that if steel is to harden, it must contain carbon. Tool steel, a high-carbon steel, is hardened by heating and immediately quenching in oil or water. An expert can tell by the color of the steel when it is at the correct temperature for chilling. You may tell quite accurately by holding a small magnet to the steel. When the magnet no longer attracts the steel, the temperature is about right.

Tempering. Hardened steel is often so hard and so brittle that it is useless. Tempering is the process of removing excess hardness so that the steel, though still hard, is tougher. Polish the part of the metal to be tempered with aluminum oxide cloth. Slowly heat this part. Remove it from the furnace frequently to watch for the colors. The first color noticeable is a yellow. As the temperature increases, this changes to a straw color, then to brown, purple, and blue. When you get the desired color for the tool you are making, quench the piece in cold water. Hammers are quenched at the straw-color stage, cold chisels and punches at purple, and screwdrivers at blue.

Annealing. Annealing is a softening process (see page 145). Since hardened steel cannot be readily machined, it is annealed, then machined, then hardened and tempered. To anneal a piece of steel, heat it to a dull red. Then cool it slowly in warm ashes, or let it cool down with the furnace. It should become as soft as it was originally.

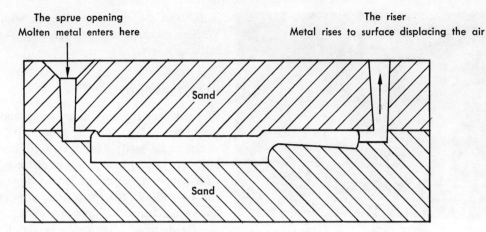

The sprue opening
Molten metal enters here

The riser
Metal rises to surface displacing the air

Sand

Sand

Fig. 4–55. Theory of casting.

Case Hardening. Since mild steel, a low-carbon steel, contains little carbon, it will not get very hard. It can be given a hard outer shell by case hardening. The steel is heated red and is then placed in a compound rich in carbon. The hot surface absorbs carbon and becomes like tool steel. At this point the steel is reheated and hardened as is tool steel. Follow the specific instructions accompanying the compound you are using.

✚ SAFETY SENSE

When you are forging or heat treating:
1. Wear a face shield while you are hammering the hot metal. The scale that flies is hot.
2. Wear gloves when necessary, and use tongs for handling hot metal.

FOUNDRY PROCESSES AND PRODUCTS

One of the first things that early man found out about metals was that they could be melted and cast into shapes in sand. Today this is a key process in the metals industries. Molding sand is still used for making molds into which large castings of molten metal are poured. Sand molds are broken up after each casting. Permanent molds of steel are used for producing great quantities of identical pieces, such as certain automobile parts of aluminum. Jewelry and dental structures are cast in plaster molds. The foundry industry has grown up around this process of shaping molten metal in molds. Today much casting is done automatically with metals melted in furnaces controlled automatically. As in all industries, certain tools and equipment are necessary.

Foundry Equipment

Patterns. Patterns are the models from which the molds are made. They are identical to the finished casting in shape except that they are enough larger to make up for the shrinkage of the metal when it cools. Patterns are usually made of wood, metal, or plaster.

Flask. The flask is the metal box that holds the pattern and the sand. It has two parts: the upper is the *cope,* and the lower, the *drag.* It has two locating pins, which keep the two parts aligned.

Sand. The special sand used in foundry work is called *molding sand.* Before it is used, it is *tempered* with water. The whole is then mixed

Fig. 4–56. A camper's outfit including grill, hamburger fryer, frying pan, and flapjack griddle. The last two were cast in aluminum. The hamburger fryer consists of two pans formed of black iron sheet, hinged together and brazed to wire handles. The legs of the grill fold to make it portable.

Fig. 4–57. The patterns for the frypan and griddle are made of wood to the exact shape desired for the castings.

thoroughly. The sand should be just damp enough to hold together. Squeeze a handful tightly and then break it in two. If it breaks clean and sharp, the sand is well tempered. Sand that is too wet is dangerous. The hot metal turns the water to steam, which may cause the mold to explode. Sand with a plastic binder mixed in takes very little water. It rams up tighter, too.

Melting Furnace. The metal to be cast is heated in a melting furnace. This burns gas or oil, or may be heated electrically. The metal is melted in a special container, a *crucible,* made of clay and graphite. Be sure to get the operating instructions from your teacher before you attempt to light the furnace.

Other Tools. The *riddle* is a sieve for the sand. Sand is packed into the flask with a *bench rammer.* The *striking board* is a straightedge for leveling the sand across the top of the flask. The *sprue pin* is a tapered, slender, wooden cone set in the sand to form the hole into which the metal is poured. The *riser pin* is similar and forms the hole for the air to escape as the metal rushes in. The *gate cutter* is used to cut channels in the sand to carry the molten metal from the sprue to the mold and out the riser.

Foundry Processes

How to Make a Sand Mold

The measure of a good sand mold is the accuracy with which the cavity is made from the pattern. Follow these suggestions as carefully as possible:

1. Place the drag on a molding board with the locating pins pointed down. Center the pattern with its flat side down on the board inside the drag. Dust the pattern with *parting compound.*

2. Riddle about a half-inch of sand over the pattern.

3. Fill the drag heaping full with unriddled sand and pack it firmly with the rammer. Strike the sand off evenly with the striking bar.

4. Place another mold board on top of the drag. Carefully turn the drag over, with the boards

Fig. 4–58. (top left) The griddle pattern is placed on the pattern board in the drag and is dusted with parting compound. Fig. 4–59. (top right) Sand is riddled over the pattern.

Fig. 4–60. (right) With the sand rammed firmly, the drag is turned over and the other side of the pattern is dusted. Fig. 4–61. (bottom right) The sprue pins are pressed into position.

held firmly in place. Remove the top board and place the cope on the drag. Set the sprue and riser pins in place, about an inch from the pattern on opposite sides. Dust the pattern with parting compound.

5. Riddle, then fill with sand as before. Ram and strike it off.

6. Remove the pins and with your fingers shape each opening like a funnel.

7. Lift off the cope and set it down gently so that the sand is not disturbed. Tap the pattern lightly from side to side so that you can lift it out. Be careful not to damage the cavity. Cut the gates.

8. Blow off any loose sand with a bellows and replace the cope on the drag. The mold is now ready to pour.

Pouring the Metal. When the metal is hot enough, lift the crucible out of the furnace with the tongs. Pour it into the sprue hole in a continuous

stream as fast as the hole will take it. When the metal comes to the surface in the riser, the cavity is full. After the metal has cooled, set the flask in the molding bench and dig out the casting.

Pattern Making. Almost any simple shape can be used as a pattern. But you must be able to remove it from the sand without damaging the cavity. The sides of the pattern must be tapered or rounded so that it can be withdrawn easily. This taper is called *draft*. Inside corners should be rounded too. They can be filled and smoothed with wax. This is called a *fillet*. Wood and plaster of paris make good patterns. Shellac and wax them well before using so they will part from the sand cleanly.

Fig. 4–62. (top left) Cutting the gates, through which the metal will flow. Fig. 4–63. (left) The mold is filled with molten aluminum. Note the protective gear that is worn to protect against hot metal that might be spilled and that might splatter from the mold if the sand were too moist.

Fig. 4–64. (bottom left) The rough castings. Fig. 4–65. (bottom right) Patterns for outboard engine crankshafts to be cast in steel. The patterns are of the two halves of two crankshafts.

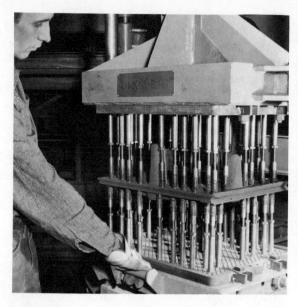

Fig. 4–66. This machine makes the mold from the patterns. A special waterless molding sand is used. The particles of sand are held together by a plastic material binder.

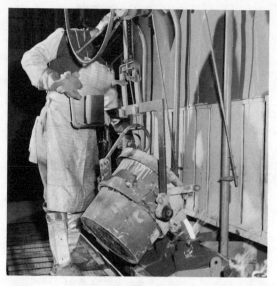

Fig. 4–68. Molten metal is poured into a shell mold for casting of the crankshafts. The pattern includes big lobes for excess metal, the "gating system," that insures that the metal gets into all parts of the mold. Notice the protective garments on the operator.

Fig. 4–67. The two halves of the mold will be closed together like a book. They will then contain the cavities to be filled with molten steel. This kind of mold is called a "shell" mold because it is thin and hard.

Fig. 4–69. Drilling is a machining process. Note the shavings. Both lips of the drill bit (twist drill) are cutting as they should. A stream of coolant keeps the tool cool. A dull drill overheats more quickly than a sharp one. Can you figure this out?

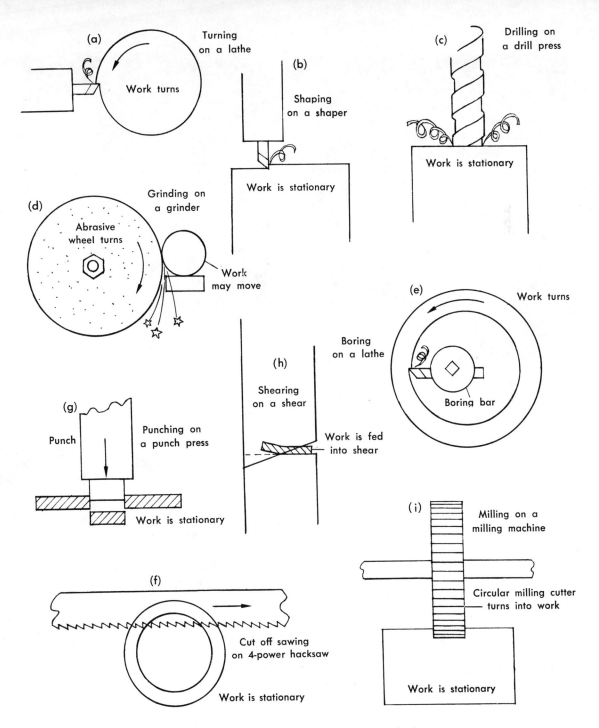

(a) Turning on a lathe — Work turns

(b) Shaping on a shaper — Work is stationary

(c) Drilling on a drill press — Work is stationary

(d) Grinding on a grinder — Abrasive wheel turns — Work may move

(e) Boring on a lathe — Work turns — Boring bar

(g) Punch — Punching on a punch press — Work is stationary

(h) Shearing on a shear — Work is fed into shear

(f) Cut off sawing on 4-power hacksaw — Work is stationary

(i) Milling on a milling machine — Circular milling cutter turns into work — Work is stationary

Fig. 4–70. Some basic processes for shaping metal. These are often called "machining metals."

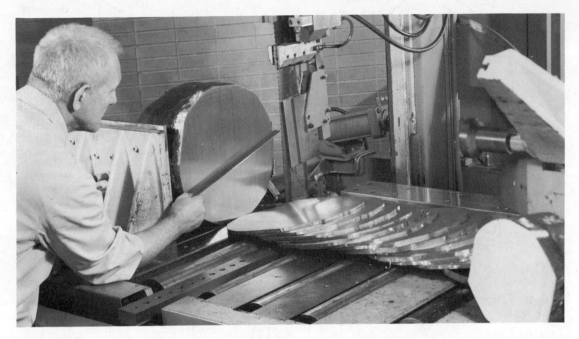

Fig. 4–70(a). A metal cutting band saw cuts plates from a huge block of metal.

✚ SAFETY SENSE

When you are working with molten metals:
1. Wear a face shield, gloves, and leggings.
2. Never add wet metal to molten metal, nor pour when the sand mold is wet. Can you figure out why?
3. Light a melting furnace with a torch, not a match.
4. Keep those persons not actually helping with the pouring at a safe distance.

METALS MACHINING

Products that originate as castings and forgings often require shaping or cutting to precise dimensions. Such processing is called *machining*. It involves a cutting away of metal, not a stretching, bending, or hammering. The department in a factory that does this work is a *machine shop*.

Fig. 4–71. A horizontal milling machine shapes the end of a bar. Coolant carries away the heat.

The skilled workers are *machinists*. The basic machines are machine lathes, drill presses, grinders, shapers, and mills. Automatic models are used for production. When automatic machines are linked together we have *automation*. Parts are processed and moved from one machine to another automatically.

Measuring Devices

Layout and Measuring. All of the machining processes require layout, measuring, and setup. Layout refers to the actual marking of the locations for the cuts. Making the setup involves the adjustment of the machine so that the correct cuts are made. Measuring includes the measuring and checking of dimensions before and after cutting. (See pages 139–143 for the tools and techniques in layout.) Measurements are made with such devices as surface plates and gauges, depth gauges, screw pitch gauges, and micrometers. (See Fig. 4–71.)

The measurement devices described here are those which you will be able to use with accuracy. They are commonly found in industrial arts laboratories. Modern metals industries use types that measure with greater accuracy, in millionths of an inch, for example. This accuracy is required in today's precision machines such as computers, jet engines, and space capsules. If you think that one one-thousandth of an inch is tiny, imagine one one-millionth of an inch! How much smaller is it?

Surface Plate and Gauge. A surface plate is a flat, true, iron surface that serves as the base line from which heights are measured on the piece being machined. The most accurate surface plates are made of black granite. A surface gauge is the tool used on the plate for marking the heights on the work.

Depth Gauge. This is a narrow steel rule with a movable crosspiece. It is used to measure depths of holes and slots.

Screw Pitch Gauge. The number of threads per inch on a bolt or nut can be accurately checked

Fig. 4–72. Some layout and measuring tools for machine work.

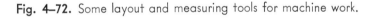
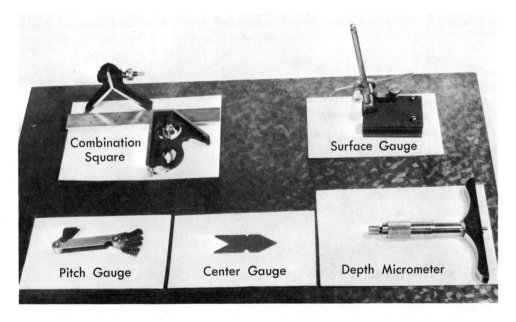

with this tool. Find the blade that fits the thread and read the number on it.

Micrometer. A micrometer is the machinist's most-used precision measuring tool. He calls it a "mike." There are styles for inside and outside measuring; all are read alike. The outside mike resembles a C clamp. It has 40 t.p.i. (teeth per inch) on the screw, so one turn moves it $\frac{1}{40}$ inch, or 0.025 inch. The thimble is marked off in 25 equal parts, each of which is $\frac{1}{1000}$ inch, or 0.001 inch. On the hub, the divisions are marked off as on a rule. The numbers and the long marks below them are $\frac{100}{1000}$ inch, or $\frac{1}{10}$ inch, or 0.1 inch apart. There are four spaces between two adjacent numbers, as between 2 and 3 and 4. Each of these spaces is $\frac{1}{4}$ of $\frac{100}{1000}$, or $\frac{25}{1000}$, or 0.025 inch.

How to Read the Micrometer

1. Hold the mike in one hand and turn the thimble until it is closed. The reading should then be zero with the 0 mark on the thimble directly over the 0 mark on the hub.

2. As you slowly back off the thimble, each mark on it that passes the horizontal line on the hub means 0.001 inch. In one complete turn there are 25 of these, or 0.025 inch. Notice that this one complete turn exposes the first short mark on the hub, or 0.025 inch.

3. To read the mike, then, the reading on the hub is added to the reading on the thimble. In Fig. 4–74 the reading on the hub is $\frac{150}{1000}$, or 0.150. This is added to the $\frac{3}{1000}$ inch or 0.003 inch on the thimble, for a total of 0.153 inch.

4. To measure with the mike, hold it in the left hand. Screw the thimble down until it lightly touches the stock. Do not screw it down tight. Then carefully slide it off and read the setting. Never mike a piece of metal while it is turning, as in a lathe. If the micrometer is dropped, the accuracy may be destroyed.

Fig. 4–73. The micrometer measures in thousandths of an inch.

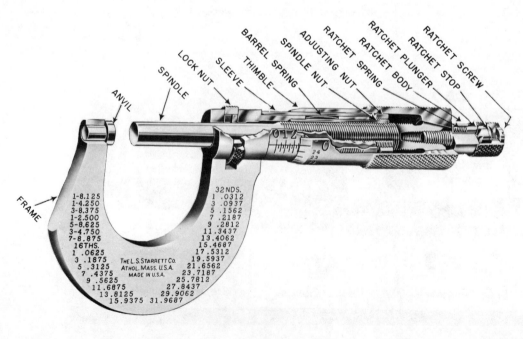

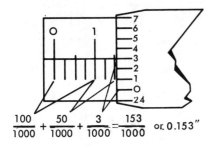

$$\frac{100}{1000} + \frac{50}{1000} + \frac{3}{1000} = \frac{153}{1000} \text{ or, } 0.153''$$

Fig. 4–74. Reading a micrometer.

Grinding

Grinding is the process of cutting away material by means of abrasives. The abrasive may be in the form of a wheel that cuts as it revolves. It may be applied to a disc or to a sheet or belt. *Surface grinders* cut flat surfaces and *cylindrical grinders* cut surfaces that are cylindrical or round. The *tool grinder* is the type with which you are probably most familiar.

Tool Grinder. This machine is intended for sharpening tools and the light grinding of stock. It usually has two wheels, one coarse and one medium or fine. Here are some tips to help you use the machine wisely and safely:

1. Always use goggles or a face shield when grinding.

2. Support the tool on the rest, which should be about ⅛ inch from the wheel. Adjust the rest before you turn on the motor.

3. Press the tool lightly against the wheel. Keep it moving over the surface to prevent the wearing of grooves in the wheel and the burning of the metal. The wheel should turn into the edge, not away from it.

Limitations on the Tool Grinder. The grinder is not intended to take the place of a hacksaw. If any appreciable amount of metal is to be removed, use the saw whenever possible. Soft metals such as copper, brass, aluminum, and lead should not be cut on the tool grinder. They clog the abrasive and make it useless. When a tool gets dull it may need only honing. For every grinding there are probably six or eight honings. Ask your teacher when you are in doubt about which process to use.

Drilling

Holes in thick metal are usually drilled. In thinner stock they may be punched. Drilling is a process of cutting away chips or shavings. Punching removes the metal in one piece the size of the hole. (See page 149.) Industry uses many types of drilling machines, or drill presses. There are models that drill one hole at a time or a dozen or more. (For instructions on the use of the drill press, see pages 104–107.)

Drill Bits. The tool that cuts the hole is the drill bit. (See page 90 for details.) Drill bits for machine shop use should be of high-speed steel rather than ordinary carbon steel.

Fig. 4–75. Grinding on a tool grinder.

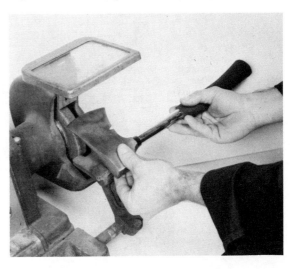

The Machine Lathe

The machine lathe is the king of machine tools. It is the most versatile because so many processes can be performed on it. These include facing, turning to diameter, cutting off, drilling, boring, tapering, and threading. Some lathes are so small that watch parts can be made on them. Others are so large that parts weighing tons can be turned.

The principal parts of a small lathe are shown in Fig. 4–77. The motor turns the *headstock*. This supports one end of the stock to be machined. The *tailstock* supports the other. The *cutting tool* is clamped in a *tool holder*. This is held in the tool post on top of the *carriage*. This is moved along the *bed* by the *hand wheel* or driven automatically by the *lead screw*. The cutting tool is moved into the stock by turning the handwheel of the *cross-feed* or that on the *compound slide* or both. The tailstock can be moved back and forth and locked

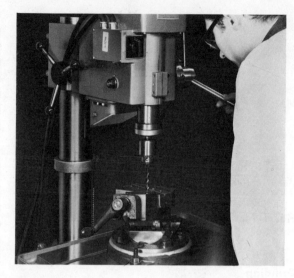

Fig. 4–76. A small drill press. Note that the work is held in a vise. This assures accuracy and safety. When it is not possible for you to clamp the work in a drill press check with your instructor. He can help you figure out a way to drill the hole.

Fig. 4–77. A machine lathe for school use.

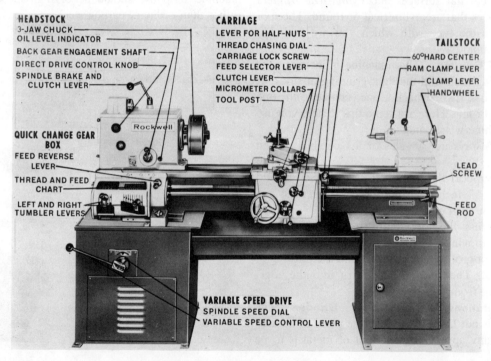

HEADSTOCK
3-JAW CHUCK
OIL LEVEL INDICATOR
BACK GEAR ENGAGEMENT SHAFT
DIRECT DRIVE CONTROL KNOB
SPINDLE BRAKE AND
CLUTCH LEVER

QUICK CHANGE GEAR
BOX
FEED REVERSE
LEVER
THREAD AND FEED
CHART
LEFT AND RIGHT
TUMBLER LEVERS

CARRIAGE
LEVER FOR HALF-NUTS
THREAD CHASING DIAL
CARRIAGE LOCK SCREW
FEED SELECTOR LEVER
CLUTCH LEVER
MICROMETER COLLARS
TOOL POST

TAILSTOCK
60° HARD CENTER
RAM CLAMP LEVER
CLAMP LEVER
HANDWHEEL

LEAD
SCREW

FEED
ROD

VARIABLE SPEED DRIVE
SPINDLE SPEED DIAL
VARIABLE SPEED CONTROL LEVER

in position. Its *ram* is moved in or out by the hand wheel.

Limitations of the Machine Lathe. The machine lathe is a precision machine and must be used carefully if it is to retain this accuracy. No hammering or other rough treatment is done on the stock in the lathe. Cutting metal is a slow process, so you must be patient. Take only as deep a cut as the lathe can adequately handle. For the most part, any adjustment and any setting on the lathe is either right or wrong. Be sure to have your teacher check you out on any setup before you turn on the motor.

Speeds and Feeds. Different speeds are required for different metals and for different sizes of stock. Soft metals can be cut faster than steel of the same size. Large-diameter stock requires slower speeds than small diameters. Speeds of the headstock spindle can be changed by shifting belts. *Back gearing* provides extra slow speeds. To obtain this, pull out the *bull pin* on the front of the headstock pulley. Then gently engage the gears by means of the *back gear lever*.

Feed is the rate at which the carriage, carrying the cutting tool, travels along the bed. When the feed is automatic, the power is transmitted to the carriage by the gears on the end of the lathe through the long *lead screw*. Some lathes have a *quick change gear box* by which several different feeds can be quickly obtained.

Chucks and Chucking. Chucks are devices that screw on the headstock spindle and hold stock without the aid of centers. The 4-jaw chuck has jaws that are tightened independently. It is used to hold any shape of stock that is centered by trial and error. For a round piece, the jaws are screwed

Fig. 4–77a. A small horizontal milling machine for school use.

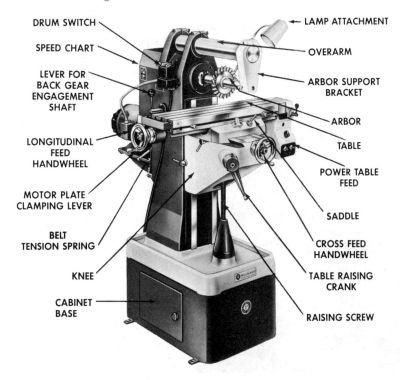

DRUM SWITCH

SPEED CHART

LEVER FOR BACK GEAR ENGAGEMENT SHAFT

LONGITUDINAL FEED HANDWHEEL

MOTOR PLATE CLAMPING LEVER

BELT TENSION SPRING

KNEE

CABINET BASE

LAMP ATTACHMENT

OVERARM

ARBOR SUPPORT BRACKET

ARBOR

TABLE

POWER TABLE FEED

SADDLE

CROSS FEED HANDWHEEL

TABLE RAISING CRANK

RAISING SCREW

down so that they are spaced equally by the concentric rings on the face of the chuck. Then the motor is started and a piece of chalk is held so that it marks the high side of the revolving metal. The high point is moved toward the center by adjusting the jaws.

The 3-jaw *universal chuck* holds round work. The jaws move at the same time when they are screwed in or out and automatically center the stock. Either type chuck holds either inside or outside as the stock permits. Drill chucks are used on the lathe for drilling holes and centers. They fit into the headstock spindle or tailstock by means of tapered *sleeves*.

Cutting Tools and Holders.   Cutting tools are small square-sectioned cutter bits. They are ground to various shapes and angles to suit cutting requirements. Tool holders are either left, right, or straight. The left-hand holder is most used. When you set up the lathe, the cutter bit should project not more than ⅜ inch from the tool holder. The holder should be back in the tool post as far as possible. The point of the bit is set to the center

Fig. 4–78. Chucking a piece in a 3-jaw chuck. It is locked securely with the chuck key.

height of the stock. To do this, align it with the point of the tailstock center.

Center Drilling.   When stock is to be turned between centers, it must have bearing surfaces cut into the ends to fit the lathe centers. These surfaces are provided by center drilling. Center the stock in the lathe chuck and hold the *center drill,* a special drill, in a drill chuck that is inserted in the tailstock. Move the tailstock up close to the work and lock it on the bed. Turn on the motor. Lubricate the drill with cutting oil as you screw it into the stock with the hand wheel. Drill to about three-fourths of the taper.

Turning Between Centers.   With centers drilled in both ends of the stock, you are ready to turn it. These are the steps:

1. Screw a *face plate* on the headstock spindle. Wipe out the spindle hole and tailstock hole. Insert the centers.

2. Attach a lathe *dog* to o   end of the stock (use the smallest dog that will fit).

3. Place a small amount of white lead lubricant in the center hole at the tailstock end. Insert the stock between the centers and screw up the tail center just snug enough to remove any play.

4. Tighten the tool bit and the straight tool holder in place with the bit on center.

5. The cut should be made from the tailstock end toward the headstock. Turn on the motor. Screw the cutter bit into the stock to take a light cut. Move the carriage to the left for ¼ inch or so to try the cut. Reset the cut if necessary. Engage the power feed and let the machine do its work.

6. If the stock must be turned even smaller, disengage the carriage at the end of the cut. Take a reading on the micrometer collar of the crossfeed knob. Then turn the knob counter-clockwise just enough to clear the stock. Return the carriage to the starting place. Screw the tool bit to the same micrometer reading. Start the lathe, screw the cutter bit in for a new cut, and proceed as before.

Facing.   Facing is the cutting of a flat surface on the ends or faces of stock. The work is held in a chuck while the cutting tool is moved across

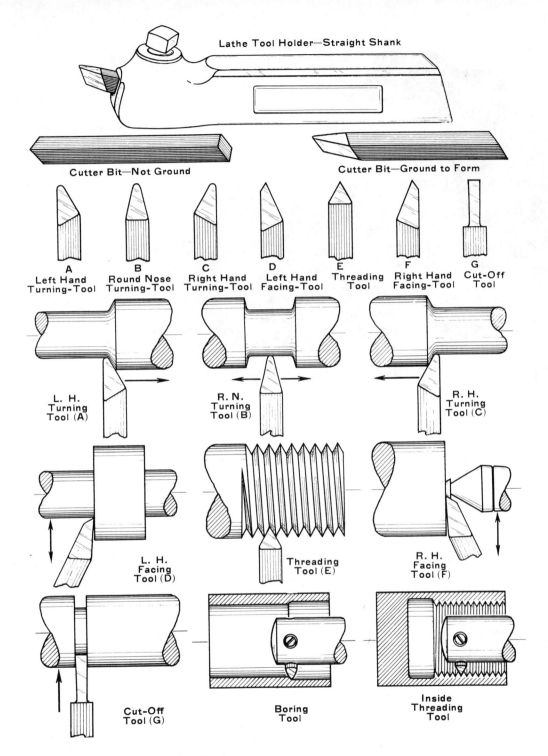

Lathe Tool Holder—Straight Shank

Cutter Bit—Not Ground   Cutter Bit—Ground to Form

A
Left Hand
Turning-Tool

B
Round Nose
Turning-Tool

C
Right Hand
Turning-Tool

D
Left Hand
Facing-Tool

E
Threading
Tool

F
Right Hand
Facing-Tool

G
Cut-Off
Tool

L. H.
Turning
Tool (A)

R. N.
Turning
Tool (B)

R. H.
Turning
Tool (C)

L. H.
Facing
Tool (D)

Threading
Tool (E)

R. H.
Facing
Tool (F)

Cut-Off
Tool (G)

Boring
Tool

Inside
Threading
Tool

Fig. 4–79. Application of lathe tools.

Fig. 4–80. Drilling a center hole in stock held in a 3-jaw chuck. Feed the drill with the tailstock hand wheel. The drill is held stationary in a chuck while the work revolves. Use cutting oil while drilling. Start with a small diameter hole and enlarge it, if necessary.

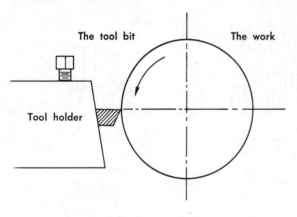

Fig. 4–80a. The tool bit is set to cut at the center line of the work.

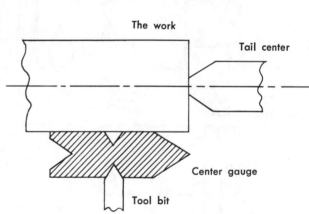

Fig. 4–80b. Setting the tool bit square with the piece being turned.

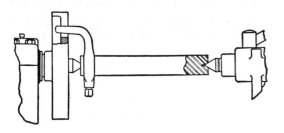

Fig. 4–81. Work properly set between the lathe centers is now ready for machining.

the surface. The tool should be set at right angles to the direction of travel. Set it for the center height and cut from the center of the stock toward yourself. Lock the carriage to keep it from moving away. The stock should not project more than an inch or two from the jaws of the chuck.

Cutting Off Stock. This is done with the cut-off tool. The stock is held in a chuck as close to the mark as possible. The tool point is set on the exact center and at 90 degrees to the work. Run the lathe at the slowest speed and feed the tool into the metal by hand at a uniform rate. Use cutting oil on it. Do not try to cut off stock held between centers.

Turning Tapers. For short tapers, as on a center punch or lathe center, hold the stock in a chuck and turn the compound rest to the degree of taper required. The tool is set on center and the carriage locked in place. Turn the feed screw to make the cut.

Long tapers are cut by means of a taper attachment or by setting the tailstock over. The latter is done by means of the small set screws in the base of the tailstock. For straight turning, the tailstock and the headstock must be in perfect alignment. This is indicated on the scale at the rear of the tailstock base. To find the amount of set-over for a taper, use this formula:

$$\text{Set-over} = \frac{\text{Taper per foot, in inches}}{2} \times \frac{\text{length of piece}}{12}$$

Thread Cutting. Threads can be cut on a machine lathe. This is precision work. It is usually done when the thread cannot be cut with taps or dies. The following procedure is for cutting external threads, as on a bolt. Your teacher may change the procedure to fit your lathe.

1. Set the quick change gear box for the desired number of t.p.i. (threads per inch).

2. Tighten a sharp, correctly ground tool bit in the tool holder.

3. Set the tool bit point square with the work. It should cut at the exact center line.

4. Set the compound tool rest to the right at 29 degrees and tighten.

5. Screw the tool bit in to just touch the work. Loosen the thumb screw on the micrometer collar on the crossfeed. Align the collar to "0" and tighten the screw.

6. Engage the thread dial indicator. Your teacher will explain this device.

7. Make a light trial cut. The cut is always made from the tailstock end.

Fig. 4–82. The flapjack griddle is held in a 4-jaw chuck for facing.

8. Stop the lathe at the end of the cut. Check the cut for the proper t.p.i.

9. Turn the cross feed screw one turn to the left. Release the thread indicator. Return the carriage to slightly beyond the starting point.

10. Turn the cross feed screw one turn to the right, to "0." Engage the thread dial. Screw the compound tool rest in for the desired depth of cut. Begin the cut.

Knurling. Knurling is the process of rolling a raised, patterned texture on the surface of stock. It makes a good hand grip on tools and adds a decorative touch. The knurling tool is clamped in the tool holder at right angles to the work. Set the lathe for a slow speed and turn the knurls into the stock until they make a distinct pattern. Screw them in until they make a deep pattern. Then engage the carriage feed. Use plenty of oil. To make a deeper cut reverse the lathe spindle when the knurl has reached the left end of the work. The

Fig. 4–83. Knurling the surface of a piece of steel in a machine lathe. Knurling is the process of embossing the surface for decorative or functional purposes. A special tool is used.

tool then automatically moves back to the start without removing the knurls from the pattern. Screw them in deeper and repeat the process.

Lubricants

Machining metals always produces heat. Special lubricants and coolants are used when needed. Use *lubricating oil* for bearings. *Cutting oil* is for threading, and *white* or *red lead* for the tailstock center.

The Shaper and the Mill

The shaper uses a long, horizontal stroke in cutting similar to the stroke of a hand plane on wood. The mill uses a revolving gearlike cutter that moves across the face of the material either horizontally or vertically.

Innovations in Processes

With the advancing technology in metals, there are innovations in processes as well as in products. Sheet and plate metals can be shaped by *explosive forming*. A flat piece is clamped over the opening in a mold. Over the top of the work an airtight chamber is clamped. The pressure wave of the explosive charge within the chamber forces the metal into the mold. (See copper core coins, page 136.)

Electrical discharge machining (EDM) is an automatic process for machining very hard metals. A high frequency electric arc between the tool and the piece being machined melts away the surface of the piece. This is done with the work submerged in a nonconducting fluid. This washes away the eroded metal. This process is especially useful in forming pieces too large for available presses.

As electroplating deposits a thin coating of one metal onto another, *electrochemical machining* (ECM) removes metal. A direct electric current passes between the tool and piece being machined. The current is carried by a stream of

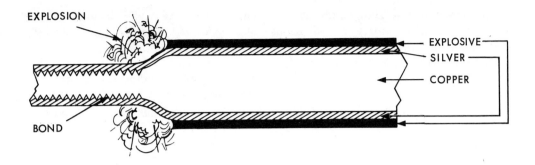

Fig. 4–84. Explosive bonding of copper and silver to produce the new coin stock. The controlled and continuous explosion across the width of the metals is done automatically in a special machine. The explosive force drives the metals together after which the silver-clad copper is rolled to coin thickness.

electrolyte rather than in a bath as in electroplating. This process makes it possible to machine sheet metals.

✚ SAFETY SENSE

When you are machining, be sure to observe these rules:
1. Take off your tie and roll up your sleeves.
2. Wear a face shield for turning, grinding, drilling, or any operation where chips may fly.
3. Never brush off chips with your hands. This is like brushing off razor blades.
4. Do not oil or adjust machines that are running.
5. Get your teacher's help before you begin using the machine lathe.

FINISHES FOR METALS

Many finishes may be used on metals. Select one that best suits the use for which the product is intended.

Buffing. A high polish is obtained by buffing. Hold the metal against a revolving cloth wheel that is charged with a buffing compound, a fine

Fig. 4–85. Tool marks can be removed by filing. The lathe is run on a speed permitting two to three revolutions of the work for each stroke of the file. Use a full stroke with a fine mill file. Too much filing can destroy the accuracy of the work. File marks can be removed by polishing with abrasive cloth and oil.

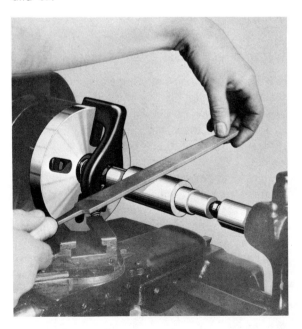

abrasive. (Common buffing compounds are tripoli and rouge.) Hold it securely in both hands below the center of the wheel.

Scratch Brush Finish. A soft, satin finish is produced by holding a soft metal such as copper or aluminum against a revolving wire brush. Be sure to wear goggles when you do this.

Spot Finish. This finish is for flat surfaces. Chuck a short piece of ½-inch dowel in the drill press. Spread a coat of fine abrasive powder and lubricating oil on the surface of the metal. Bring the spinning dowel down against the metal to make spots of shiny rings. Overlap them uniformly.

Copper Finishes. Copper finishes require chemicals for antique effects. Many prepared finishes are available. Liver of sulphur (potassium sulfide or sulphurated potash) mixed with water will do this, too.

Lacquers. To preserve the bright polish on copper and brass after buffing, brush or spray on a coating of clear lacquer. Transparent colored lacquers tint copper, brass, and aluminum.

Enamels. First use a metal primer over the bare clean metal. Flat black is appropriate on many iron and steel projects.

WELDING PROCESSES

Welding is the process of joining metals by melting them together. In his day the village blacksmith was the local welder, too. He could heat two pieces of steel in his forge and then hammer them together on his anvil. Since then better methods of welding have been perfected. Skyscrapers, ships, bridges, automobiles, airplanes, and many other products are held together by welding. Today there are two basic types of welding. The gas method uses heat resulting from the burning of acetylene gas with oxygen, producing temperatures as high as 6000°F. The other type uses electricity in various ways to produce heat; for example, there are electric arc welding, resistance welding, and induction welding. All types are done automatically in industry. Oxyacetylene and electric welding are described here because you can perform them safely and successfully.

Oxyacetylene Welding Equipment

This equipment includes two metal cylinders or tanks, one of which contains oxygen, and the other, acetylene (a-*set*-e-lin). A *regulator* is used to reduce the pressure and maintain an even flow from each tank. Each of these regulators has two *gauges*. One tells how much gas is in the tank. The other shows the working pressure for welding. In the center of each regulator is a hand screw by which the working pressure is adjusted.

The green hose carries the oxygen, and the red hose, the acetylene. The torch has two valves for adjusting the flow of each gas. Different *tips* are used on the torch, depending on the size of the flame needed. A cutting head may be attached to the torch for burning cuts through heavy metals. The *ignitor* is the lighter for the torch. Do not use a match. The flame is very bright, so special goggles must be worn when you weld or watch welding. Asbestos gloves will protect your hands.

Fig. 4–86. The welding gauges and regulators on acetylene and oxygen tanks.

How to Weld with Oxyacetylene

You should be able to make satisfactory welds after a few trials. The idea is simply to melt two pieces of metal together. Follow these steps:

1. First, make sure that the regulator hand screws are backed off so that no gas can get through. Close both valves on the torch. Open the oxygen tank valve slowly to allow pressure to build up gradually on the high-pressure gauge. When the hand stops, open the tank valve as far as it will go. Turn the hand screw in until the low-pressure gauge shows the correct working pressure for the tip you will use. See your teacher about this.

2. Turn on the acetylene equipment in the same manner except that the tank valve should be opened only 1½ turns. When you open tank valves, stand on the side opposite the regulators as a safety precaution.

3. Slip the goggles on your head and put on the gloves. Hold the ignitor in one hand. Open the acetylene valve on the torch a part of a turn. Light the gas by holding the tip close to the ignitor and directed into the flash cup. Open the valve until you have a large, sooty flame, without blowing. The flame should touch the tip. Quickly turn on the oxygen until the flame is neutral. This is your welding flame.

To close down the equipment, first shut off the acetylene, then the oxygen at the torch. Now turn off each tank valve. Open the torch valves and drain the hoses. Back off the regulator hand screws.

4. Making puddles is the first step in welding. Get a piece of ⅛-inch steel, a few inches long to practice on. Light the torch and hold it so that it balances easily in your hand. Bring the tip down to the steel with the flame at a 45-degree angle. Hold the cone of the flame about ⅛ inch from the metal. As soon as it begins to melt a puddle, move the torch slowly along, revolving the tip in small circles at the same time. These puddles should be about 3/16 inch wide. Practice this several times. Try to keep the puddles uniform, with even ripples. When you move the torch too slowly, the metal burns through. When it moves too fast, there is no melting.

5. When you can run a good puddle on a flat surface, place two pieces of the metal edge to edge. Puddle each end together, then puddle the edges.

Laying Beads. The next thing to try after puddling is the laying of a *bead* on a flat surface. A welding rod is used for filler. Hold the torch as before, in one hand, and a piece of 3/32-inch welding rod in the other. Keep the tip of the rod just inside the outer layer of the flame so that it will be almost melted. As the puddle forms, touch the rod to the puddle to deposit some metal there. Then withdraw the rod slightly. Continue puddling and adding filler rod as needed to build up a bead. A bead should be similar to a puddle except that it is built up above the surface about 1/32 of an inch.

Now place two pieces of stock edge to edge. Tack each end as before, then weld them. This is a *butt weld* and it should penetrate to the other side of the metal. To make a *lap weld* lay one piece of stock overlapping the other. The heat must be directed to the lower piece to keep the other piece from burning.

Brazing. Brazing is a combination of welding and hard soldering. A bronze filler rod is used with a *flux*. You can more easily braze cast iron than you can weld it. It will be fully as strong. The process is also used on steel, copper, brass, and bronze.

First grind or file the surfaces clean or the bronze will not stick. Bring them to a red heat, not a melting heat. Heat the end of the rod and dip it in the flux. Touch the rod to the hot metal and play the flame over the surfaces. When the metal is at the right heat, the bronze will flow over it. If it is too hot, the bronze bubbles and burns off. If it is not hot enough, the bronze rolls up into globs. Repeat this heating and filling to build up thickness and strength. Let the stock cool until the red has disappeared before moving it.

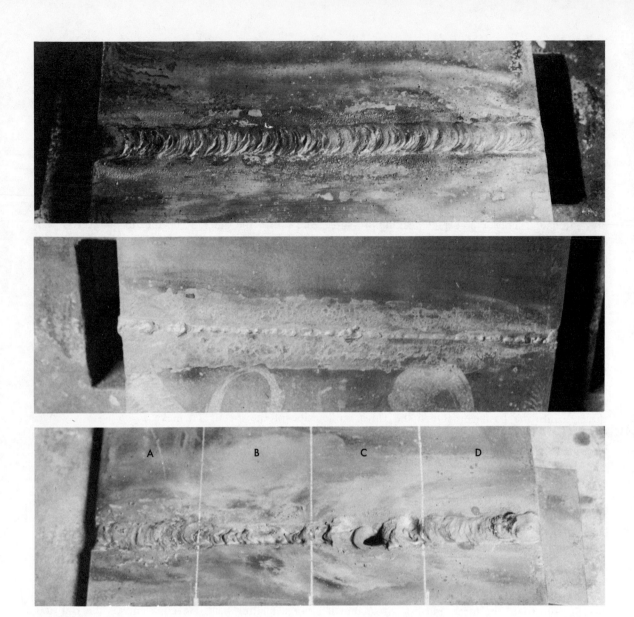

Fig. 4–87. (top) A good fusion weld. The bead stands only slightly above the metal being welded. The ripples are uniform and properly spaced. **Fig. 4–88.** (middle) The underside of a good fusion weld. The small beads indicate that the penetration of the weld through the metal is complete. **Fig. 4–89.** (bottom) Four different welds. Section A shows a poor weld caused by the use of too much oxygen. Section B shows a poor weld due to insufficient heat. Section C has holes burned through the metal because of too much heat. Section D is a good weld.

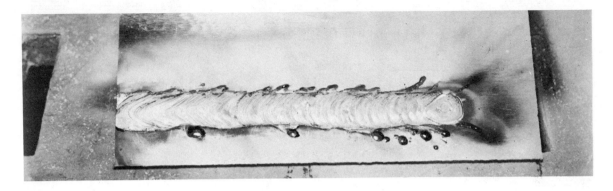

Fig. 4–90. Here is an example of a properly done braze weld. During braze welding, the torch and the filler rod should be held at angles of about 45 degrees to the surface of the plate, with the flame at the forward edge of the puddle.

Fig. 4–91. The intense light, heat, ultra violet rays, as well as sparks emitted during electric arc welding make protective clothing and face shields necessary. Be sure that no one watches you while you weld unless he is adequately protected.

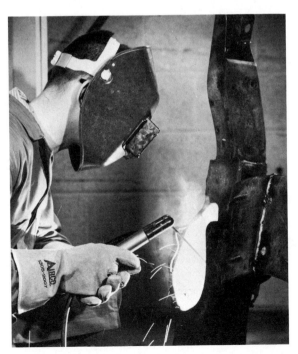

Electric Arc Welding

Arc welding uses the heat of an electric arc to melt the metal and the rod at the same time. It has the advantage of speed and economy and is especially suitable to the welding of thick metal.

Arc-welding Equipment. There are two main types of arc welders, the alternating and the direct current. The first generally uses a transformer, which steps down the voltage and steps up the amperage. The second uses a motor-driven generator to supply the welding current. Two heavy cables carry the current. The electrode (welding rod) holder is attached to one of them. The other is clamped to the work or to the welding table. (See Fig. 4–92.) A special helmet with built-in filter glass is used for protection of the face and eyes.

How to Arc Weld

Arc welding is fascinating and is entirely safe when properly done. You will get accustomed to the flash, which at first may cause you to jump. As you watch the flame closely through your shield, imagine that you are peering into the crater of a spewing volcano. Actually, arc-welding tem-

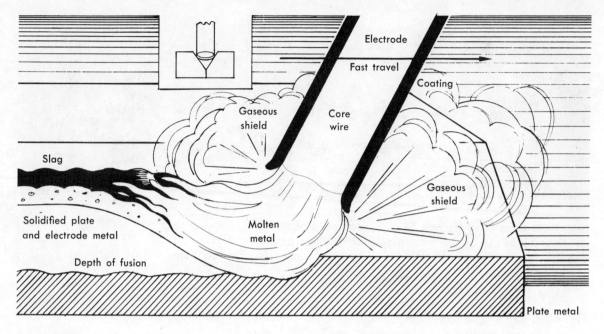

Fig. 4–92. Diagram of fusion of metals in electric arc welding.

peratures are hotter than the interior of a volcano; they exceed 6000°F. Here are the steps:

1. Use a piece of scrap metal not less than ⅛ inch thick and then check a welding chart to find the correct size of electrode. Your instructor will show you how to set the machine for the proper welding current.

2. Lay the metal on the welding table and clamp the ground cable to it or to the welding table. Adjust the helmet to fit. Turn on the ventilating fan, and draw the curtains around the booth. Insert the electrode in the holder and turn on the welder.

3. Drop the helmet over your face. Bring the end of the rod down to the metal in a short, sweeping motion as in striking a match. As soon as the arc forms, raise the electrode to about 1/16 inch to ⅛ inch from the metal to hold the arc. When the arc is struck, metal and electrode melt together.

4. Move the rod from side to side in small half-moons about ⅜ inch wide as you move it along the metal. The rod melts away, so you must

Fig. 4–93. By beveling butt weld edges, the coated electrode can get down to the bottom of the joint and a weld can be made through the entire thickness of the plates. This makes for a very strong weld.

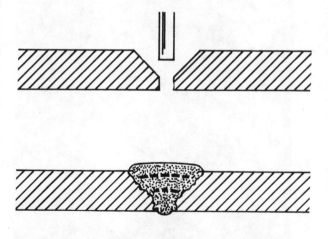

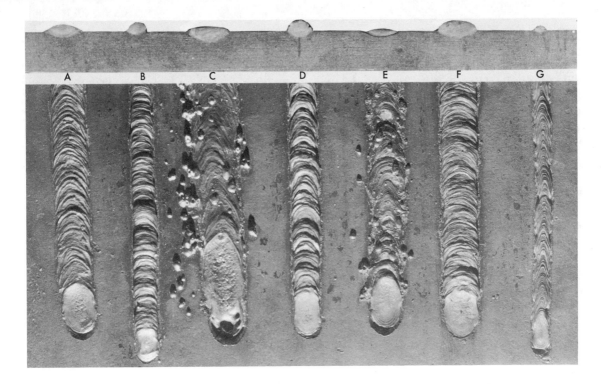

Fig. 4–94. Some variations in arc welding beads. Note the cross sections of these welds at the top of the picture. (A) A good weld with current, voltage, and speed of electrode normal. (B) Current too low. (C) Current too high. (D) Voltage too low. (E) Voltage too high. (F) Speed too slow. (G) Speed too high.

keep moving it into the metal. When you have an inch or so of bead, stop and inspect it. Chip away the scale with the chipping hammer so that the weld is exposed, then show it to your teacher.

5. When you can run a bead that passes your teacher's inspection, try to butt-weld. Lay two pieces of stock edge to edge and run the bead between them.

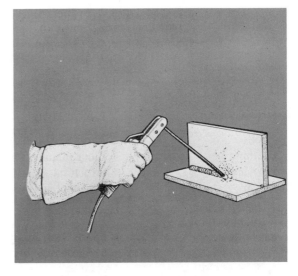

Fig. 4–95. Holding the electrode for fillet welding. Tack-weld two 3/16" or 1/4" plates at all four corners to form an upside-down T. Use 1/8" E-6010 or E-6011 electrode.

Either type of welding must be done carefully to avoid fire and burns. Consider these suggestions for oxyacetylene welding:

1. Use only enough pressure to do the work.
2. If you smell unburned gas, turn off the torch and call your instructor. The gas is not only inflammable but it is explosive if not controlled.
3. Always shut off the torch as soon as you are finished.
4. Don't watch the flame without welding goggles. However, these are not adequate for arc welding.
5. Be careful where you point a lighted torch.
6. Stand so that no molten metal can fall into your trouser cuffs or shoes. Isn't that an interesting thought? It has happened.

7. Remember that the metal may still be too hot to handle without gloves even though it is no longer red.
8. There should be adequate ventilation.

These suggestions are for arc welding:

1. The ultraviolet rays produced by the flash can burn just as the sun can. Wear gloves and keep your sleeves rolled down. Never watch the arc without a shield. Your eyes can sunburn too. You can imagine how it would feel to have your eyeballs peel. Shield your welding so that others won't watch it.
2. The welding fumes are very objectionable. Be sure there is adequate ventilation.
3. If your shoe soles are wet or if the concrete floor is damp, don't weld unless you can stand on a dry board or some other good insulator. You could get a shock.

METALS AND OCCUPATIONS

Metals are used so widely in all areas of our living that you can expect many occupations to be involved. There are the white and the blue collar jobs. White usually refers to office work. Blue refers to shop and factory work. Many jobs are classed as skilled. Some are semiskilled and others are unskilled. Craftsman includes several types of work that usually require apprenticeship. Technician positions are part engineering and part craftsman. Technical institutes train technicians. Engineers and product designers require a college background. The following outline shows the great variety of work in the broad field of metals.

Areas of Living	Typical Usage of Metals
Foods	Farm machinery, fencing, food processing
Clothing	Fiber manufacturing, weaving, printing, sewing
Housing	Structural steel, heating, cooling, water, sewer
Power	Power plants, transmission, distribution, servicing
Transportation	Vehicles, tracks, conveyors, ships, boats
Communication	Electronic, printed, photographic
Recreation	Sports and hobby equipment, toys
Security	Medical, military, fire, police equipment

Some Typical Occupations

Semi-skilled
Machine operators
Assemblers
Shippers
Packers
Clerks
Attendants

Skilled
Machinists
Tool and die makers
Mold makers
Welders
Structural steel workers
Foundrymen
Patternmakers
Electricians
Mechanics

Technicians
Draftsmen
Electricians
Electronics
Computer
Hydraulic
Pneumatic
Medical
Inspectors
Servicing

Engineers
Structural
Mechanical
Electrical
Manufacturing
Metallurgists
Chemical
Aerospace
Marine

Product Designers
Sports equipment
Appliances
Jewelry
Watches
Tableware
Automobiles
Tools
Cameras
Artist-craftsmen

METALS AND RECREATION

With your experience in working with metals you probably have discovered that they are rather easily formed. There are ways to make the forming easy. But tools and machines are usually necessary. Some hobbies with metals require only hand tools. Others require both hand and machine tools. The following hobby ideas require few tools.

Art metalwork in copper, brass, aluminum, silver. For example: bowls, trays, ladles, forks, lamps, boxes, jewelry.

Ornamental ironwork: lamps, furniture, plant stands, fireplace tools, camping gear, grills, bird cages.

Jewelry: enameled copper pendants, etched aluminum, and cast silver or bronze.

Welded sculpture: scrap steel and miscellaneous junk parts welded or brazed together.

Metal casting: low melting metals can be melted with a liquid propane torch.

Fishing lures: combine metals with plastic, cork, wood.

Antiques: polishing and restoring antique brass, copper, pewter, wrought iron.

Reclaiming scrap copper and brass and aluminum.

If you have access to such machine tools as a machine lathe, drill press, and grinder, you can make quite a different line of items. For example, you could design and build a wood lathe, potter's wheel, jigsaw, steam engine.

METALS AND THE ENVIRONMENT

Metals are essential to our technology and economy. The two largest manufacturing industries in the country are steel and automobiles. Tools and machines basic to technology are chiefly of iron and steel. You could probably fill an entire notebook with a list of products made of or with metals.

The effects of metals' mining and smelting on the environment are well-known. Great areas of some states have been converted into huge craterlike pits by open pit mining. Metals are smelted and refined with heat. This is obtained from coal. Coal mining, too, has made vast sections of farm and forest land uninhabitable. Smoke and gases from steel mills and coke ovens blanket large sections of the country. Rivers are contaminated with wastes from mines and mills.

You see we depend heavily on metals for products, machines, jobs, and income. Yet we know that we must clean up and preserve the environment. This makes for really difficult national problems. You should study these problems. Try to come up with sensible answers. Keep alert to what is being done to solve them. Find out all you can about the following: strip mining, open-pit mining, deep mining, coke ovens, fly ash, sludge, smog.

TO EXPLORE AND DISCOVER

1. What are diamond dies?
2. Why steel can be magnetized.
3. How thin can gold be hammered?
4. How metal spinning is done.
5. How uranium is refined.
6. Why metals become hard with hammering.
7. How a blacksmith fastens shoes on a horse.
8. How your bicycle frame is fastened together.
9. Why some coins are made of silver.
10. Why a left-hand thread is used.
11. How copper and chrome plating are done.
12. Why oil makes a machine run easier.
13. Why copper wire is not suited for the element in your toaster.

14. How ball bearings are made.
15. The different metals used in an automobile.
16. What rust is; what tarnish is.
17. Why annealing makes metals soft.
18. The comparative tensile strengths of some common metals.
19. How your mother's pressure cooker was made.
20. How window screen is made.
21. What kind of metal makes the flash in a photographic flashbulb?
22. How tin-can metal (tin plate) is coated with tin.
23. Why some watches are nonmagnetic.
24. How a bellows works. What were they formerly used for?
25. How many different metals are known to man.

FOR RESEARCH AND DEVELOPMENT

1. How many aluminum soft drinks cans will make one pound of ingot? Can you convert this ingot into a useful project? This is an example of recycling.
2. Does your teacher need an automatic paddling machine for unruly students? Design and build a small working model. Consider all possible means of powering it; for example, gravity, spring, motor, magnetic.
3. What is the difference between spring steel and mild steel? Which would be better for an automobile bumper, if the bumper were actually to be that? Can you design a better bumper? How could it be better than the usual one?
4. Why do metals fatigue? How do they change in nature when fatigued? Conduct comparative tests among samples of aluminum and mild steel of the same cross-sectional area. What is work-hardening in metals?
5. Can you design a sports car chassis so that it can have interchangeable bodies? Try for a roadster and a convertible. Make scale models and drawings.
6. What differences are there among emery, aluminum oxide, and silicon carbide abrasives? How are they manufactured? Is sand used on sandpaper?
7. What do the color and shape of the sparks from a grinding wheel tell about the iron and steel being ground? Make some tests. Can you make drawings of the sparks? Could they be photographed in color?
8. Design a wood lathe for a home workshop using the arbor shown on page 113 as the headstock. Could this arbor be converted to a tailstock?
9. Why do metals expand when heated? Which expands more with equal heating, aluminum or steel? Test the expansion of two bars of these metals each 10 inches long. Measure them accurately at room temperature and then at 100 degree intervals up to 700°F. Plot the figures on coordinate paper and draw the curves. Compare your findings with those in an engineer's handbook.
10. How is wire made? Can you design and make a simple drawing die to demonstrate the principle? Why is it necessary to anneal the wire before drawing?

FOR GROUP ACTION

1. As an exhibit, make a model steel mill and rolling mill, showing steps and processes from ore to steel.
2. Another exhibit: The story of aluminum. Include a display of common aluminum products found in the home.
3. As a production project, make some jewelry by forming copper wire in jigs. First names or nicknames, and animals would be popular.
4. Small trays of aluminum can be pressed out in dies constructed of hardwood with pressure in a machinist's vise.
5. Make some planters or flowerpots for the local hospital.
6. Make a portable outdoor fireplace grill. Design it for mass production.

FOR MORE IDEAS AND INFORMATION

Books

1. Boyd, T. Gardner, *Metalworking*. S. Holland, Ill.: Goodheart-Willcox Co., Inc., 1961.
2. Feirer, John L., and J. R. Lindbeck. *Industrial Arts Metalwork*. Peoria, Ill.: Chas. A. Bennett Co., Inc., 1965.
3. Fraser, Roland R., and Earl E. Bedell. *General Metal*. Englewood Cliffs, N. J.: Prentice-Hall, Inc., 1961.
4. Johnson, Harold V. *General Industrial Machine Shop*. Peoria, Ill.: Chas. A. Bennett Co., Inc., 1970.
5. Walker, John R. *Modern Metalworking*. S. Holland, Ill.: Goodheart-Willcox Co., Inc., 1968.

Booklets

1. *American Standard Safety Code for the Use, Care, and Protection of Abrasive Wheels*. USA Standards Institute, 10 E. 40th St., New York, N. Y. 10016
2. *The Care and Operation of a Lathe*. Sheldon Machine Co., Inc., 4258 N. Knox Ave., Chicago, Ill. 60641
3. *Hack Sawing Helps*. The Henry G. Thompson Co., P.O. Box 1304, New Haven, Conn. 06505
4. *How to Run a Lathe*. South Bend Lathe, Inc., South Bend, Ind. 46623

5. *The Making of Steel; The World of Steel.* United States Steel Corp., Public Relations Dept., Room 1800, 71 Broadway, New York, N. Y. 10006
6. *The Oxyacetylene Handbook.* The Linde Co., Division of Union Carbide Corp., 270 Park Ave., New York, N. Y. 10017
7. *The Picture Story of Steel.* American Iron and Steel Institute, Public Relations Dept., 150 E. 42nd St., New York, N. Y. 10017
8. *Solders and Soldering Information.* Advertising Dept., Chase Brass and Copper Co., 20600 Chagrin Blvd., Cleveland, Ohio 44122
9. *Welding, Soldering, Brazing Aluminum—Different not Difficult.* Public Relations Dept., Reynolds Metals Co., P.O. Box 2346, Richmond, Va. 23218
10. *The Story of Aluminum.* Kaiser Aluminum and Chemical Corp., Technical Publication, 300 Lakeside Drive, Oakland, Calif. 94612

Charts

1. *Choosing and Using Disston-Porter Files; Disston-Porter Hack Saw Chart.* Disston Tools—H. K. Porter Co., Inc., Porter Bldg., Pittsburgh, Pa. 15219
2. *How Steel Is Made.* (Wall chart; teacher's kit of samples) United States Steel Corp., Public Relations Dept., Room 1800, 71 Broadway, New York, N. Y. 10006
3. *Steelmaking Flow Charts.* American Iron and Steel Institute, Public Relations Dept., 150 E. 42nd St., New York, N. Y. 10017

Films

1. *Alcoa Informational Aids.* (catalog of films and literature) Motion Picture Section, Aluminum Company of America, 1246 Alcoa Bldg., Pittsburgh, Pa. 15219
2. *Educational Films Catalog.* Free. 16 mm, sound films. Modern Talking Picture Service, Inc., 1212 Avenue of the Americas, New York, N. Y. 10036
3. *Film Catalog.* United States Steel Corporation, Public Relations Dept., Room 1800, 71 Broadway, New York, N. Y. 10006
4. *Free Film News.* (catalog) Sterling Movies Inc., 43 W. 61st St., New York, N. Y. 10023
5. *Motion Picture Catalog.* Ford Motor Company, Educational Affairs Dept., The American Road, Dearborn, Mich. 48120
6. *Refining Copper from Sudbury Nickel Ores.* 16 mm, sound, color, 39 minutes. Rothacker, Inc., 241 W. 17th St., New York, N. Y. 10011
7. *Refining Precious Metals from Sudbury Ores.* 16 mm, sound, color, 29 minutes. Rothacker, Inc., 241 W. 17th St., New York, N. Y. 10011
8. *The Story of Stainless.* 27 mm Modern Talking Picture Service, Inc., 1212 Avenue of the Americas, New York, N. Y. 10036

GLOSSARY—METALS

Acetylene highly inflammable gas, C_2H_2 produced by action of water on calcium carbide.

Alloy a metal that is a mixture of two or more metals.

Anneal to soften metals by heating.

Bauxite the ore from which aluminum is refined.

Bead the raised ridge made when welding; also a raised round ridge in sheet metal.

Bellows a device for producing a volume of low pressure air, using a hinged action.

Bessemer furnace in which steel is made from iron, named for its inventor.

Buff cloth wheel used for polishing metals and stones.

Calipers tool for measuring inside or outside diameters.

Carat unit of measure of purity of gold, also unit of weight of gem carats.

Case harden to surface harden by immersing heated steel in a compound of carbon.

Cast iron iron produced by refining pig iron.

Chase form decorative designs on art metal with specially shaped punches.

Clads sandwiches of silver and copper from which coins are made.

Cope upper half of a foundry flask.

Crucible clay-graphite container in which metals are melted.

Dap round end punch used with a die to form designs in art metal.

Die tool for cutting threads on a bolt, also the steel block form for punching out shapes of sheet metal.

Dividers a double-pointed adjustable tool for stepping off or dividing distances.

Draft the taper on a foundry pattern that permits it to be withdrawn from the sand.

Drag lower half of a foundry flask.

Draw-file to pull a file at right angles to the work, as on the edge of metal.

Drawing process of pulling metal wire and rod through openings in a die block to reduce its diameter.

Ductility capability of a metal to be drawn into a wire.

Etch to produce a decorative design on metal by means of chemical action.

Flask the box used to contain the molding sand and cavity for holding molten metal.

Forging process of shaping metal by hammering.

Galvanize to coat steel with a layer of molten zinc.

Gate opening that connects the sprue and the cavity in a mold.

Hardy a cold chisel set in an anvil.

Hem a folded joint in sheet metal.

Ignitor mechanical lighter for gas torches.

Ingot a huge chunk or billet of metal as cast from a smelting furnace.

Jig a device in which identical parts can be formed.

Knurl a series of ridges on a metal surface as a handle or control.

Malleability capability of a metal to be hammered into a sheet.

Mallet a soft-faced hammer.

Micrometer a precision instrument for measuring diameters.

Mild steel steel of low carbon content.

Metallurgy field of study on the chemical and physical properties of metals.

Mold form block over or in which metal is shaped.

Needle file a fine file used in jewelry making.

Oxygen chemical element used to obtain higher temperatures in welding and metal melting.

Patterns full sized models used in making sand molds, also full size outlines to be traced.

Peen working face of a hammer, for example, a ball peen.

Pig iron iron produced directly from ore.

Planish to smooth the surface of formed art metal shapes by hammering.

Pritchel a punch tool set in an anvil.

Protractor a precision instrument for measuring angles.

Riddle the sieve used for screening molding sand.

Rivet a permanent holding device furnished with one head. The second is applied after installation.

Rolling mill series of heavy rolls for reducing ingots to sheet and bars.

Scriber steel scratch tool for marking on metal.

Sintering process of bonding powdered metal into a solid by heat and pressure.

Spinning shaping sheet metal by pressing over a revolving mold.

Sprue vertical hole in a sand mold into which the metal is poured.

Stake anvil over which sheet metal is formed by hammering.

Taconite low grade of iron ore.

Tap tool for cutting inside threads, also to cut such threads.

Taper gradual decrease in diameter width or thickness.

Temper to heat treat metal by removing excess hardness.

Thread spiral wrapped around a bolt to engage the nut, also the inside spiral of the nut.

Tin to deposit a layer of solder on a part before soldering; also the chemical element SN.

Tinplate tin-coated sheet steel as used in cans.

Weld to fuse two pieces of metal by means of heat.

Electrical and Electronics Technology **5**

Electricity is the wonder worker in our homes and industries. It not only works wonders, but it is a wonder how well it can work for us. If the electricity were cut off from your home for a day, how many things would you have to do without? Imagine how inconvenient it would be if the current were off for a whole year. But worse than that would be the effects if electricity were not available to our industries. They would have to go back to steam, water, and wind power, which would throw millions of workers out of jobs. Electricity is indispensable in the United States, in fact in any modern, technological country.

A doctor uses electricity to treat his patients, but when it is used unwisely electricity can cause injury and death. It lights a tiny flashlight, and the whole sky in a storm. It makes your home cheery with light, but it can also set fire to a building. It can ring a doorbell and guide a rocket to the moon. It can coat metal with plating less than a thousandth of an inch thick and it can melt tons of steel. Electricity is truly a wonder worker, so we should become well acquainted with it. All of the possible uses for it have not yet been found. It could be that you or one of your friends will find a new way to make it useful.

Primitive man cringed and hid when lightning flashed. For centuries people were afraid of lightning because they could not understand it. Today we know about electricity and how to control it, so we need not fear it. As you study and work with electricity it is hoped that you will come to understand and appreciate it.

Electric Shock

Most people fear electric shock. You may know why from experience. It is the high voltage that produces the characteristic jolt. But a high amperage causes burns. The two together are very dangerous. Even the 110/115 volt house current can produce a fatal shock. Operating an ungrounded electric tool while you stand on a damp basement floor is inviting danger. Be sure you know how to handle electricity safely before you begin your experimentation. With your teacher's assistance you can draw up a list of "Safety Senses."

ABOUT ELECTRICITY

Static Electricity

There are two kinds of electricity, *static* and *current*. When you rub a cat's fur and you hear a snap and crackle, that is static. Shuffling across a thick wool rug will sometimes build up a charge of static that sparks when you touch metal or another person. The Greek wise man, Thales, is said to have discovered static electricity about 600 B.C. while working with amber. The Greek word for amber is "elektron," from which our word *electricity* comes.

A lightning flash is a discharge of static electricity that we cannot restrain or put to use. Static is often more bothersome than useful. It creates radio and television interference and damages power lines and buildings. However, scientists are learning to harness static electricity to make it work for us. It is used in smoke eliminators in cities and to coat paper with abrasive grit in making sandpaper.

Fig. 5–1. Thomas Alva Edison in his first electric generating station. This is a model of the Pearl Street station in downtown New York City. With six steam-driven generators, the station began operation on September 4, 1882, serving 59 customers. Each generator had a capacity of 120 kilowatts of direct current. Electric power could be transmitted less than a mile from the station.

Current Electricity

Electricity that flows along a conductor is called *current*. The scientist knows electricity as a flow of electrons along a conductor, which is usually a wire. Because electricity flows, it is useful to us. It is sent from the power station to your house and to industries over miles and miles of wires.

Kinds of Current. There are two kinds of electric current. You probably use both of them. One is *direct current* (D.C.) and the other is *alternating current* (A.C.). The current produced in your flashlight and auto batteries is D.C. It flows in one direction only. The generator in a car generates direct current. Industry uses D.C. for such purposes as electroplating, battery charging, electric welding, electric locomotives, and streetcars.

Alternating current flows forward and backward through the circuit. The current for your home is probably 110/120 volts, 60 cycles, A.C. The *cycle* tells us that the current flows 60 times forward and 60 times backward through the circuit in one second. Most of the electricity consumed today is A.C. It has an advantage in that its voltage can be stepped up or down by means of transformers. (See page 201.) This makes it possible for the power station to send it for hundreds of miles over wires. This cannot be done with D.C.

The automobile generator has been replaced by an alternator. This is an A.C. generator. It has the advantage of greater voltage output at low engine speeds. But the battery and ignition require

MILESTONES IN THE DEVELOPMENT OF ELECTRICAL-ELECTRONICS TECHNOLOGY

B.C.

2600   The Chinese used a compass.

600   Thales discovered static electricity.

A.D.

1269   Peregrinus discovered that like magnetic poles repel, unlike poles attract.

1747   Benjamin Franklin believed that there was both positive and negative electricity.

1791   Luigi Galvani discovered electric current.

1820   Hans Christian Oersted discovered that electricity can produce magnetism.

1820   André Marie Ampère found that a coil of wire carrying a current is a magnet.

1827   George Simon Ohm published his book explaining "Ohm's Law."

1831   Michael Faraday and Joseph Henry discovered electromagnetic induction.

1844   Samuel Morse sent the first public telegram.

1858   Cyrus Field completed the first transatlantic cable.

1860   Gaston Plante invented the storage battery.

1875   Alexander Graham Bell invented the telephone.

1879   Thomas A. Edison invented the electric light and set up the first public light and power station.

1895   Guglielmo Marconi invented the wireless telegraph.

1895   Wilhelm Roentgen invented the X ray.

1897   Joseph Thomson discovered electrons.

1920   First regular radio broadcasting in the United States, Station KDKA, Pittsburgh.

1938   Regular television broadcasts began in New York. They had begun two years earlier in London.

1945   First electronic computer. Designed by J. W. Mauchly and J. P. Eckert, Jr., at University of Pennsylvania.

1946   First radar contact with the moon.

1948   First transistor. Developed in Bell Telephone Laboratory.

1953   Regular color television broadcasts began in the United States.

1960   First nuclear power plant serving communities put in operation in the United States.

1969   Neil A. Armstrong, first man to set foot on the moon.

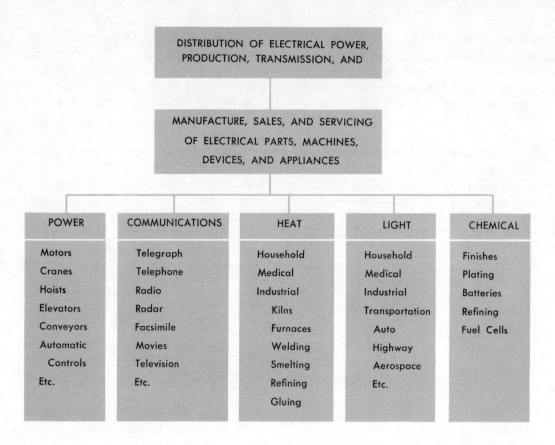

DISTRIBUTION OF ELECTRICAL POWER, PRODUCTION, TRANSMISSION, AND				
MANUFACTURE, SALES, AND SERVICING OF ELECTRICAL PARTS, MACHINES, DEVICES, AND APPLIANCES				
POWER	COMMUNICATIONS	HEAT	LIGHT	CHEMICAL
Motors	Telegraph	Household	Household	Finishes
Cranes	Telephone	Medical	Medical	Plating
Hoists	Radio	Industrial	Industrial	Batteries
Elevators	Radar	Kilns	Transportation	Refining
Conveyors	Facsimile	Furnaces	Auto	Fuel Cells
Automatic	Movies	Welding	Highway	
Controls	Television	Smelting	Aerospace	
Etc.	Etc.	Refining	Etc.	
		Gluing		

Fig. 5–2. Typical electrical-electronic applications and industries.

D.C. Consequently the A.C. current is passed through a *rectifier* (see page 203). This permits the current to pass in one direction only. It becomes direct current.

Why Current Flows. Water flows from a faucet only because pressure forces it through the pipes. Electricity flows too, because of pressure. This is called *voltage* (after Alessandro Volta). Voltage, also known as Electromotive Force (EMF), pushes the electrons along the conductors, the wires. We are told that electrons are so tiny that it takes 25 trillion of them placed end to end to equal an inch.

Current doesn't actually flow through a conductor as water through a pipe. Instead current flow is thought to be a wave movement of electrons. To illustrate this, put several marbles in a cardboard trough just wide enough to hold them in a line. They should be touching and free to move. Now move a separate marble along the trough to strike the end one. Almost instantly the far marble moves out. Each electron is assumed to push the one next to it. It is the pushing that moves at the speed of light.

Current flows from the terminal of a battery or of a generator having the most electrons to the one with the fewest. It is the difference in the EMF that forces the excess of electrons of the *negative* terminal (—) to flow to the *positive* (+) when a switch is turned on.

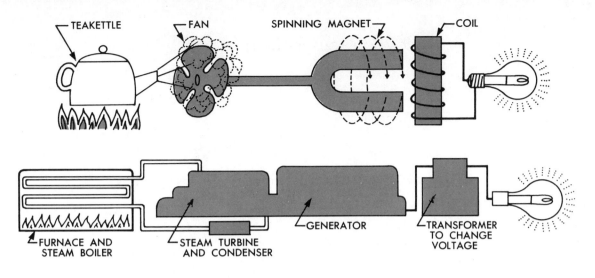

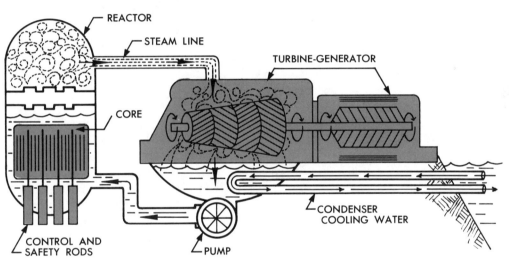

Fig. 5-3 (top). Heat is used to produce electricity. While electricity can be produced in several different ways, heat is the most common source of energy for it. Here is shown the basic principle of a steam power system. Fire boils water in a teakettle, producing steam pressure to turn a fan attached to a magnet. The lines of force of the spinning magnet are cut by the turns of wire in the coil. This produces an electric current.

Fig. 5-3 (middle). The teakettle has been replaced by a steam boiler heated with coal. The fan becomes a turbine driving the generator. A transformer has been added to change the voltage.

Fig. 5-3 (bottom). The atomic reactor has replaced the steam boiler. The main difference between an atomic power plant and a coal-burning plant is in the method of obtaining the steam. Heat from atomic fission produces steam as heat does from the burning of coal.

COMMON SOURCES OF ELECTRICITY

Source	Current	Application
Friction	Produces static electricity.	Has no commercial value.
Generator	Produces direct or alternating current.	Electric power plant; automobile engine.
Dry Cell	Chemical action produces direct current; self-contained.	Flashlight; transistor radio.
Wet Cell	Chemical action produces direct current; self-contained.	Automobile storage battery.
Fuel Cell	Chemical action produces direct current; chemical fuel is added.	Space vehicles; experimental.
Solar Cell	Produces direct current from light.	Space vehicles; satellites; experimental.
Crystal	Certain crystals, like quartz, when squeezed, produce direct current.	Experimental.
Thermocouple	A pair of unlike metals, joined at one end and heated, produce direct current.	Heat-control instruments, as a kiln shut-off.

In your home you probably have a voltage of from 110 to 120 for lights and appliances. You may, however, have a motor on a saw or a pump that uses 220 to 230 v (volts). An appliance wired for the lower voltage will burn out when connected to the higher. A motor wired for 220/

Fig. 5–4. A voltmeter is connected across the two wires; an ammeter, in one wire.

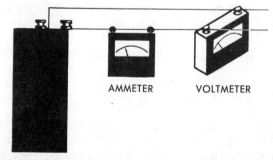

CURRENT SOURCE · AMMETER · VOLTMETER

230 v will run slowly and with little power on the 110/120 v. You see, it is important that the correct voltage be used.

Voltage is measured with a *voltmeter,* which is connected across the two wires conducting the current in a circuit.

Current and Amperage. To an electrician, current means *amperage.* The amount of current consumed by a light bulb or an appliance is given in *amperes* (after André Marie Ampère). Both voltage and amperage are usually indicated on a plate fastened to a motor or other appliance. See if you can find them on a motor. Both voltage and amperage are necessary for the flow of electricity. When the voltage is high and the amperage low, electricity easily jumps an air gap. In an automobile spark plug 20,000 v or more are required. With low voltage and high amperage, great heat is possible, as in electric welding. The combination may be about 40 v with 200 to 300 amp.

Amperage is measured by an *ammeter,* which is connected in one wire of the circuit. An electric clock consumes about $\frac{1}{30}$ amp; a ¼ hp (horse-

Electrical and Electronics Technology

power) motor, about 1¾ amp; and a trolley bus, about 50 amp.

Resistance. There is always opposition to the flow of a current over a conductor. This is called its *resistance*. The amount varies with the kind of material, its length, diameter (cross-sectional area), and its temperature. The greater the length of a wire, the smaller its diameter, and the higher its temperature, the greater is its resistance. Resistance is measured in *ohms* (after George Simon Ohm).

Copper is most often used for wire conductors because it has the least resistance of the plentiful metals. Silver has even less resistance, but it is too costly. Materials that conduct current with little resistance are called *conductors*. Those that have great resistance, for example, rubber, plastics, porcelain, and glass, are called *insulators*. In between these two groups of materials are those that are both poor conductors and poor insulators. Some of these are iron, steel, tungsten, and various alloys used for heating elements. Some high-temperature elements are nonmetallic. The term, *semiconductor,* is commonly used with the materials from which transistors are made.

Heat is given off in a wire when the voltage overcomes the resistance. The electrons resist moving and, in their being forced to move, heat is created.

Voltage, Amperage, and Resistance. These three qualities are always found together in a *circuit* (the path over which the current flows). They behave according to Ohm's Law:

$$I = \frac{E}{R} \text{ This means: Amperes} = \frac{\text{Volts}}{\text{Ohms}}$$

If you know any two of the three, you can find the third by arithmetic.

Circuits. An electrical circuit is a closed path over which the current flows. When you turn on the light switch, the circuit is complete and the light burns. Turning it off breaks the circuit. There are two common types of circuits with which you can easily work, the *series* and the *parallel*.

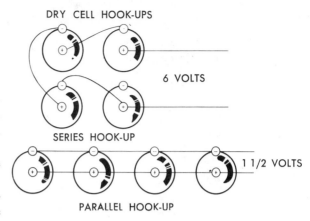

Fig. 5–5. Series and parallel hookups.

Series Circuits. When several lights are connected in series they are all connected into the same wire. You may have seen Christmas tree lights connected in this way. When one burns out, they all go out. Connecting several dry cells in series increases the voltage, but the amperage remains unchanged. Suppose you have four dry cells. Each has 1½ v. In series, the total voltage is 4 times 1½, or 6 v.

Parallel Circuits. The other kind of Christmas tree lights are connected in parallel. Each light is connected to two wires. A bulb may burn out without affecting the rest. This is the type of circuit used in wiring your home and in the transmission lines that send electric power over the country. Four dry cells connected in parallel provide four times the amperage of one. The voltage remains unchanged.

Short Circuits. In any well-planned circuit there is just enough resistance among all the parts to balance the voltage and amperage. When the current is routed over a shorter path, the current is too great for the resistance. Sparks fly, heat is formed, and a fuse burns out. Faulty insulation on a lamp cord permits the wires to touch and build up excessive current, and a short circuit results.

Fig. 5–6. Three common fuses. The center fuse is for a main switch; the one on the right is a house circuit fuse. The small fuse is used in an automobile electrical system.

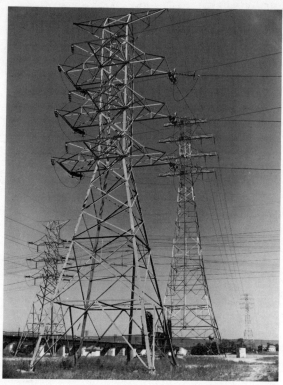

Fig. 5–7. TVA power is distributed to 2 million customers by more than 150 locally owned municipal and cooperative distribution systems buying electricity from TVA at wholesale.

Fuses. Fuses are silent sentries standing watch over the current flowing in a circuit. If too much flows, as it does when too many lights and appliances are turned on, the wiring becomes overheated. Instantly the fuse burns out, thus breaking the circuit and preventing a fire. When two bare wires touch, as they can when the insulation is broken, the current becomes excessive. Again the fuse instantly burns out, thus preventing further damage. New fuses will continue to burn out until the trouble is remedied.

The size of the fuse is determined by the size of the wiring in the circuit. A 15-amp fuse is proper in a house circuit using a 14-gauge wire. A 20-amp fuse is used with 12-gauge wire. If the larger fuse is used in place of the smaller, it does not provide the necessary protection.

Circuit breakers are commonly used instead of fuses in homes. A switch is automatically shut off when the circuit is overloaded. It can be reset after the cause of the excess has been removed and the switch has cooled.

Underwriters Laboratories

On some appliances, lamp cords, and other electrical goods you will find a tag bearing the label "U.L." This means that the product has passed the tests used on it by the Underwriters Laboratories, Inc. It tells the consumer that the product or part will operate safely. Look for this when you buy any electrical materials.

ABOUT WIRE

Many sizes and shapes of wire are used in conducting electricity. Copper is preferred for most wiring. Aluminum is often used in cross-country power lines. The lightness of aluminum makes possible the great spans between the steel towers. When ice forms on the wires the power is stepped up through them, warming the wire and melting the ice. That is the reason high-tension lines are often bare rather than insulated.

Types of Wire. The copper wire used for winding motors, transformers, magnets, and in radio circuitry is *magnet wire*. It is available bare, with tin coating, and with different kinds of insulation such as enamel, plastic, and cotton. It is usually sold by the pound.

COPPER WIRE TABLE

American Wire Gauge	Diameter in In.	Capacity in Amp.	Resistance in Ohms per 1000 feet at 68°F.
1	0.289	100	0.123
10	0.102	25	0.998
12	0.081	20	1.588
14	0.064	15	2.525
16	0.051	6	4.016
18	0.040	3	6.385
20	0.032		10.15
22	0.025		16.14
24	0.020		25.67
26	0.016		40.81
28	0.013		64.90
30	0.010		103.2

Resistance wire is an alloy and a poor conductor. Because of this quality it is useful both in the home and in industry. It is used for heating purposes as in stoves, corn poppers, waffle irons, heat-treating furnaces, ceramic kilns. It is also used in rheostats and similar devices for the control of current flow in a circuit, such as the volume control on a radio.

Nichrome, Chromel, and Kanthal are some brand names for resistance wire. Types are available for use at low temperatures as well as at those of 2200-2300°F.

Electrical *power wire* includes the many types used for power transmission in power lines, residences, schools, and factories. Such conductors may be single solid wire, stranded, or in cables of two or more conductors. The insulation is color coded to assure proper connections in circuits. Your teacher will have samples of these.

CHARACTERISTICS OF CHROMEL "A" RESISTANCE WIRE
Maximum Temperature 2200°F

Watts	A. W. Gauge	Ohms/ft	No. of ft Req'd.	Ft/lb
100	30	6.50	19.0	3495
200	28	4.09	15.1	2200
300	26	2.57	16.0	1380
500	22	1.017	24.3	545
700	20	0.635	27.8	341
1000	18	0.406	30.4	219

Making Electrical Connections. Good electrical connections are necessary if your projects are to work well and safely. Some joints should

Fig. 5–8. The wire must be put under a terminal screw, pointing clockwise so that it will draw down tight.

Fig. 5-9. The underwriter's knot used in the same attachment takes the strain, rather than the terminals, if the cord is pulled.

be soldered and others fastened with screws. In your radio construction, soldering is used for quick, permanent connections of low resistance. Remove any insulation at the area of contact. Use a light, clean, well-tinned electric soldering iron and rosin-core solder. Hold the iron on the connection to heat the wires. Touch the joint with the solder and melt off a small drop. Very little solder is needed to make a good connection.

Connections within the radio circuitry are not taped. But electrical connections that are exposed to any possible short circuiting should be. Use plastic electrical tape.

When a screw connection is made with wire, first make a hook to fit the screw. Then insert it under the head so that the end of the hook points in the direction that the screw will be turned for tightening. Can you figure out why? Try it the other way and you will have the answer.

MAGNETS AND MAGNETISM

The discovery of magnets and magnetism was the first step in the long chain of discoveries and inventions that has made electricity commonplace

today. The first magnets were *lodestone,* a natural rock. The early Chinese noticed the peculiar nature of this rock. It attracted iron. They used it for compasses 5000 years ago.

The best deposits of lodestone, actually an iron ore, were found near the city of Magnesia in Asia Minor. The ore was called *magnetite* and its quality of attracting and repelling was called *magnetism.* Early sailors used lodestone for compasses until it was found that a piece of iron could be magnetized by rubbing it with the ore.

Your compass has a needle in it that is a *permanent* magnet. As you have noticed, one end always seeks the north and the other, the south. If you suspend a bar magnet by a string so that it is free to move, it will point north and south by itself unless it is attracted by other magnetic material. Here are some facts about magnets:

1. Magnets attract many materials, especially those bearing iron.

2. Each magnet, no matter how tiny or large, has two *poles:* the north-seeking, or north pole, (N), and the south-seeking, or south pole, (S).

3. Iron and steel can be magnetized by rubbing with a magnet. Stroke several times in the same direction.

4. Soft iron does not retain magnetism long. Hard steel retains it permanently.

5. Among magnets, the unlike poles attract and the like poles repel each other. North attracts south and repels north.

6. Each magnet has a *field of force,* or a magnetic field. You can prove this. Lay a piece of clear plastic (glass or cardboard will do, too) over a permanent magnet. On this sift some iron filings. Tap the plastic lightly and watch the particles arrange themselves between the poles. (See Fig. 5-10.)

Permanent magnets are used in such devices as scientific instruments, motors, radio speakers, telephones, and generators. New magnetic alloys have been developed that make tiny magnets very powerful.

 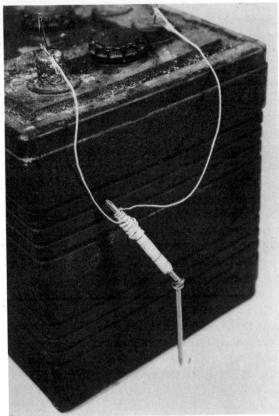

Fig. 5–10. (left) The magnetic field between two magnetic poles forms a distinct pattern with iron filings. **Fig. 5–11.** (right) A nail wrapped with magnet wire becomes a strong magnet when current is passed through it.

Electromagnets. Electromagnets are magnets produced by electric current. Connect one end of a 2-foot length of No. 24 or 26 copper wire to a dry cell. Hold a compass below the wire. Touch the other end of the wire to the other terminal and watch the compass needle. The closer the compass is to the wire, the stronger the attraction. This experiment proves that there is a magnetic field about a current-carrying wire. Coil this wire like a spring and it will be a stronger magnet. Coil it about an 8d nail and it will be even stronger. With this nail as the core, it will pick up another nail.

Make the iron-filings test with this magnet. Hold a compass near each end of the nail. One will be the north pole and the other the south. Reverse the battery connections and watch the compass. This shows that changing the direction of the current changes the *polarity*. The more turns of wire you put on the core, the stronger the magnet.

Electromagnets have more uses than has the permanent type. They are the hearts of electric motors and generators. Because of them the generation of electricity as we have it today is possible.

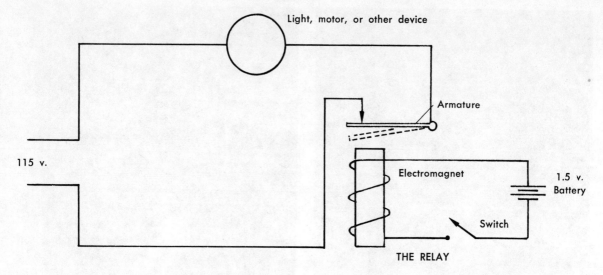

Light, motor, or other device

Armature

115 v.

Electromagnet

1.5 v. Battery

Switch

THE RELAY

Fig. 5–12. A relay is a magnet switch. Here the battery supplies a small current when the switch is closed. This causes the electromagnet to attract the armature. It can be made to open or close the circuit.

Fig. 5–13. Electricity produces mechanical action in a solenoid. When the current flows the magnetic field produced in the coil draws the iron core into its center. When the current is cut off the spring pulls it back. This is sometimes called a "sucking" coil.

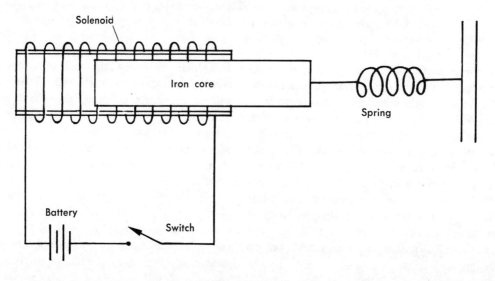

Solenoid

Iron core

Spring

Battery

Switch

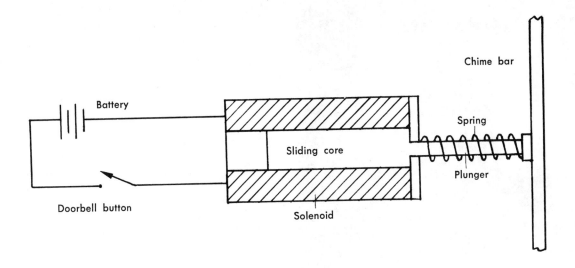

Battery

Chime bar

Sliding core

Spring

Plunger

Doorbell button

Solenoid

Fig. 5–14. Door chime circuit. Pushing door-bell button closes circuit and causes a magnetic field in the solenoid. This pulls the core into the center, compressing the spring. Releasing the button cuts off the current. The spring drives the plunger against the chime. When the front door has two notes and the rear, one, another unit is added. How would these then be wired?

Solenoid. A solenoid is a coil of magnet wire with a freely moving core of soft iron or other magnetic material. When current flows in the coil, the core moves in the direction of the lines of force. Try this with a coil of fifty turns of wire. Wrap it around a small diameter of paper or plastic tube. Use a piece of ¼″ round iron rod for the core. Connect the magnet to a dry cell.

Solenoids are used for operating switches and valves, as in an automatic washer and doorbell chimes.

Relays. A relay is an electrically-operated switch. It consists of a solenoid and a movable iron armature. The current produces a magnetic field in the solenoid, which attracts and moves the armature. The latter is connected to a switch. A field trip through your telephone company will show hundreds of relays switching automatically to the dialing of numbers. They are opening and closing circuits.

THE GENERATION OF ELECTRICITY

Faraday discovered that when a wire conductor is moved through a magnetic field (see page 197) a current is *induced* in the wire. This current will flow only if the ends of the wire are connected into a circuit.

The device for generating an electric current is a *generator*. At the power station, which supplies the homes and industries in your area, are huge generators driven by either steam turbines or water power. When water power is used, the station is called a *hydroelectric* plant. Atomic power plants are becoming common. Atomic energy furnishes heat to make steam, which drives the turbine that turns the generator. Study the diagram in Fig. 5–18 to see how a power station produces and transmits electric power.

Electrical and Electronics Technology

Fig. 5–15a. (top) A giant motor under construction. The rotor of the world's most powerful direct current motor is being lowered into place. Capable of 18,000 horsepower, the motor drives a centrifuge. This is used for the training of astronauts by simulating the forces experienced during the launch and reentry of a space capsule. **Fig. 5–15b.** (bottom) The drawing below shows how the huge motor drives the arm to which is attached the capsule in which the astronauts train. The 50-foot arm turns at such a rate that a force of 50G (fifty times gravity) can be produced.

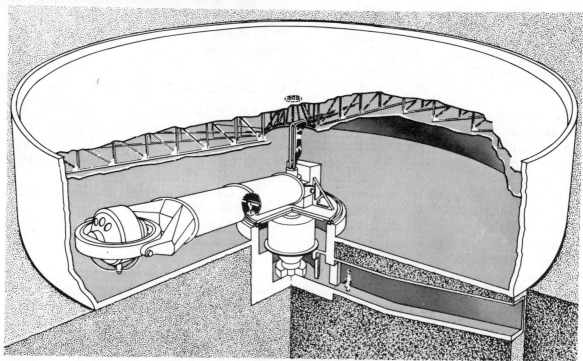

Fig. 5–16. (top left) The world's most powerful Diesel locomotive. Diesel power is used to drive electric generators to furnish electricity to the motors that drive the wheels. This locomotive develops 6600 hp. Such compact and efficient locomotives have made the steam locomotive a museum piece. **Fig. 5–17.** (top right) Assembly of the Diesel engines for the 6600 hp locomotive pictured.

THE TRANSFORMER

In Fig. 5–18, as you trace the power lines from the station to the house, you find that whenever the voltage is stepped up or down, there is a transformer. This device works automatically, using alternating current. It does not work with direct current.

Because the voltage can be stepped up, electric current can be sent for hundreds of miles with very little loss. In a pipeline carrying gasoline from the oil fields to bulk stations across the country, pumping stations along the way are required to keep the gasoline flowing. Transformers do this for electric current. If a wire conductor is long enough it will have resistance enough to overcome the flow of current so that hardly any will flow. A step-up transformer can take a small voltage and increase it so that current will again flow. A step-

down transformer works in the reverse; it reduces voltage to whatever is needed for a particular use.

How a Transformer Works. In our explanation of electromagnetism it was pointed out that a coil of wire carrying a current has a magnetic field set up about it. When another coil of wire is brought close to it, the magnetic field is cut and a current is caused to flow in the second coil. We say that a current is *induced* in the second coil. There must be an actual cutting of the lines of force. Remember that alternating current changes direction frequently. Because it does, the coil does not have to be moved through the magnetic field. The ebbing and flowing of the current takes care of the cutting and causes the field to build up and then collapse. This provides the necessary cutting of the field by the wire.

Induction Coil. An induction coil consists of two separate coils of magnet wire around a core.

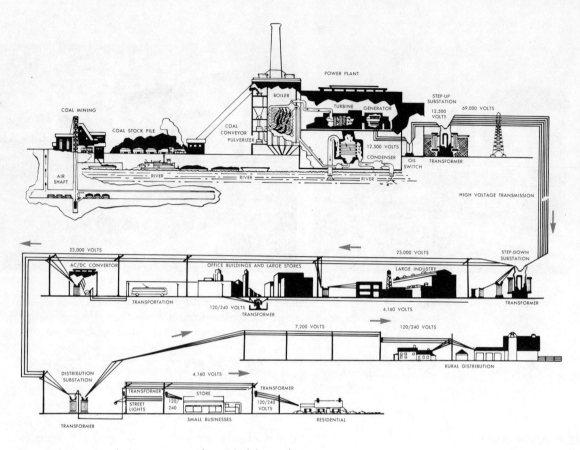

Fig. 5–18. How electricity is made and delivered.

The core is usually of soft iron wires or laminated sheet iron. The coils have different numbers of turns. When a current is passed through the first, the primary, it becomes an electromagnet. When its magnetic field is cut by the turns of the secondary, a current is induced in it.

Step-Up and Step-Down. A transformer is an induction coil. A step-up transformer has more turns of wire in the second coil than in the first. We say more in the *secondary* than in the *primary*. If the secondary has one hundred times as many turns as the primary, it will have one hundred times the volts.

In the step-down transformer the primary has the more turns. If the secondary has only one-tenth as many turns, it will have only a tenth as many volts. An electric train set is operated from a transformer. When you plug it in, 110/120 v go into the primary winding and from 10 to 20 v come out of the secondary to operate the train. The various voltages are obtained as you move the control dial. The secondary coil was tapped and connections made at different numbers of turns. So it operates as though there were several different coils.

Transforming Direct Current. Direct current, when it is made to start and stop rapidly, can be transformed. The timer on a model airplane engine does this, as do the distributor points on an automobile engine. As the current starts and stops, with

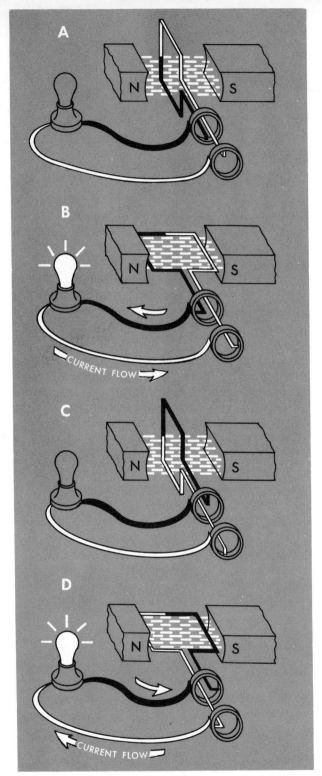

the closing and opening of the points, the magnetic field in the primary winding of the ignition coil is cut. A high voltage current is induced in the secondary.

ELECTRICITY AND POWER

It has been pointed out that an electric current is a flow of electrons along a conductor. As the electrons flow, work is done; for example, resistance is overcome, heat is given off, and a magnetic field is set up. How rapidly work is done by the moving electrons depends on how many there are (amperage) and how much pressure is pushing them along (voltage). We know that

$$\text{Power} = \text{Current} \times \text{Voltage}$$

and since the unit of electrical power is the *watt,* the formula becomes

$$\text{Watts} = \text{Amperes} \times \text{Volts}$$

This means that 1 amp times 1 volt equals 1 watt.

The ton as a unit of weight is more convenient to use than the pound when large quantities are being weighed. Likewise, the *kilowatt,* a thousand watts, is the more convenient unit of electrical measure for large quantities. When you visit your local power plant you can find out how much

Fig. 5–19. How a generator produces electric current. In A, no current is flowing, because the magnetic field between the north and south poles is not cut through by the revolving wire. In B, the wire cuts through the magnetic field as it turns, and generates a current. In C, no current is flowing. In D, current is flowing, but in the opposite direction. The revolving wire is called an armature. In a real generator, the armature has a great number of wires. The more wires, the greater the current.

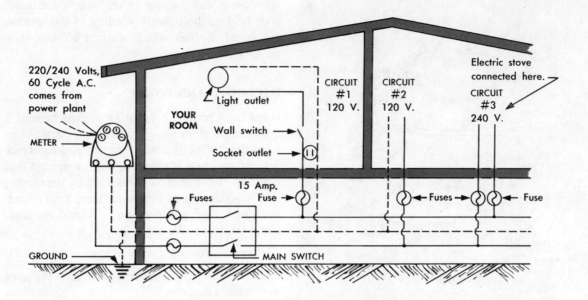

Fig. 5–20. Electricity in your home. Notice how the incoming voltage, 220/240, is divided into two 110/120v circuits and one 220/240v circuit. The center wire, indicated by broken lines, is the ground. This wire with either outside wire gives 110/120v. The two outside wires, indicated by solid lines, are "hot" wires and it is to these that an electric stove, or other appliance that requires a 240v circuit, must be connected. Be safe; let a licensed electrician do your house wiring!

power is being produced there. It may generate only about 25,000 kw—or several times that many.

Current Consumption. The amount of current consumed by an electrical device is measured in watts. The power consumed by an electric motor, for example, is found by the above formula: Watts = Amperes × Volts. On light bulbs and many household appliances the wattage is indicated. Check a toaster and an electric iron.

In figuring your electric bill the power company uses *kilowatt hours* (KWH) instead of watts. For example, if your rate is 5¢ per KWH you could use one thousand watts for one hour for a nickel. In actual practice this would permit you

to burn, say, ten 100-watt lamps for an hour, or five of them for two hours, and so on.

You can figure the cost of operating an electrical device by this formula:

$$\text{Cost} = \frac{\text{Watts used}}{1000} \times \text{Rate per KWH}$$

What, then, does it cost to use your radio if it consumes 100 watts and the rate is 5¢ per KWH? The answer is $\frac{100}{1000} \times 5¢$, or ½¢ per hour.

When the wattage is not given, as is common with motors, use this formula:

$$\text{Cost} = \frac{\text{Volts} \times \text{Amperes}}{1000} \times \text{Rate per KWH}$$

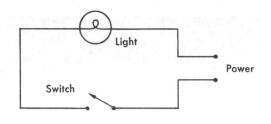

Light

Power

Switch

Three-way switch circuit.
Two switches control one light

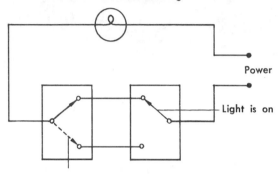

Power

Light is on

Light is off when one switch breaks the circuit

Fig. 5–20a. Single pole switch circuit as in a flashlight or in house lighting.

Fig. 5–21. Transformers step up the voltage of electric power for transmission over distances. These shown deliver 765,000 volts. Compare their size with the man at the center.

Reading the Kilowatt Hour Meter

A meter keeps a record of the number of kilowatt hours used in your home, just as the speedometer on an auto records the number of miles traveled. It is a clever instrument, since it multiplies volts by amperes, divides this by 1000, and multiplies the result by the hours. Here's how to read the meter (see Fig. 5–23a).

1. Start at the right and read each dial in order.

2. Find the last number passed by the pointer in each dial.

3. Jot down each number from right to left.

IGNITION SYSTEMS FOR ENGINES

Current from the battery flows through the primary winding of the coil when the *timer* closes. When the timer is opened, the flow of current is interrupted. This causes a current of high voltage to be induced in the secondary winding. This current flows to the *spark plug*. Here it jumps the gap at the *points,* making a spark that ignites the fuel. The *condenser* is placed across the timer points to keep them from becoming burned and pitted.

Because an automobile engine has several cylinders and spark plugs, there is a rotary switch, called a *distributor*. This directs the high-voltage

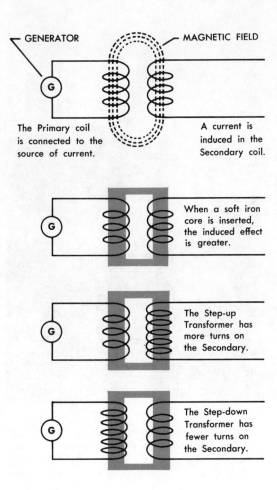

GENERATOR — MAGNETIC FIELD

The Primary coil is connected to the source of current.

A current is induced in the Secondary coil.

When a soft iron core is inserted, the induced effect is greater.

The Step-up Transformer has more turns on the Secondary.

The Step-down Transformer has fewer turns on the Secondary.

Fig. 5–22. How a transformer works.

Fig. 5–23. An electric watt-hour meter such as is found on the side of a house.

current to the correct plugs. When 6 v or 12 v go into the primary, 15,000 to 20,000 leave the secondary.

Some internal combustion engines, such as those on lawn mowers and motor bikes, use a *magneto ignition system* instead of a battery. A magneto is really a generator. It uses permanent magnets instead of electromagnets to set up the magnetic field.

Piezoelectrics. Crystal quartz has the property of producing electric current when it is compressed. The principle has been successfully applied to the ignition systems of small electric engines. Instead of a magneto, the engine has a piezo unit. It is mechanically operated to produce a current at the right time. The voltage is stepped up by an induction coil to fire at the spark plug.

ELECTRICITY AND COMMUNICATION

Take electricity out of our methods of communicating with one another and there would be nothing left but handwriting, printing, and speaking. Ships, airplanes, and trucks would have to carry our messages. There was a time, as you know, when the Pony Express system tried to keep one side of our country in touch with the other. Think what the old-timers missed in not having the telegraph, telephone, radio, television, and facsimile (fac-*sim*-i-lee). Nowadays you can know what is happening anywhere in the world almost as soon as it happens. Do you take advantage of this opportunity to keep up with the news?

Reading No. 1

Reading No. 2

Fig. 5–23a. Reading the watt-hour meter. What is the electric bill when the rate is 5¢ for the first 1000 and 4% for the remainder?

Fig. 5–24. (right) An ignition system for a model airplane engine. Some model airplane engines use a glow plug. This operates as in a Diesel engine and eliminates the ignition system. A dry cell is connected to the plug to start the engine. It heats a tiny heating element to ignite the fuel. Once started, the element stays hot to ignite the fuel. The battery is then disconnected. **Fig. 5–24a.** (bottom) Wiring diagram of an ignition system for a 4-cylinder engine.

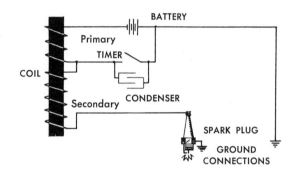

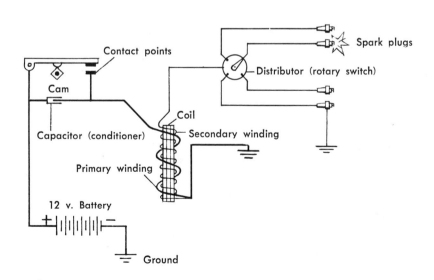

The Telegraph

The telegraph was the first successful electrical device for sending communications over great distances. Invented by Samuel Morse in 1840, it was followed by a chain of communications inventions. Most of these Mr. Morse probably never dreamed of.

In the early telegraph systems, wire was used to carry the current between stations. Marconi's wireless telegraph, called *radio*, eliminated the need for connecting wire.

A code system of dots and dashes is used in sending messages. It is necessary for an operator to memorize the code so that he can send as well as receive. The story is told that when Steinmetz, the great electrical wizard, and Edison would meet, they used the code in place of talking. Edison was

Fig. 5–26. (top) A telegraph key and oscillator. This combination permits you to practice and learn the code as is required for a ham radio license. When tapped, the key closes and opens the circuit of the oscillator. This produces the familiar "beep" signal of the telegraph. **Fig. 5–26a.** (bottom) The schematic wiring diagram is for the oscillator.

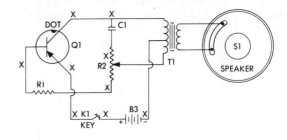

Fig. 5–25. Continental International Code.

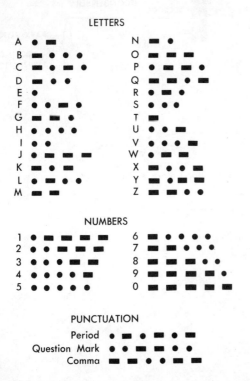

deaf and Steinmetz would tap out his message on his friend's knee.

If you have listened to shortwave radio you have probably heard the high-pitched tones of dots and dashes being transmitted by a "ham," an amateur radio operator. He uses the Continental International Code. "A" sounds like "dit-dah," and "C" like "dah-dit-dah-dit." Hams need a Federal license to broadcast.

Teletype

Teletype is a form of telegraph. The operator copies the message on a special typewriter that transmits it over wire to its destination. There the receiving typewriter types the message automatically on paper, ready to read. Newspapers use teletype to receive news reports from national and international sources.

The Telephone

It was on October 9, 1876, that Alexander Graham Bell carried on the first successful two-way telephone conversation. This was between Boston and Cambridge, Massachusetts, two miles away. Since then the telephone industry has become a most important one.

Today it is possible to talk around the world by radio telephone (a combination of wire and wireless telephone). Two-way telephones are used in automobiles. Ship-to-shore telephones have been used for years. The telephone wires and cables laid across the nation are also used to carry telegraph messages and radio and television programs. Telephone cables are laid across the floor of the Atlantic Ocean.

The Telephone Transmitter. The mouthpiece of a telephone has a tiny transmitter in it. There is the same kind of device in a radio microphone. It changes sound waves into electrical current.

Sound waves from the mouth strike a sensitive steel diaphragm, which is a thin, circular disc. The diaphragm springs in, packing the tiny carbon particles together. Since carbon is a conductor, current from the battery can then flow around the circuit. The louder the voice, the more the particles are compressed and the more current flows. Actually the diaphragm vibrates with the variations of the voice. This causes a vibrating current to flow.

In the earpiece of the receiver at the other end of the line, is an electromagnet operated by this

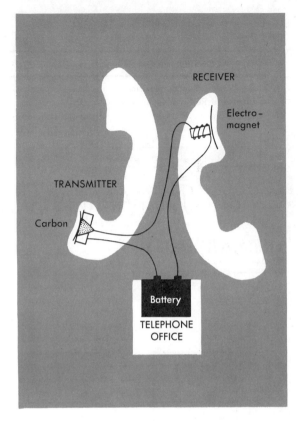

Fig. 5–27. Telephone circuit.

current. A diaphragm springs in and out as the current vibrates. This causes the electromagnet to attract and release it. As this diaphragm vibrates, it sets the air in front of the ear in motion. This creates sound waves like those spoken into the mouthpiece of the telephone.

When You Dial a Number. Between your telephone and another is the telephone company's central office. It is here that your call is connected to the telephone of the party to whom you wish to speak. As you lift the receiver, a switch in your telephone closes and permits a current from the batteries in the central office to flow.

This flow of current set a switching mechanism at the office to work, hunting for your line.

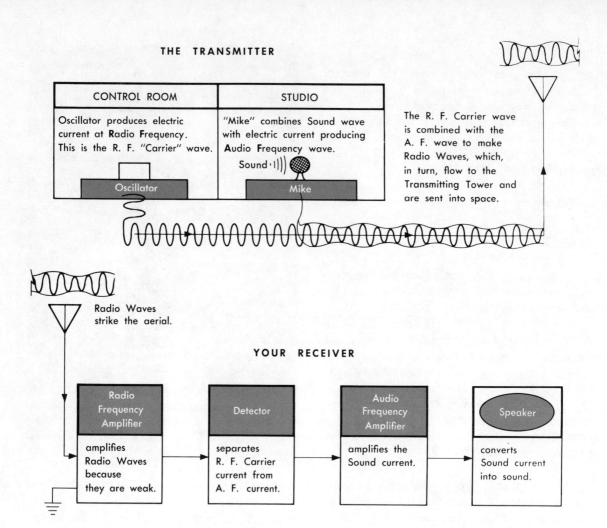

THE TRANSMITTER

CONTROL ROOM	STUDIO
Oscillator produces electric current at **Radio Frequency.** This is the R. F. "Carrier" wave.	"Mike" combines Sound wave with electric current producing **Audio Frequency** wave.

Sound ·)))

Oscillator

Mike

The R. F. Carrier wave is combined with the A. F. wave to make Radio Waves, which, in turn, flow to the Transmitting Tower and are sent into space.

Radio Waves strike the aerial.

YOUR RECEIVER

Radio Frequency Amplifier	Detector	Audio Frequency Amplifier	Speaker
amplifies Radio Waves because they are weak.	separates R. F. Carrier current from A. F. current.	amplifies the Sound current.	converts Sound current into sound.

Fig. 5–28. Radio transmitting and receiving.

When it finds the line with the current in it, it stops and makes a connection. Then you hear the dial tone which means, "Number, please."

When you dial the number, each click as the dial returns to its starting place sends a pulse of current to the switchboard in the office. The board keeps track of the number of clicks and connects your party. Then the current from your telephone makes a circuit with the other telephone as you

talk with the person there. This is all done automatically, without the aid of a telephone operator, as was formerly necessary. Dials are being replaced with push buttons for greater convenience.

Radio

Radio transmitting begins at the studio *microphone.* Here the sounds of the program strike the

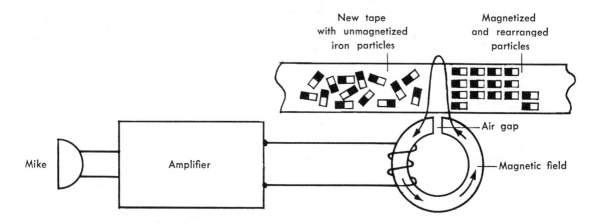

New tape
with unmagnetized
iron particles

Magnetized
and rearranged
particles

Air gap

Magnetic field

Mike

Amplifier

Fig. 5–29. Tape recording. To play back the process is reversed with sound coming from the speaker.

diaphragm of the "mike" and cause the current within it to pulsate, just as in the mouthpiece of a telephone. (See Fig. 5–27.) These pulsations are electrical waves at an audio frequency (AF). At the same time, the *oscillator* in the station produces an electric current at a radio frequency (RF). The two currents are now passed through the *modulator,* which combines them into radio waves. The RF wave is called the carrier wave for the AF wave. The radio waves are then sent out from the antenna as an electrical current. They travel in all directions and some strike your radio antenna. When the radio is tuned to a station, the radio waves from that station first flow through the RF *amplifier,* where they are increased in strength. Then they pass through the *detector* stage of the set where the carrier wave is separated from the AF current. The audio frequency amplifier increases the strength of the current. It then goes to the speaker, which converts it into sound. (See Fig. 5–28.)

Radar

The term RADAR is an abbreviation of *Ra*dio *D*etection *A*nd *R*anging. It is a device which, by means of radio waves, locates objects and tells how far away they are and how fast they are moving. PPI (Plan Position Indicator) is a form of radar. A ship or airplane having this equipment can "see" in the dark and through fog and clouds. A rotating antenna sends out the radar beam in all directions. When an object is struck by the beam, it is reflected back to the radar receiver. This shows the object on a screen similar to a television screen.

Radar is employed in automobile traffic control. Mounted in a patrol car, it can instantaneously report the speed of a moving auto at which it is beamed.

Sonar

Sonar does under water what radar does in air. Installed in a ship or submarine, it sends out *ultrasonic* signals (very high frequency sound waves) which are reflected from an object they strike back to the source. Sonar analyzes the signals electronically and converts them into information on the nature of the object, its location and speed.

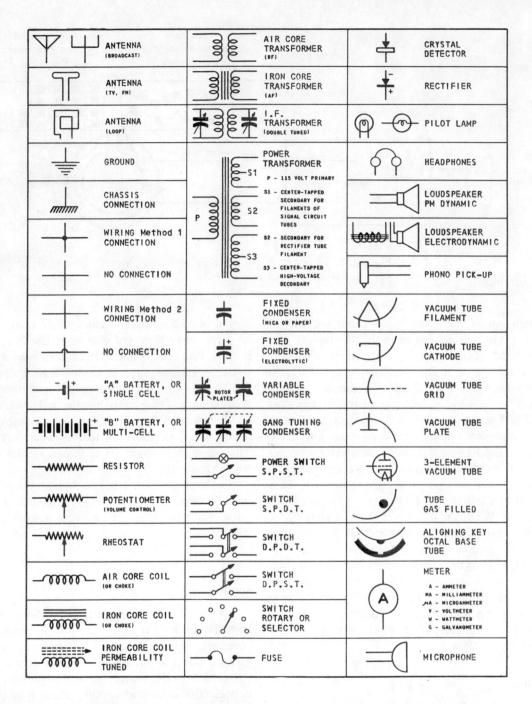

Fig. 5–30. Schematic symbols for radio. Use them to help you read wiring diagrams.

Fig. 5–31. This radar is used on the Navy's Pacific Missile Range. This is an extremely compact model, using microminiature components extensively.

Fig. 5–32. (top) Televising a football game. The cameras take long shots of the whole field or closeups of details. **Fig. 5–32a.** (bottom) Men work at top speed to make selections from the monitors for the audience.

Facsimile

With *facsimile* a photograph can be converted into an electric current and radioed or wired to a receiver that reverses the process and reproduces the photo. This process is used by newspapers in covering events around the world. It is also useful to the police. Officers in one city can work closely with police in another on the trail of a wanted person if they have his photo.

Television

Television transmission begins at the TV camera that records the scene. This camera does not shoot a picture as does an ordinary film camera. It traces what it sees in horizontal lines back and forth, 525 lines for each picture. It takes 30 pictures per second. The camera converts the picture into electrical current. This current is combined with a radio carrier wave and is transmitted from the station antenna. Television waves have a

shorter range than radio waves. The station can send them out for about 100 miles. For longer distances they are picked up by relay towers, amplified, and sent on. Or they are sent out by *coaxial* cables from the station.

At your receiver the TV waves are collected by the antenna. The picture and sound signals pass through the tuner, where the right ones for your channel are selected. These signals are then amplified, and the two waves are separated. The sound waves are amplified and fed to the speaker, which converts them into sound. The picture signals are fed to an electronic beam in the picture tube. This beam *scans* the inside face of the tube (the screen) with the 525 lines as transmitted by the camera. A coating on the inside of the tube is sensitive to the electrons being shot at it. With a great number of electrons, it glows white. Fewer produce the grays and black. These tones make up the picture.

Color Television. A color TV camera photographs the scene through three color filters, red, blue, and green, forming three picture images. All are changed into picture signals and are sent out from the TV station. In a color receiver three different electron guns receive the signals. Each bombards the screen with its own color beam. The screen is coated with equal number of tiny phosphor dots sensitive to the three colors of light.

The dots are arranged in triangles including one of each color. Each electron gun bombards only one color of dot. Together the three guns produce three pictures at the same time on the receiver screen. The eye sees these as one image in full color.

ELECTRICITY AND HEAT

Electricity makes heat in various ways. The heat in a stove, toaster, corn popper, and the like, is caused by a resistance to the flow of the current. The heating element is a poor conductor. When the current is forced through it in spite of this, it

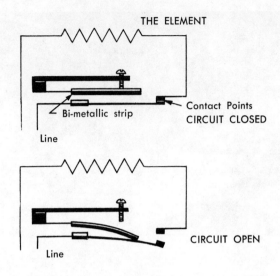

Fig. 5–33. A thermostat is an automatic switch activated by heat.

gets hot. The more it is forced, the hotter it gets. (See page 192.) Such appliances consume considerable power, so be sure not to connect too many into the same circuit.

Resistance Wire Devices. An *electric blanket* gives off heat because it contains a network of very fine resistance wires. It must be handled with care to prevent their breaking. An *automatic toaster* contains resistance wire or ribbon as the heating element. A timer determines the length of time the bread is toasted before it pops up. An *electric dryer* for household laundry has a resistance wire heating element and a fan to circulate the heated air to the tumbling wash. A timer determines the length of the drying time.

Infrared Rays. These are used to provide a deeply penetrating heat, in fact, they are called *heat rays*. They dry the paint on automobiles from the inside out. A doctor uses them to relieve pain. Infrared photo film is sensitive to infrared rays.

Electric Arc. Heat from an electric arc is used in welding metals. A current of very heavy

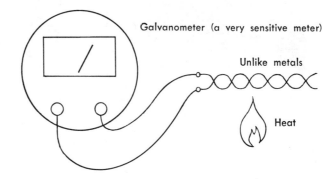

Fig. 5–33a. A thermocouple. Two wires of unlike metals are twisted together. When heated a small voltage develops.

Galvanometer (a very sensitive meter)

Unlike metals

Heat

amperage and low voltage melts the metal and fuses it together. (See pages 175–177.)

Carbon Arc. Carbon arc melting furnaces are used in the smelting of high-grade steel, such as stainless. One electrode is of heavy carbon. The metal to be melted is the other electrode. The carbon is moved close to the metal to make contact and is then withdrawn a bit to keep the electric arc flowing. Temperatures of 6000°F are obtained.

Thermostatic Controls. Most electric heating devices for home use have thermostatic controls to regulate the heat. The heart of a thermostat is a bimetallic strip. It has two different kinds of metals fastened together. When the strip is heated, one of the metals expands more than the other. This causes the strip to bend. The contact points open, shutting off the current. Upon cooling, the strip straightens and closes the circuit, allowing current to flow again. There may be a heat adjustment on the appliance. Turning the screw changes the distance between the contact points so that more or less heat is required to bend the strip and control the current.

Thermoelectricity

Electricity can be produced directly from heat. This occurs in a *thermocouple*. Two wires of unlike metals are joined by twisting and welding. The free ends are connected to a voltmeter reading in degrees. When the joined end is heated, a

Fig. 5–34. The parts of an incandescent lamp. The lamp produces light by heating the metallic filament until it glows brightly. The element has very high resistance to the flow of current. This kind of light gets very hot.

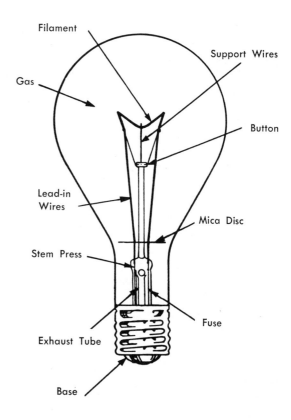

Filament

Support Wires

Gas

Button

Lead-in Wires

Mica Disc

Stem Press

Fuse

Exhaust Tube

Base

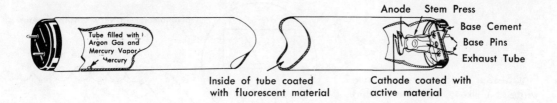

Inside of tube coated
with fluorescent material

Cathode coated with
active material

Anode   Stem Press

Base Cement

Base Pins

Exhaust Tube

Tube filled with
Argon Gas and
Mercury Vapor
Mercury

Fig. 5–35. The parts of a fluorescent lamp. The inside of the tube is coated with fluorescent material, the cathode is coated with active material, and the tube is filled with argon gas and mercury vapor.

current is produced that is translated into temperature readings. Thermocouples are used for high temperature readings such as in furnaces and kilns.

ELECTRICITY AND LIGHT

Man-made light today is almost entirely electrical. Candles and oil lamps are used mostly for decorative purposes now. Edison's incandescent lamp showed the possibilities of electric illumination. Since then, many types of lighting have been developed.

The Incandescent Lamp. We have learned that when a current flows through a resistance wire, heat is formed. When enough current flows, the wire gets white hot and glows. It becomes *incandescent.* This piece of resistance wire in a lamp is called the *filament.* It is made of the metal *tungsten,* also called *wolfram.* This has a very high melting temperature. To keep the filament from burning up, the air is removed from the bulb and replaced with either nitrogen or argon gas. (See Fig. 5–35.)

The Fluorescent Lamp. The fluorescent lamp makes another type of lighting possible. It is soft, glareless, cooler, and less costly to operate than is the incandescent lamp. It gives about three times as much light on the same amount of current. Inside the tube at each end is a small heating coil

coated with chemicals. When the switch is turned on, these coils heat up and send out electrons that react with the mercury vapor. Ultraviolet rays, which are given off, cause the phosphor coating on the inside of the tube to glow.

How Light Is Measured. The lamp bulbs you use at home are stamped with the number of watts they consume. Since the consumer pays for electric power according to watts per hour (kwh, see page 205), this marking is convenient. The intensity of electric light is not measured in watts but in *candlepower.* A standard candle, which burns according to certain specifications, is used as the measure, just as a pound is used in measuring weight. A headlight lamp on an auto may be 32 cp (candlepower). This means it is 32 times as powerful as the standard candle. The great beacons and searchlights at airports may have a million or more candlepower.

Electric Eye. An electric eye opens the door for you at the supermarket. When you approach the door you interrupt a beam of light directed to a photo tube. This tube contains a light-sensitive metal. When light strikes one terminal of the tube, the cathode, electrons are forced off. They are attracted to the anode and so set up a current flow. This is very small. It is amplified by means of vacuum tubes to the point where it can operate a relay. In turn, the relay operates a switch, which turns on the motor to open the door.

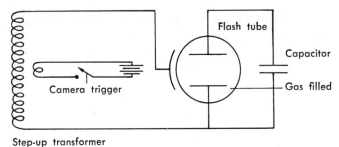

Step-up transformer

Glowing for approximately 1/1000 second

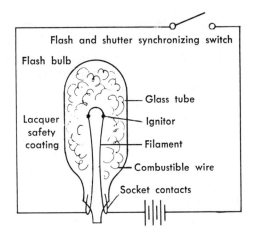

Fig. 5–36a. A photo flash gun. Pressing the camera trigger closes the primary circuit of the step-up transformer. This produces a surge of some 10,000 volts in the secondary. The gas is ionized, producing a glow like a neon tube. The charge stored in the capacitor is released to keep it glowing for approximately 1/1000 of a second.

Fig. 5–36b. Circuit for a battery operated photo flash using bulbs. Battery current heats the ignitor material, which ignites the combustible wire. This is usually aluminum-magnesium or zirconium metal in the presence of oxygen.

Fig. 5–37. (bottom left) The first use of laser beams in mass production was in the making of diamond dies. These are small pieces of diamond through which a laser beam pierces a tiny hole. Through this, fine wire is drawn to reduce its diameter. **Fig. 5–38.** (bottom right) The laser beam strikes the diamond in the making of a die and produces a plumed effect as it instantaneously pierces the hardest of materials.

Light exposure meters, such as those used by photographers to measure the intensity of reflected light, operate on the same principle. Since the only work performed is moving a pointer, little or no amplification of current is used. A light cell of metal such as cadmium or selenium is used in place of the phototube.

Solar Cell. Certain materials such as cadmium, selenium, sodium, and potassium produce small electric charges when exposed to light. They have been commonly used in photoelectric cells. The light-sensitive material is coated on a metal plate. These plates are connected to produce the desired quantity of current. Such cells are used in satellites. It is estimated that there is enough light falling on the roof of a house on a sunny day to supply more than three times the amount of electricity needed in the house. The problem is storing it for use when the sun is not shining. Got any ideas?

Use Good Lighting

Adequate lighting in your school shop means lighting that permits you to work accurately and safely without glare and eyestrain. You need better lighting when you are drawing or working on fine details than when you are planing a board. Your teacher can get a lighting specialist from the local power company to check your shop lighting. Ask him about good lighting in the home, too. You will have but one pair of eyes all your life.

ELECTRICITY AND CHEMISTRY

When electricity and chemistry are combined, new uses and products result. Two common uses are batteries and electroplating.

The Dry Cell. A dry cell, such as is in your flashlight, is a portable source of electric current produced by a chemical action. (Study Fig. 5–40.) The center terminal is positive (+) and the outer, negative (−). When these are connected with a wire, current flows. A dry cell, no matter how large or small, always has 1.5 v. The larger the cell, the greater the amperage. A *battery* is a group of cells. In a 9 v radio battery, 9 divided by 1.5 means 6 dry cells.

Cells connected in series multiply the voltage, 1.5, by the number of cells. (See page 193.) When

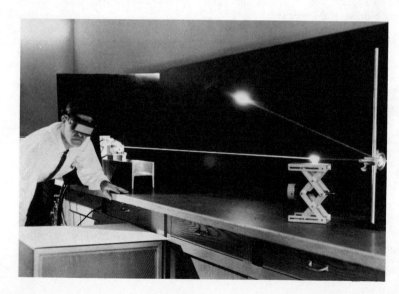

Fig. 5–39. The slender beam of light shown is being produced in a laser. It is known as coherent light because it has only one wavelength. Natural light in contrast is made up of all colors of light, each with a different wavelength. Laser light can be pinpointed rather than dispersed like that from a lamp. There are many types of lasers. Some produce continuous beams, others give out short bursts of light. The argon beam shown here is used in bloodless surgery.

in parallel, the voltage remains the same, but the amperage is increased. In a flashlight the cells are in series because the extra voltage is needed to make a bright light. For electroplating, low voltage and high amperage are needed.

Dry cells are often stamped with a date to insure freshness when you buy them. They lose their power with age as well as with use.

The Storage Battery. A storage battery is a group of *wet cells* connected together. An auto battery has 3 or 6 cells depending on whether 6 v or 12 v are required. Each cell, large or small, has 2 v. (See Fig. 5–41.) A storage cell does not produce its own current as does a dry cell. It must be *charged* before it can be used. Electric current is passed through it until its capacity is reached. Then the battery is fully charged and ready for use. The automobile generator or alternator keeps the battery charged. Watch the ammeter on the instrument panel. It tells whether the battery is charging or discharging.

The sulfuric acid will eat holes in cloth, except woolens, so when you work around a storage battery, remember to keep the acid away from your clothes.

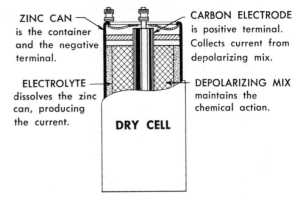

ZINC CAN is the container and the negative terminal.

CARBON ELECTRODE is positive terminal. Collects current from depolarizing mix.

ELECTROLYTE dissolves the zinc can, producing the current.

DEPOLARIZING MIX maintains the chemical action.

DRY CELL

Fig. 5–40. A section through a dry cell.

Fuel Cell. This cell produces electricity from chemical reaction. It consumes fuel in the process and requires no charging. The fuel is oxygen and hydrogen, which combine within the cell. When oxygen is obtained from the air, only a supply of hydrogen is required. The by-product of fuel cells is water. The fuel cell may be the power source for the auto of the future. Such an auto

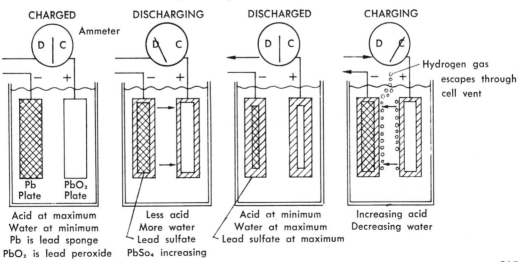

Fig. 5–40a. The cycle of a lead-acid storage cell. This is the type of cell in an auto storage battery.

CHARGED
Ammeter
D | C
− +
Pb Plate
PbO₂ Plate
Acid at maximum
Water at minimum
Pb is lead sponge
PbO₂ is lead peroxide

DISCHARGING
D | C
− +
Less acid
More water
Lead sulfate
PbSo₄ increasing

DISCHARGED
D | C
− +
Acid at minimum
Water at maximum
Lead sulfate at maximum

CHARGING
D | C
− +
Hydrogen gas escapes through cell vent
Increasing acid
Decreasing water

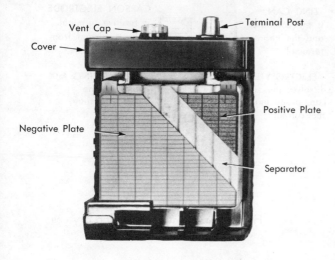

Fig. 5–41. A section through a storage cell.

would be quiet, have no exhaust fumes, and no clutch or transmission. Fuel cells provide current to operate the instruments in satellites and the water produced is used by astronauts for drinking.

Electroplating. Electroplating is a process of removing metal from one object and depositing it on another. Plating is used to protect the object, which is usually of steel, and to make it more attractive. The common types are brass, copper, cadmium, nickel, and chromium. Bicycle wheel rims are often cadmium plated. Handle bars are chromium plated.

Direct current of low voltage and high amperage is used. You can easily copperplate a tool that you have made on the machine lathe. Clean it thoroughly in a strong detergent to remove all grease or oil. Rinse in water. Pick it up with tongs or pliers, not fingers, and place it in the solution as shown in Fig. 5–37.

ELECTRICITY AND ELECTRONICS

Electronics is that field of electricity based on applications of the flow of electrons, as in the vacuum tube or transistor. It includes radio; television; radar; radio-telegraphy; electronic control systems used in aircraft, ships, submarines, missiles, and rockets; and electronic computers used to control automated machines in industry. Applications of electronics are increasing so rapidly that there appears to be no limit to the possibilities.

The Vacuum Tube. The vacuum tube has made possible the development of radio, television, and the entire field of electronics. Edison is credited with the discovery that later led to the perfection of the vacuum tube. He noticed that when he connected a second element in an incandescent lamp, a current could be drawn from the hot filament to this element, even though they were not in contact. The filament is now called the *cathode* and the element is called the *anode,* or plate. When a positive voltage is applied to the anode, a current flows. With a negative voltage, no current flows. This discovery is called the "Edison Effect."

In 1896 J. A. Fleming placed the cathode and the anode in a glass tube and removed the air. This was the first vacuum tube. Since it had two electrodes, it was called a *diode* (*di* means "two"). This was used as a detector in the first radios.

Lee DeForest later placed a fine wire mesh between the cathode and the anode. He found that the amount of current flowing to the plate could now be controlled. This mesh is called the *grid*. It is supplied with a voltage for controlling the plate current. Having these three electrodes, the tube is called a *triode*. This was the original amplifier tube. It could increase the strength of radio signals. Many different tubes are now available. They all depend upon a source of electrons within the tube.

Crystal Detector. Certain minerals have been found to have peculiar characteristics useful in radio. Galena and germanium, for example, will permit current to flow in only one direction. Like the diode tube, they serve as detectors.

Transistors. Transistors are crystal-type amplifiers. They are made of *semiconductors* such as

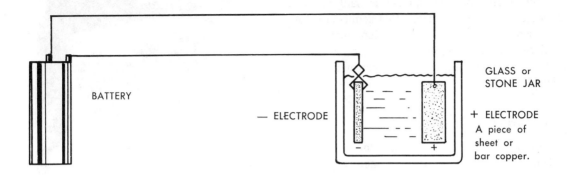

BATTERY

GLASS or
STONE JAR

— ELECTRODE

+ ELECTRODE
A piece of
sheet or
bar copper.

Fig. 5–42. (top) Electroplating with copper. The object being plated is a —electrode. It must be perfectly cleaned. Use detergent in hot water to do this. The voltage is low, amperage high. Use as much as six volts. The solution is ½ cup copper sulfate in two quarts of distilled water. **Fig. 5–43.** (bottom) Inside a vacuum tube.

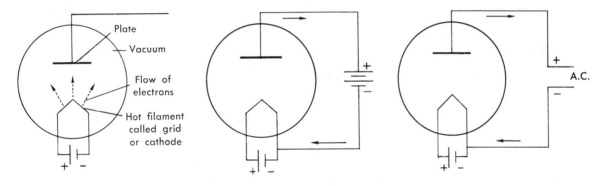

Plate

Vacuum

Flow of electrons

Hot filament called grid or cathode

A.C.

An alternating current in place of the direct current causes a different reaction. Current flows only when the plate is positive. When it is negative current cannot flow because the negative plate repels the electrons. (Remember, like charges repel; unlike charges attract.) This tube operates as does a crystal diode in a crystal radio. Current can flow in only one direction.

germanium and silicon. These are called semiconductors because they can serve as either conductors or insulators, depending on the conditions. Actually they are neither good conductors nor insulators. A transistor is made up of three layers of materials bonded together with heat. One layer is the emitter, one is the collector, and one is the base. The emitter serves as the cathode in a triode tube. The collector is the plate, and the base is the grid. Transistors require much less current for their operation than do tubes. They give off much less heat and react instantaneously. Unlike tubes, they do not require a warm-up before emitting electrons. Their small size is also an advantage.

When electronic equipment is said to have *solid state* circuitry, it means that semiconductors are used rather than tubes. Semiconductors are used as detectors, amplifiers, rectifiers, resistors, and circuits for radio, television, instruments, and controls.

ELECTRONICS AND AUTOMATION

Developments in electronics have made possible today's automation in the manufacturing industries. Electronics systems take over the controlling of the machines. They activate mechanisms causing the machines to do their work. Electronic devices control the rate of production, moving the part or product from machine to machine with exact timing. They see to it that parts are correctly positioned in the machines, tell them when to work, and inspect the quality of the work. And, too, they stop any machine that is not performing properly. In some systems the controls can correct for error as soon as it occurs, rejecting the faulty product.

Fig. 5–45a. A pocket-size transistor radio typical of many kits available for assembly.

Fig. 5–44. Transistor components.

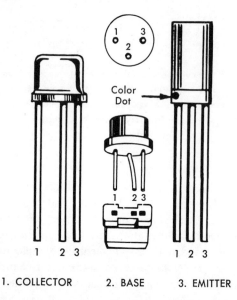

1. COLLECTOR   2. BASE   3. EMITTER

Many of today's manufactured products would not be possible without the precision that electronic controls make possible. Electricity traveling at the speed of light makes this control practically instantaneous. Gasoline, autos, jet airplanes, watches, appliances, manufactured food, and many other items are produced by means of electronics.

Automation at its highest level of development appears to be "almost human" in many of its functions. It can make decisions, correct its mistakes, and evaluate its work. The present highest level is known as *heuristics*. This involves a computer into which goals can be programmed. The computer then directs the machine to seek out and accomplish the goals. The further automation develops, the more humanlike it becomes.

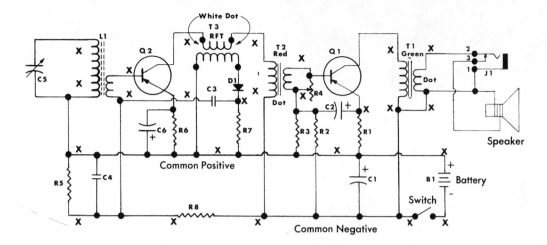

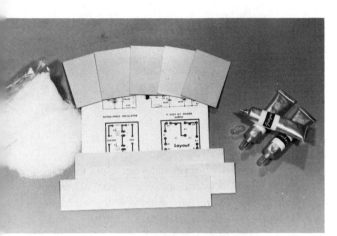

Fig. 5–45b. (top) The schematic wiring diagram for it. **Fig. 5–46.** (middle) You can make your own printed circuit boards with a kit of materials such as these. A special ballpoint pen is used to lay out the circuit with an acid-resistant fluid on copper-clad boards. These are placed in an etching bath to remove the unwanted copper. **Fig. 5–47.** (bottom) A space-age watch. Using a miniature power cell and a magnetic circuit, this is an electronic watch. The concept of miniaturization is applied here. Because it is possible to make parts or components in miniature or sub-miniature sizes, many new products are possible.

✚ SAFETY SENSE

Electrical current is not particularly dangerous unless both voltage and amperage are high. You can get a jolt from a spark plug on an automobile engine, but you don't get burned, because the amperage is very low. The "wise guy" who sticks his finger into a house lamp socket to show how tough he is, is just asking for trouble. That's like shaking hands with an octopus. Study these suggestions. Can you add others?

1. Do not try to do your own house wiring. This is for an expert electrician. Cities have laws that prohibit anyone but a licensed electrician from doing wiring.

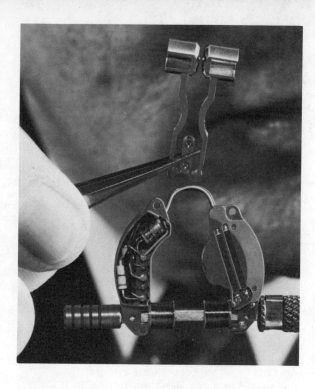

2. Replace a blown fuse only with the same size fuse (no pennies, please). If the fuse continues to blow, the trouble is still there.
3. Why doesn't it make safety sense to touch an electrical appliance or switch while taking a bath in the tub?
4. If your shoes are wet, do not stand on the concrete floor while operating a machine in your shop.
5. Don't do your electric welding on a damp floor.

Fig. 5–48. The heart of the electronic watch is this tiny tuning fork, which is kept vibrating in a magnetic field. The vibrations activate the mechanism that moves the hands.

ELECTRICITY-ELECTRONICS AND OCCUPATIONS

The chart, Fig. 5–2, shows the nature and extent of electricity-electronics applications and industries. In all of these there are occupations. The production of electrical power includes the occupations in the operation of power plants. Most of our electrical power is produced by what are called electrical utilities. There are occupations in the design, construction, and maintenance of power lines and substations.

The manufacturing industries produce electrical-electronic components, machines, and equipment. These are used in "E-E" industries, power generation, and distribution. There are occupations in the fields of communications, heating, lighting, chemical, medical, and security.

There are more occupations in the servicing of E-E equipment than in the manufacture of it. New uses for electricity are continually being developed. Service technicians are needed for all types of this equipment from home appliances to computers.

The following are typical of the **occupations** found in the electrical-electronic fields. There are not many that do not require either special training or college studies. Large corporations often have their own training programs.

Accountants
Appliance service
Aircraft electrical-electronic installation and service

Attorneys
Automobile ignition specialists
Computer operators
Computer technicians
Draftsmen
Electrical engineers
Electricians
Industrial equipment service
Machine operators
Machinists
Medical equipment service
Motor rebuilding
Office personnel

Operating engineers
Plant engineers
Plant maintenance
Power line crews
Product designers
Production engineers
Production line workers
Radio and TV service
Research scientists
Sales engineers
Supervisors
Welders

ELECTRICITY-ELECTRONICS AND RECREATION

You can imagine that there are many hobby possibilities with electrical things. The "double E's" are natural areas for experimenting. Moreover, this is a good way to learn about them providing you do the necessary studying, too. Trial and error experimenting with electricity is rather stupid. It is also dangerous.

The following suggestions are only a few of the hobby possibilities. See if you can add more. Be sure that in whatever you do, you handle electricity with respect. Until you are qualified to handle dangerous currents, you will be wise to use low voltage battery power.

Electrical Hobby Possibilities

Electroplating, copper and brass
Construct apparatus to demonstrate basic laws
Electromagnet design and construction
Remote control outfit for a camera
Design and build several types of electric motors
Tesla coil experiments and demonstrations

Hobbies with Electronics

Build a code oscillator
Learn the Morse Code
Build electronics gadgets from old radio parts
Repair radios
Civilian Band Communications
Build a radio control system for a model airplane
Ham radio

ELECTRICITY-ELECTRONICS AND THE ENVIRONMENT

The demand for electricity continues to increase. This is due to the increasing variety of uses for it. Most of the electrical power produced in the United States is by steam turbine. Coal is the common fuel to produce the steam. From this comes smoke and gases to pollute the air. Increased efficiency in steam power plants has reduced these pollutants considerably in recent years. Most boilers use powdered coal, which is blown in. This gives a more complete burning and gets more of the heat units (British Thermal Units) from the coal.

Nuclear power plants produce heat to generate steam to drive turbines. They discharge no smoke or gas. They do require great quantities of cool water for the condensers. This water, after circulating through the condensers, is returned to a river or lake. This can raise the temperature of the water enough to destroy the oxygen and kill off the fish. Until a better way is found to produce electric power we have a very difficult environmental problem. At this time successful experiments in producing electricity directly from nuclear energy without the need for turbines have been conducted. Watch for further developments along this line. This will be a giant breakthrough when it becomes a reality.

TO EXPLORE AND DISCOVER

1. How many KW's of electrical power are consumed in one day in your city? In the U. S.?
2. What is a "brownout"? This has happened in cities along the East Coast. What causes it? What are the remedies?
3. Why can automobile headlights shine high and low?
4. Explain how a traffic light operates.
5. Estimate the number of electrical devices in your home. Now make an actual count. Did you miss it?
6. Why does an airplane battery have 24 volts but an auto battery only 12?
7. How is electricity used in the manufacture of aluminum?
8. Explain the circuitry in door chimes. The rear door has one "bong" and the front two.
9. Why should you not be in swimming during an electrical storm?
10. What is the scientific explanation for lightning flashes and thunder? How did primitive man explain it?
11. Draw a circuit diagram for a bike lighting system.
12. What is the difference between an open and a closed circuit?
13. Explain the principle of a sun lamp such as can be used in the home. Point out the possible dangers in its use.
14. Add up the wattage of all the electric lights in your home. What does it cost to operate all of them for one hour when the power costs 5¢ per KWH?
15. How is sound converted to electricity and back again in a telephone? In a radio?

16. Connect four dry cells in series and then in parallel. Compare the voltage and the amperage using the proper meters.
17. How does electric shock cause death? Burning?
18. A toy train transformer operates on 120 volts. The locomotive operates on 12. What is the ratio of turns in the primary and secondary coils?
19. Why won't copper wire work for the heating element in an electric toaster? Why won't resistance wire serve well as a conductor? Could you set up a demonstration of these using dry cells or auto battery for power?
20. For sight saving what is the best lighting for reading and for drawing?

FOR RESEARCH AND DEVELOPMENT

1. Compare the conductivity of salt water and tap water. Set up an experiment using two stone crocks or glass jars. Take five readings in the salt water. Each time double the amount of salt. Start with one ounce per quart of water. Show the readings on a graph.
2. What is a theory? A principle? A law? Explain each and illustrate with applications from electricity.
3. How does faulty wiring cause fires? Set up a demonstration using an auto battery for power. In what different ways can dangerous amounts of heat build up in circuit?
4. The oceans are our largest sources of water but the water is not potable. Can it be converted to drinking water? Design and construct a demonstration model.
5. Produce a printed circuit board for a crystal diode radio. Does it function as well as a wired set? What are the advantages and disadvantages of printed circuits that a consumer should know about?
6. Wanted: an automatic device to close a window when it rains.
7. Wanted: a device that will announce the arrival of a visitor in your home without his knowing.
8. Wanted: collision-proof vehicles. Some 55,000 persons are killed each year in auto accidents. Is it possible to design cars so that they cannot collide?
9. Wanted: a bicycle burglar alarm. They are available for autos.
10. Compare the processes of electroplating with copper, brass, zinc, nickel, and chromium. Can any of these be done safely in your laboratory? Figure out what is needed to demonstrate the one you recommend. Conduct tests experimentally.

FOR GROUP ACTIVITY

1. Arrange for a demonstration before your class on rescue and first aid for electrical accidents. Contact the local Fire Chief or a Red Cross first aid instructor.

2. Make a field trip to your local power station. What source of energy or heat is used? Compare the processes of the production of electricity: steam, atomic, and hydro. What effects do each have on the ecology of natural environment?

3. Form a ham radio club. Let each member construct a code oscillator. Get a licensed ham to instruct the group in code practice. What are the licensing regulations? Can you pass the test?

4. Get several buddies to build simple crystal radios. Put on a contest for the most stations received and for the farthest. What evidence or proof will be required?

5. Several of you could team up on the design and construction of an electric-powered mass transportation system. Build to scale a three-dimension model showing it serving both a city and the suburbs.

FOR MORE IDEAS AND INFORMATION

Books

1. Delpit, George H. *Electronics in Action.* Peoria, Ill.: Chas. A. Bennett Co., Inc., 1969.
2. Gerrish, Howard H. *Electricity and Electronics.* S. Holland, Ill.: Goodheart-Willcox Co., Inc., 1964.
3. Lush, Clifford K., and G. E. Engle. *Industrial Arts Electricity.* Peoria, Ill.: Chas. A. Bennett Co., Inc., 1965.
4. Miller, Rex, and Fred W. Culpepper, Jr. *Energy, Electricity and Electronics.* Bloomington, Ill.: McKnight & McKnight Publishing Co., 1964.
5. Vergara, William C. *Electronics in Everyday Things.* New York: Harper and Row, Publishers, 1961.
6. Woodward, Robert L., and J. Lyman Goldsmith. *An Introduction to Applied Electricity-Electronics.* Englewood Cliffs, N. J.: Prentice-Hall, Inc., 1963.

Booklets

1. *The ABC's of Radio; The Story of X-Ray; You and the Computer.* Educational Publications, General Electric Co., One River Road, Schenectady, N. Y. 12305
2. *Instructors Guide for Basic Electricity and Electronics.* Electronics Industries Assn., Service Coordinator, 2001 I St., N. W., Washington, D. C. 20006
3. *Magic Behind Your Dial.* American Telephone and Telegraph Co. Obtain from nearest Bell Telephone Co. office. Ask also for list of booklets and films.
4. *Mr. Bell Invents the Telephone.* Western Electric Co., Inc., Technical Information and Public Relations, 195 Broadway, New York, N. Y. 10007
5. *Portable Power Handbook.* Batteries. Union Carbide Co., Consumer Products Division, 270 Park Ave., New York, N. Y. 10017
6. *Power and Progress.* Edison Electric Institute, Editorial Dept., 750 Third Ave., New York, N. Y. 10017

7. *RCA-NBC Firsts in Television.* National Broadcasting Co., Dept. of Information, 30 Rockefeller Plaza, New York, N. Y. 10020
8. *The Story of the Bell Solar Battery.* Available from your local Bell Telephone Co.
9. *The Story of the Transistor.* Available from your local Bell Telephone Co.

Films

1. *Free Educational Films from Industry.* (catalog) Modern Talking Picture Service, Inc., 1212 Avenue of the Americas, New York, N. Y. 10036
2. *New Developments in Communications.*
3. *Optical Masers and Lasers.*
4. *Telstar.*
   (The last three films may be obtained through your Bell Telephone Co. office. Ask for complete list of films available.)

Teacher's Kit

1. *Electricity.* Teacher's Kit of instructional material on wiring in the home. Contact your local power company or write: National Wiring Bureau, 155 E. 44th St., New York, N. Y. 10017

GLOSSARY—ELECTRICITY AND ELECTRONICS

Alternating Current electric current that flows forward and back in a circuit; designated AC.

Alternator a mechanical rotating device that produces alternating current, as on an automobile engine.

Ammeter an instrument for measuring amperage.

Ampere unit of electric current.

Amplifier a device for increasing the strength of RF waves.

Anode positive terminal or plate in a vacuum tube.

Candlepower the illuminating power of a standard candle.

Capacitor condenser.

Cathode negative terminal that emits electrons in a vacuum tube.

Cat's whisker the movable wire that makes contact with a crystal.

Circuit breaker a safety device to prevent overloading of a circuit; can be reset and reused.

Coaxial cable a transmission line in which both outer tube and inner wire are conductors and insulated from each other.

Condenser a device for holding an electrical charge between metal plates separated by insulation. Newer term is capacitor.

Conductor a material that conducts electricity with little resistance.

Crystal detector a small diode consisting of a semiconductor (crystal) and cat's whisker.

Current a flow of electrons along a conductor.

Cycle a single complete period of alternating current flow from negative to positive and back again.

Diode a vacuum tube with two (di) elements: cathode and plate. Also a semiconductor with the same function.

Direct current electrical current that flows in one direction only, from negative to positive.

Dry battery a group of several dry cells connected together.

Dry cell a self-contained cell that produces electricity by chemical action.

Edison effect Edison first noticed that in a vacuum the emitted electrons were attracted to the positive plate.

Electric arc heavy discharge or sparking of current used for producing heat or light.

Electric eye a device in which a light sensitive cell produces a voltage to activate a relay. Also called photoelectric cell.

Electronics the technical-scientific field based on application of the flow of electrons in vacuum tubes and semiconductors.

Electromagnet a magnet produced by an electric current.

Electroplate the process of removing metal from one source and depositing it on an object by means of electric current.

EMF electromotive force, also voltage.

Facsimile a device for transmitting pictures by radio or telephone.

Filament the resistance wire in a vacuum tube that, when heated, emits electrons; also the element in an incandescent bulb.

Fluorescent the property of producing light when acted upon by a stream of electrons.

Fuel cell a cell that produces electricity by means of chemical action. Requires continual feeding as with hydrogen and oxygen.

Generator a mechanical rotating device that produces electric current.

Heuristics the level of automation using computers capable of goal seeking.

Hydroelectric plant an electric power generating facility using water power to drive the generators.

Incandescent glowing with intense heat, white hot.

Induced current current produced in a coil whose turns cut the magnetic field of another conductor.

Induction coil a coil employing the principle of induction.

Infrared rays heat rays produced by electricity and the sun.

Insulator material having great resistance to the flow of electrons.

Kilowatt KW. One thousand watts.

Kilowatt hour KWH. A unit for the sale of electric power equivalent to one thousand watts for one hour.

Magnet a material, usually iron, that attracts other materials having magnetic qualities.

Magnetism the condition of being magnetic.

Magneto a machine that produces electricity as used for ignition on some internal combustion engines.

Magnet wire copper wire used for magnets and coils.

Microphone a device for converting sound (audio) waves to electrical impulses.

Modulator in radio broadcasting, a device that combines the audio frequency (AF) wave with the radio frequency (RF) wave.

Ohm unit of measurement for electrical resistance.

Ohm's Law the mathematical relationship between amperes, volts, and ohms, as

$$I \text{ (amperes)} = \frac{E \text{ (voltage)}}{R \text{ (resistance in ohms)}}$$

Oscillator a vacuum tube generator of AC voltage.

Negative a terminal also known as the cathode.

Parallel circuit a circuit that provides two or more routes for current flow at the same voltage.

Permanent magnet a material body that tends to retain its magnetism over a long period.

Photoelectric cell see electric eye.

Piezoelectric property of a material to produce electric current when subjected to mechanical pressure.

Polarity the state of the ends of a magnet as being positive or negative.

Positive a terminal also known as the anode.

Power the product of current and voltage, as watts = amperes $\times$ volts.

Primary the first coil or winding of a transformer and/or induction coil through which the current flows.

Radar radio detection and ranging system.

Rectifier a device for changing alternating current to a pulsating direct current.

Relay an electrically operated magnetic switch.

Resistance opposition to the flow of electrons, measured in ohms.

Resistance wire a poor conducting wire used to produce heat, as in a stove.

Secondary the coil in a transformer or induction coil in which current is induced.

Semiconductor a material that can both conduct and restrict current flow, as in a transistor.

Series circuit a circuit providing a single route for current flow.

Short circuit a circuit in which the normal path of the current has been shortened. Since the resistance is lowered, excess heat and sparking may result.

Solar cell a metal plate coated with a light-sensitive material, which produces an electrical charge when exposed to the sun.

Solenoid an electromagnet with a freely moving core.

Solid state circuitry circuits in which semi-conductors rather than vacuum tubes are used.

Sonar underwater radar.

Spark plug the electrode in an internal combustion engine which ignites the fuel.

Static electricity electricity generated by friction between two unlike materials.

Step-down transformer a transformer that lowers the incoming voltage.

Step-up transformer a transformer that increases the incoming voltage.

Storage battery a group of wet cells connected together.

Teletype a telegraphic device that receives and types the message on paper.

Thales Greek scientist (600 B.C.) who discovered static electricity

Thermocouple two wires of unlike metals joined at one end. When the joint is heated a DC voltage is generated.

Thermoelectricity electricity that is produced by heat.

Thermostat a heat-sensing switch for automatically opening and closing a circuit.

Timer a mechanical device that permits current to flow at the right time.

Transformer a device for changing voltage by application of the principle of magnetic induction.

Transistor a semiconductor that functions both as a conductor and insulator.

Transmitter the device for radio broadcasting.

Triode a vacuum tube with three elements: cathode, plate, and grid.

Turbine an engine driven by pressure of steam, water, or air against blades or vanes.

Vacuum tube a closed device for regulating the flow of electrons in a circuit; operates in a vacuum.

Volt the unit of electrical pressure.

Voltage the EMF in a circuit.

Voltmeter an instrument for measuring voltage.

Underwriters Laboratory (UL) a private testing company for electrical products.

Wet cell a storage cell for containing an electrical charge.

Wolfram tungsten.

Graphic Arts Technology 6

People had been writing and drawing for many centuries before anyone figured out how copies could be made quickly and easily. Until this was discovered, each copy was written or lettered by hand, letter by letter. Imagine how long it would take you to make a hand-lettered copy of the Bible. Back in the Middle Ages, copies of the Bible were made this way by monks who spent their lifetimes at copying with quill, ink, and parchment of their own making.

You know the Bible story of how the Ten Commandments were cut into stone tablets. The Ark of the Covenant was built to house these tablets because they were the only copies of the law by which the people lived. Only a few people, called *scribes,* could write in those days. Scribes kept records and made copies of the laws, lettering them by hand.

The first use made of printing seems to have been on government seals in Egypt, Syria, and Babylonia, probably about 5000 years ago. The designs were cut into stones, which when inked made impressions on other material. The Chinese found out how to carve their language characters into wood blocks and to print from them as early as A.D. 700–800. They actually were manufacturing paper a thousand years before other people. They printed paper money about the year A.D. 807.

Printing from Movable Type

Johann Gutenberg, a German silversmith, got the idea that each letter of the alphabet could be made separately on a block. He first made these letters of wood, but being a metal worker, he soon made them of metal. This was *movable type,* which could then be set up, printed from, and then taken down to be used again. Gutenberg is credited with being the first European to use movable type.

With movable type there was the problem of sufficient quantities of the characters. Each had to be carved from hand. Gutenberg's greatest contribution is said to have been the process for casting them in molds. He had to use wood for his molds. These did not last long, since melted lead was poured into the mold. Nevertheless the casting of metal type characters has been done ever since.

In 1452 Gutenberg produced the famous 42-line Bible, printed from movable type. It was Gutenberg who saw the great possibility of cultural, religious, and commercial advancement through such printing. This proved to be true. The Bible and other printed books were made available to more people than ever before.

Gutenberg's press was a very simple device based on the principle of wine presses and linen presses. The paper was pressed down on the inked type with the aid of a wooden screw and a lever. Even so, it was a very slow process. It required eleven different operations and two operators, one to ink and the other to remove the sheet. For the next three and a half centuries this kind of press was standard, although improvements included the use of metal in place of wood.

Paper and Ink. Paper and ink are as important in printing as the type and the press. The early scribes used papyrus (pap-*eye*-rus), a plant that grew in the Nile valley. The long fibers of the leaves were soaked and pressed into sheets. Parchment and later vellum, made from the skins of

Fig. 6–1. Printing was a hand process in colonial America. The printer is turning on a heavy screw to put pressure on the paper so that each type character will print. This is a restoration in Colonial Williamsburg of the shop in which William Parks published the first issue of "The Virginia Gazette" on August 6, 1736.

sheep and goats were more durable. The skins were scraped thin and cut into sections that were used as scrolls. Paper, as we know it, was first made about A.D. 100 by the Chinese and Japanese from bamboo, silk, and linen. In Europe by the time of Gutenberg, paper was being made from rags. Today most of our paper is manufactured from wood pulp.

THE GRAPHIC ARTS INDUSTRIES

From such simple beginnings the graphic arts industries have grown. They are the industries that

produce our newspapers, magazines, books, and pamphlets. They reproduce in quantity drawings, photographs, and other illustrations.

An example of a big-scale printing operation is the New York *Daily News*. This has the largest circulation of any newspaper in the United States. Its presses print 65,000,000 pages of newspaper an hour. To do this, they use an average of 5,600 tons of *newsprint* (newspaper paper) a week. This amounts to 29,090 miles of paper five feet wide. Let's see, it is only 25,000 miles around the earth, isn't it? So it figures out that these presses print a mile of such paper about every three minutes. Besides, they consume 9,340,000 pounds, or 1,153,090 gallons, of black ink every year. These figures exclude the printing of the Sunday color comics and the rotogravure sections.

Fig. 6–2. (top) The printer has pulled the first print from his press. He is proofreading it. Note that his press is made largely of wood.
Fig. 6–3. (bottom) The main kinds of printing explained by method, uses, and the plates each requires.

Kind of Printing	*Method*	*Uses*	*Kind of Plates*
Letterpress	Printing from a *raised* surface.	Newspapers Magazines Books Letterheads	Block cuts Line etchings Halftone etchings Stereotypes Electrotypes Rubber plates and stamps
Planography	Printing from a *flat* surface.	Magazines Books Advertisements Billboards Posters	Crayon-stone lithography Photolithography Photogelatin

Intaglio	Printing from a *sunken* surface.	Art prints
		Wallpaper
		Stamps
		Paper Money
		Gravure section of
		  Sunday paper
			For lines:
			  Line etchings
			  Line engravings
			  Dry point
			  Steel engraving
			  Copper plates
			For tones:
			  Soft ground etching
			  Aquatint
			  Mezzotint
			Photographic:
			  Photogravure
			  Rotogravure

Other	Silk screen Duplicating Blueprinting	Etchings Folded boxes and packaging	

Fig. 6–4. A modern automatic letterpress. The paper is cut to size and is stacked in the machine. One sheet is printed at a time.

Printing Processes

There are several different printing processes. Each requires different kinds of plates, presses, and equipment, and each has certain characteristics that make it best for the given job.

Printing Inks

There are many kinds of printing inks. Each is intended for particular printing jobs. For example, there is *news ink* for printing newspapers, *job ink* for general purposes, *lithograph ink,* and the like. Inks are made for printing on metal, glass, plastics, and other materials. You will probably use job ink, lino-block ink, and silk-screen ink.

LETTERPRESS PRINTING

Letterpress printing means printing from *raised* surfaces. It is sometimes called *relief* printing. As in the case of a type *character,* all the material has been cut away except that which is to make the impression on the paper. (See Figs. 6–5 and 6–21.) The raised surfaces are inked with a roller and are then pressed against the paper to make the impression. This is done in a *press* into which paper is fed by hand or automatically.

Type

Type is made up of individual characters, one for each letter, numeral, or punctuation mark. These are assembled into words and sentences, in lines of a given length. *Type height* was standardized in America almost a hundred years ago. All type measures 0.918 inches from the base of the feet to the printing surface (see Fig. 6–5). Various type faces can thus be interchanged in a particular job or among different printers. They fit on the many kinds of typecasting machines.

In the letterpress process, the printing is done from standing type and cuts and from *slugs* cast from a *matrix* (*may*-tricks), or small mold of each character. They are made up into words and lines.

Sizes of Type. Type is made in several sizes according to a standardized *point* system:

72 points equal 1 inch
12 points (⅙ inch) equal 1 pica (*py*-ca)
 6 picas equal 1 inch
 6 points (½ pica) equal 1 nonpareil (non-par-*rel*)
5½ points equal 1 agate (used in newspaper measurements).

These measurements refer to the height of the letters from the top of the tallest, such as "h" which has an *ascender,* to the bottom of those with *descenders,* such as "p" or "y." (See *point body* in Fig. 6–5.)

Common sizes of type for school use are: 6, 8, 10, 12, 14, 16, 18, 24, 30, and 36 point. In

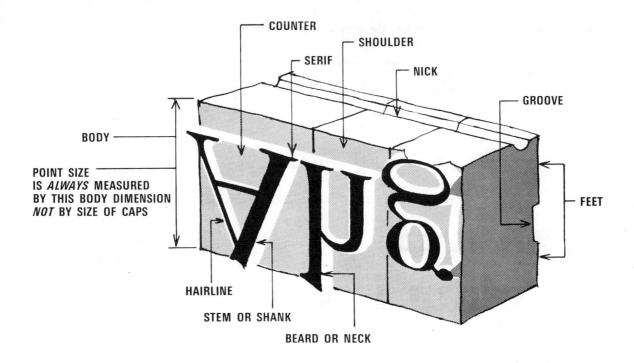

Labels on figure:
COUNTER
SHOULDER
SERIF
NICK
GROOVE
BODY
POINT SIZE IS *ALWAYS* MEASURED BY THIS BODY DIMENSION *NOT* BY SIZE OF CAPS
FEET
HAIRLINE
STEM OR SHANK
BEARD OR NECK

Fig. 6–5. The parts of printing type. The height to paper (printing surface to foot of type) is always .918 inches. Nicks vary in number and position to distinguish sizes and faces and to show if type is right side up. On a Linotype machine, these also serve to sort matrices after the slug is cast.

a printing shop, additional sizes to 120-point might be used. Larger sizes are sometimes cut in wood or linoleum block. This book is printed in 10-point type.

Serifs. The cross bars that finish off the ends of letters are *serifs*. A *sans serif* type face is one without serifs. There are many styles of both serif and sans serif types.

Type Faces. The designing of letters is an art. This is especially true in the designing of a *type face,* or distinctive style of alphabet of type. The forms of the letters must be easy to read and pleasing to the eye when composed into pages. They should be appropriate for the job.

A type face is often named for its designer.

Some of the early type designers were: Nicholas Jenson (a Frenchman working in Venice) and William Caxton (English), both in the fifteenth century; Claude Garamond (French), sixteenth century; William Caslon (English) and Giambattista Bodoni (Italian), eighteenth century. More recent master type designers were Frederic W. Goudy (American) and Eric Gill (English). Compare the styles of these designers and others of their period as shown in type books.

Nearly all types are made also in **dark, heavy bold face** and in *italic,* the latter being slanted and resembling handwriting. These are used sparingly to emphasize or call attention to words or short passages. Note their use on this page.

Fig. 6–6. Suggested type faces to show the main groups of type.

QUINQUAGE

LARGO OPEN

ROMAN TYPES. Classical models. The upper type is Bauer Text Initials. The lower is Largo Open.

PROFIL

SAPHIR 1234

DECORATED LETTERS. Sometimes too much so. Must be used with restraint.

Composing Mac

Information Service

SANS SERIF TYPES. Standard Extralight Extended and Commercial Grotesk Demibold Elongated.

Craw Modern

MYSTERIOUS RENEGADES

BOOK TYPES. Can be well used for display. They are Craw Modern above and Century Expanded Italic below.

MUSIC IN THE AIR

new york city

SCRIPT. Affects the feeling of writing or hand printing. Post Title Light and Charme.

6 't the type! It makes all the difference
sake. If you were speaking your words
ıst speak them for you—type, with its
gnificance. What types will most effec
. orator in planning his speech tries to
al speeches heard in th industry today.
3, CAREFULLY SELECT 1234567890

8 After you have selected your words, carefully
select the type! It makes all the difference in
the world, the form in which you cast your
message. If you were to speak your words you
would know what tone of voice to use. But the
type must speak them for you—type, with its
AFTER YOU SELECT THE WORD 67890

10 *After you have selected*
carefully select the type!
the difference in the world
which you cast your mes
were speaking your words
what tone of voice to use.
AFTER YOU HAVE SEl

12 AFTER you have selected your words, carefully select
the type! It makes all the difference in the world, the
form in which you cast your message. If you were to
speak your words you'd be sure to know what tone of
voice to use. But type must speak them for you—type
with its thousand and one variations and its subtle sh
AFTER YOU HAVE SELECTED YO 1234567890

Fig. 6–7. Various sizes of the same type face, Century Expanded. The 6 pt. and 12 pt. samples are roman, the 8 pt. is bold face, and the 10 pt. is italic.

Type Fonts. Type is purchased by the pound and by the font. A font includes a complete assortment of letters (capitals and lower case), punctuation, and spacing material of one size and one style of type. For the letters used most often, such as *a, e,* and *i,* more type is given than for *x*'s and *z*'s.

Spacing Devices

Spaces are needed between words, between lines, and sometimes between letters to make the print easy to look at and easy to read. *Quads, spaces, leads,* and *slugs* are used for these purposes.

Quads and Spaces. These are blocks of type metal of the same point size as the font of type.

They have nicks just as the other matrices do. The *em* quad is the square of the type size. For example, if 12-point type is being used, the em of that type will be 12 points wide. The other quads and spaces are wider or thinner than the em. (See Fig. 6–9.)

Leads and Slugs. Spacers in thin, long strips for use between lines are *leads* and *slugs.* Sometimes they are called coppers and brasses because they are made of those metals. Leads are commonly 2 point, and slugs 6 point. They can be cut to desired lengths. This book has two extra points of leading between lines.

Furniture and Reglets. *Furniture* is filler material of wood or metal. It is used to fill the space around the type in the metal frame called a

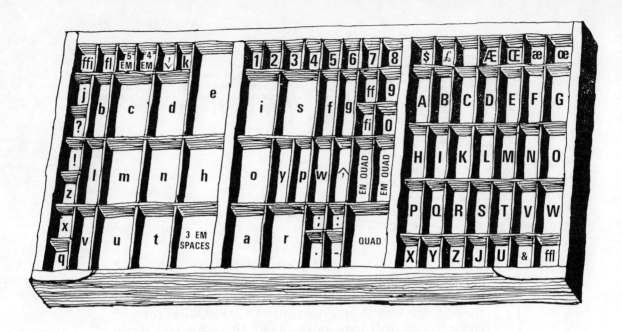

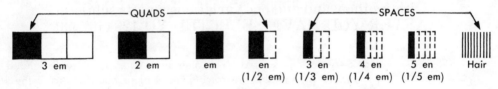

Fig. 6–8. (top) Layout of California job case. **Fig. 6–9.** (bottom) Spacing devices.

chase so that the type can be locked in securely.

Reglets are thin strips of wood used for furniture as well as for spacing in typesetting.

How to Set Type

Follow these steps for setting your name and address. Perhaps you could use this type on some stationery.

1. Select the size and type face to suit the purpose you have in mind.

2. Holding the composing stick as shown in Fig. 6–10, insert a 6-point slug, as long as or longer than the longest line of type.

3. Set the type for the first line on the slug. Start at the left end of the stick and make sure the nicks are all up. The words should read from left to right, with the letters upside down.

4. Place a lead over the first line to provide a space between the lines. Use more space if you wish.

5. Add the next lines the same way.

6. *Justify* the lines. This means to tighten or spread the line in the composing stick by adding ems, and ens, and thin spaces so that lines will be of equal length.

7. Add spacing material to even the ends of the lines so that the type forms a block.

8. Carefully slide the whole block into a *galley* and tie it securely with a string. A galley is a metal tray open at one end. Wrap the first turn

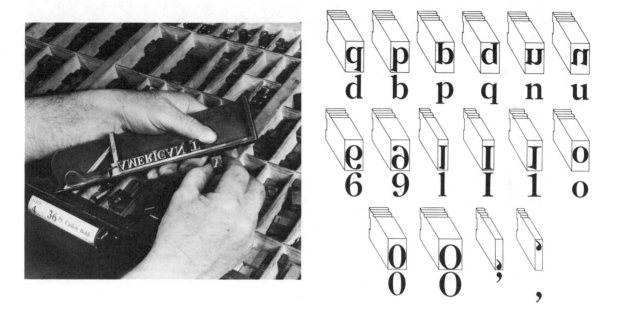

Fig. 6–10. (left) The first line of type is set against a 6-point slug in the composing stick. The stick is set to the width of the type measure or column.
Fig. 6–11. (right) Type characters difficult to identify. Do not read type in any other way than upside down, from left to right. You will need practice to do this but it will soon be easy.

over the loose end to hold it so that you can hold the block of type with one hand. Wrap it with five or six snug turns. After you have just passed a corner, tuck the end under several strands. Spread the strands up and down over the height of the type.

9. Make a proof print. With a brayer, roll a dab of ink out on a piece of plate glass. When the brayer is evenly covered, roll it over the type. Put the type and galley in a proof press for the impression. If you have no proof press, lay the paper on the type and press down on it with a piece of cardboard.

10. Check the proof for accuracy. If corrections are necessary, move the block or the line back into the composing stick.

11. Now clean the type with a cloth pad, a type brush, and some type cleaner. Clean the brayer and glass too.

12. Next comes the *lock-up*. Lock the type in the iron frame (called the *chase*) centered as high as possible. Use whatever furniture is necessary. With the *quoins* (coins) or set screws, in the hand press chase, wedge the type and furniture in securely. No piece of type should be loose. The locked-up chase is now a *form*.

13. *Plane* the type with the planing block and mallet. Place the chase and type on the imposing stone. Move the planing block over the type, tapping it lightly. This should make the type surface level so that all letters will print evenly. In commercial printing this is called *makeready*.

How to Print
With the type locked in the chase, the press must be made ready to print.

1. Place a small dab of printing ink on the lower left corner of the inking plate. Operate the

Style	Characteristics	Use
Roman	Thick and thin widths in the same letter.	It is formal and dignified; an all purpose style for books, newspapers, etc. Easy to read.
Gothic	Uniform width.	This is simple, bold, loud, and informal. It is used on posters, labels, advertisements, etc.
Italic	Slanted letters. (In Roman or Gothic)	Commands attention, gives emphasis.
Script	Resembles handwriting. (In Roman or Gothic)	Script is graceful; it commands attention. It can be formal or informal.
Text	Resembles forms of hand lettering done by scribes centuries ago.	This is very formal, and legal-looking. It is difficult to read, so use it sparingly.

Fig. 6–12. (top) Main styles of type, their characteristics and uses.
Fig. 6–13. (bottom left) When the type has been set, it is tied with string.
Fig. 6–14. (bottom right) The type is locked up in the chase. Then it is planed, or is tapped down with a block and a mallet. This sets each piece of type to the same height.

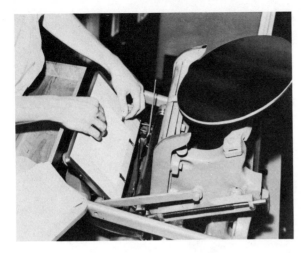

Fig. 6–15. After the chase has been placed in the press, gauge pins for holding the paper are inserted in the tympan sheet. Make a print on the tympan paper to help you locate the gauge pins.

press a few times so that the rollers can spread the ink evenly over themselves and the plate.

2. Insert the chase—be sure it is locked in place—and make an impression on the *tympan* paper.

3. Insert the *gauge pins* in the tympan paper. These will hold your paper in the proper position for printing. Three are usually needed. Adjust the *grippers*—the long slender fingers that keep the paper from sticking to the type—to clear the type and the pins. If the form does not have much inked surface, the grippers are not needed.

4. Make a print and check it for position and inking. If it appears that more pressure is needed, slip a sheet or two of paper under the tympan paper.

5. Now print as many copies as you want. Add ink when needed, but remember that beginners tend to use too much.

6. When the run is complete, remove the chase and clean the type.

7. Remove the ink rolls and clean them. Put them in their special rack. Rolls can be spoiled by leaving them on the plate or on the type. They flatten out and stay that way.

8. Clean the inking plate. Put used rags in a fire-safe container.

9. Distribute the type. Slip the form back into the composing stick or the galley and return one letter at a time to its proper compartment in the proper case. There is one person whom we can get along without in a printing shop. He is the one who leaves his type for someone else to distribute, or who mixes up the type in the case. When you meet this person you'll feel this way, too.

✚ SAFETY SENSE

Your experiences with graphic arts can be even more enjoyable if you use *safety sense*. Remember:

1. The cleaner you use on type and on the presses is inflammable. Keep the used rags in a fireproof container.

2. There isn't room for your hand between the type and the paper in the press. Keep it out of there or you will spoil the printing (and what else?).

3. If you use power presses in your shop, make sure the belts and gears are well guarded.

4. Don't talk with a person who is feeding a power press. Give him a chance to enjoy his work.

5. A paper cutter will cut fingers, too. Keep your eyes on them at all times.

COMMERCIAL TYPE COMPOSITION

Commercial Typesetting. Although type is set by hand for some specific uses, most commercial typesetting is done by machine. Type metal is an alloy of lead, tin, and antimony. The metal is melted and cast in molds to form the characters. Foundry type is cast in single characters. Slug castings, such as those made in Linotype and Intertype

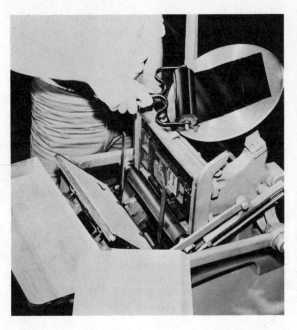 

Fig. 6–16. (left) When you are ready to print, the ink disc is inked. Use a brayer to roll the ink on uniformly. **Fig. 6–17.** (right) Proofread the first print, checking it for accuracy in spelling, spacing, inking, and the like. When the corrections are made, you are ready to print.

machines, are whole words and lines in single blocks.

Linotype

The Linotype operator sits at a keyboard tapping out the story, or *copy*. As he strikes a key, each matrix or mold falls into the equivalent of a composing stick of predetermined width. A space band key inserts a wedge between words. When the line is sufficiently full, the space bands are pushed upwards all at the same time. This spreads the type to the full width of the line and equalizes the space between the words. This is called *justifying*. Hot type metal is forced against this "line o' type," which forms a slug. After the type has cooled, the machine transports the matrices back up to the magazine of type. Keyed notches on each piece allow the sorting device to drop the character into its own compartment, ready again for use. The slugs are assembled, a whole line at a time, to form the column or page.

Monotype

The Monotype machine operates similarly except for the fact that two machines, a keyboard and a caster, are required. The operator perforates holes in a tape. When fed into the caster, this activates it to form molds of individual characters rather than a whole line.

Ludlow

In this process the composition is set by hand in a special composing stick and is then cast by machine in a line slug. Ludlow is used for larger size

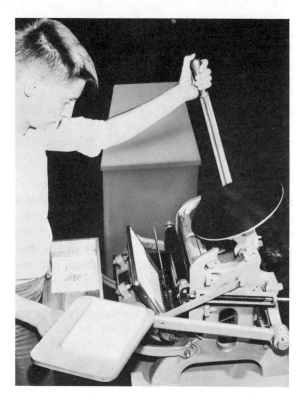

Fig. 6–18. Printing with a hand letterpress.

PLATES

The chart on printing processes on the next page will give you an idea of the three most-used methods and the kinds of plates needed for each. *Plates* provide a flat or curved surface from which individual pages or whole sheets of pages are printed. They may be metal, plastic, or rubber. In newspaper printing, the plate for a page is one solid metal casting, called a *stereotype*. (See page 256.) It includes type as well as the cuts for illustrations.

The processes used in printing pictures are more complicated than are those used in printing words. Words are printed from *type* and/or *plates;* pictures from *plates* and *cuts.* A separate plate is required for each color used. A natural color photograph, for example, can be reproduced with three color plates, a yellow, a red, and a blue; the addition of a fourth plate, black, gives even better detail.

Fig. 6–19. Copy being machine-set. There is a key for every uppercase (capital) and lower case (small) letter, plus punctuation.

type requirements such as in advertising and display as well as for tabular work.

Photo Composition

In recent years several types of machines have been developed which set copy directly on film. No hot metal is required. The matrices contain a small piece of film instead of the raised character. Instead of casting the character, a photograph of it is taken. One "font" of characters can be enlarged or reduced to give many sizes of type in the same face. The process bypasses several operations. Printing is usually done by offset lithography (which we will discuss later).

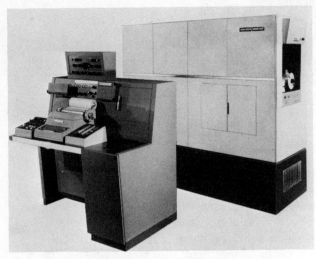

Fig. 6–20. An automatic photo tape printing machine. The operator types the copy on the console unit. This activates the printing unit which produces the tape.

PLATEMAKING FOR LETTERPRESS PRINTING

Relief Line Etching. This is a platemaking process for use with illustrations or copy that is in solid black or white, with no middle tones or grays. The design is photographed and the negative developed. A contact print (see page 270) is made on a sensitized zinc plate, which is then developed, fixed, and washed as though it were a snapshot.

Powdered rosin, called *dragon's blood,* is dusted over the plate. It sticks only to the printed areas and protects them. This rosin is melted over a flame. The plate is then put in a weak nitric acid bath that etches away the bare metal areas. The rosin-covered areas are left high and in *relief.* The plate is then trimmed and mounted on a block so that it is type-high. The finished plate is often called a *cut.*

Halftone Etchings. This process is similar to that for relief line etching except that the middle tones, known as *halftones,* are reproduced.

The design is photographed through a very fine screen that breaks it up into tiny dots. Newspapers use screens of 55–65 lines per square inch, with about 3,600 dots for this area. Magazines and books use 110 to 120-line screens, with a dot pattern of about 15,000 dots to the square inch.

With a magnifying glass, compare a photograph in a newspaper with those in this book. The smaller the dot, the finer the reproduction.

Stereotypes. Stereotypes are plates for printing newspapers. The type and cuts for an entire page are locked up in a chase. A mat, or matrix, of moist paper is laid over the type and cuts. When this is run through a press, an impression is thus squeezed into the mat. The mat is dried, put into a casting box, and molten metal is poured in. Upon cooling, the impression is formed in the metal. This plate may be flat or curved, depending on the printing press in which it will be used. Dry mats, molded by pressure, are now favored, since they do not shrink.

Electrotypes. Electrotype plates, or *electrotypes,* as they are called, are made with an electroplating process. They are more durable, give better impressions than do stereotypes, and are used chiefly for book and magazine printing.

A thin metal plate is coated with beeswax and the type or cut is pressed into it to leave the impression. The wax is then coated with graphite, which is a conductor of electricity. Then the plate is put into a tank and is copper-plated. When the plating is thick enough, it is removed from the wax as a thin metal sheet bearing the impression. A layer of lead is poured over the back to give support to the thin metal. The printing is done from the copper.

Rubber Plates and Stamps. Rubber plates and stamps can be used to print not only on flat surfaces but on curved ones as well. Intricate designs on dishes, for example, are often printed with rubber stamps. They can be made from type, linoleum block cuts, electrotypes, and the like.

	LETTERPRESS	OFFSET-LITHOGRAPHY	GRAVURE
Rotary Methods of each Process	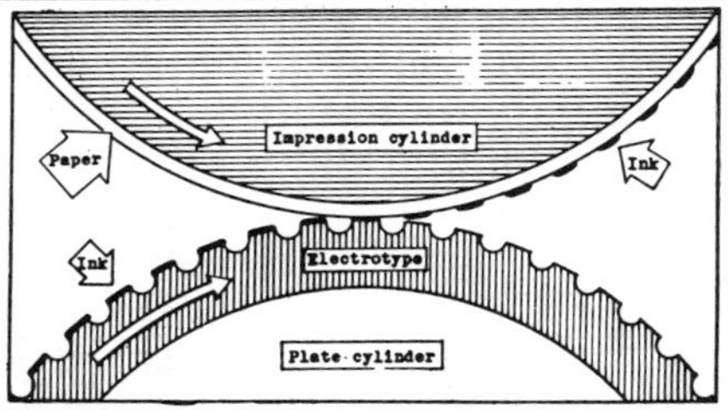	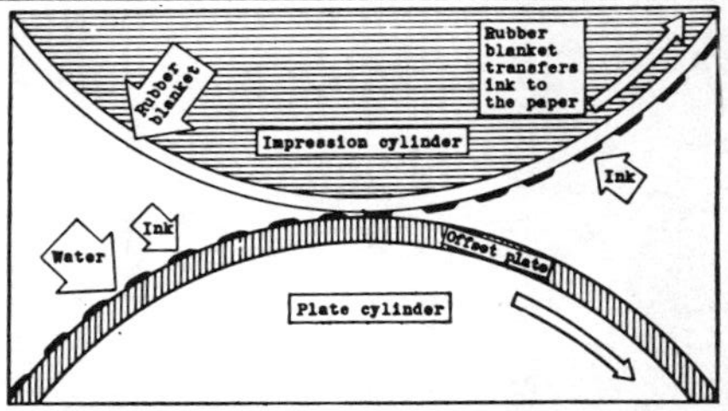	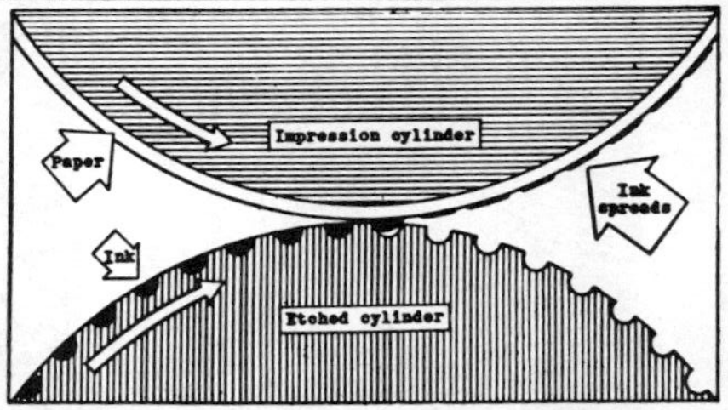
Examine Type	Often shows a slight impression on back of sheet—hold at an angle to light	Never shows an impression on the back of sheet—hold at an angle to light	All type is screened the same as illustrations—examine with a magnifying glass
Examine Line Engravings	Impression may show on back of sheet—hold at an angle to light	Never shows impression on back of sheet	Illustrations which appear as line are actually screened
Examine Ink	Usually appears as an intense black. Detail is sharp	Solid blacks usually appear gray when compared with letterpress. Detail usually less sharp than letterpress	Smooth, blending tones. Detail not as good as in letterpress or lithography
Examine Halftone Dots	Highlight dot / Shadow dot 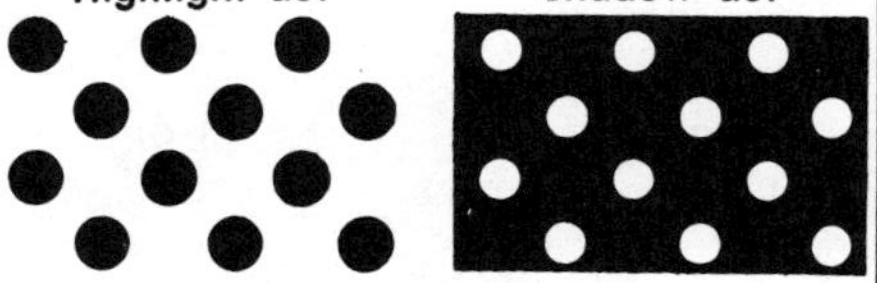 Usually 120–133 line	Highlight dot / Shadow dot 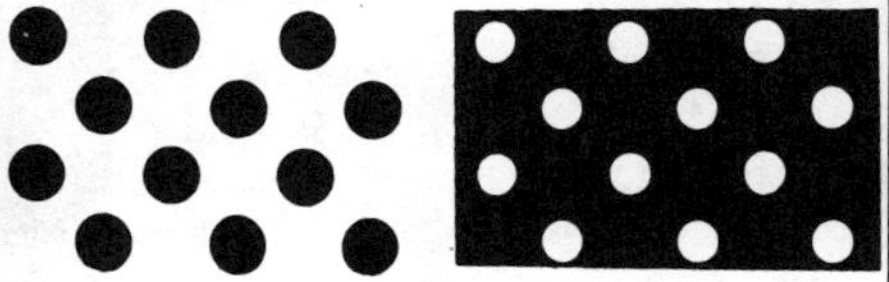 Usually 80–133 line	Highlight and shadow screen 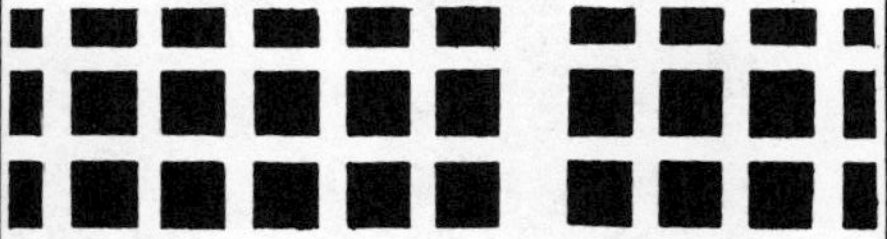 Usually 150 line; ink tends to fuse and obliterate screen
	Because of the extreme differences in quality of work done in letterpress and offset-lithography,	it is often difficult to distinguish one process from another. Offset often looks like letterpress.	

Fig. 6–21. Comparison chart of the main graphic arts processes. (From Karch: *How to Plan and Buy Printing*, Englewood Cliffs, N. J.: Prentice-Hall, Inc.)

Fig. 6–22. (top) Checking curved press plate against foundry proof and engraver's proof.
Fig. 6–23. (bottom) A papier-mâché (paper ma-shay) matrix of a full newspaper page. This "mat" is used as the form from which the metal printing plate, the stereotype, is made.

Fig. 6–24. Stereotypes, the curved printing plates, are placed on the press cylinder.

A mold is first made, using a special molding compound. The type is pressed into the compound and when removed, the impression remains, just like tire tracks in the snow or mud. A rubber compound is placed over the mold. In a heated molding press, it is forced into the impression, where it is vulcanized. The press and process are similar to those used in service stations for patching inner tubes. The rubber is then removed, trimmed, and mounted on a block. You can easily make rubber stamps. To start out, get a kit of materials from one of the many companies supplying them. Follow their instructions.

LINOLEUM BLOCK PRINTING

Linoleum block printing is printing from a raised surface with the design cut into heavy linoleum. Thousands of prints can be made from such a block.

Equipment Needed. Use ready-made blocks, which have heavy linoleum glued to thick plywood blocks. Some are type high. Either water-mixed block-printing inks or regular printing ink can be used. Special cutting tools are preferred—either the push type or the pull type. The latter is the safer. A brayer of soft rubber or plastic is needed for inking the block. The ink is rolled out on a heavy glass slab.

Methods. There are two methods (see Fig. 6–26 below). The first is easier.

1. Trace the pattern with the veining tool, making a narrow V cut about $\frac{1}{16}$ inch wide and deep. The ink prints on each side of the cut, leaving a line on a colored background.

2. The second method leaves a raised line with the background cut away, and only the line prints. A separate block is cut for each color.

How to Make a Block Print

1. First make a suitable design in ink on tracing paper. Ink those areas where you want the printing ink to be. Transfer this design *in reverse* with carbon paper onto the block.

2. With the veining tool, cut the outline into the linoleum. Cut away any unwanted background with a wide tool.

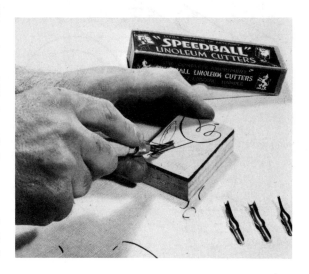

Fig. 6–25. (top) Cutting a linoleum block.
Fig. 6–26. (bottom) Linoleum-block cutting methods.

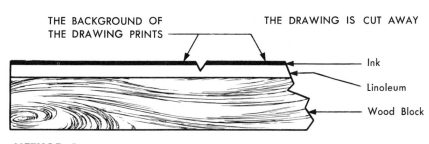

THE BACKGROUND OF THE DRAWING PRINTS

THE DRAWING IS CUT AWAY

Ink

Linoleum

Wood Block

METHOD 1

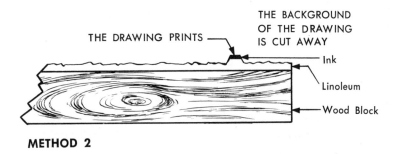

THE DRAWING PRINTS

THE BACKGROUND OF THE DRAWING IS CUT AWAY

Ink

Linoleum

Wood Block

METHOD 2

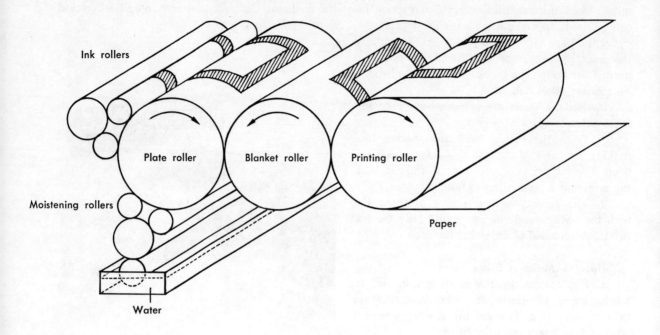

Ink rollers

Plate roller

Blanket roller

Printing roller

Moistening rollers

Paper

Water

Fig. 6–27. Offset printing.

3. Squeeze about a half inch of ink onto the glass and roll it out uniformly thin with the brayer. When evenly coated, roll the brayer over the design until it is well inked.

4. Lay your printing paper on a thick pad of newspapers and place the block where you want it. Step on the block to make the impression. From this proof you can tell whether any changes are needed in the design. Printing can also be done in a printing press or a vise.

✚ SAFETY SENSE

When using a push-type cutting tool, keep your fingers behind it, never in front. Sometimes the tool skids. It is best to put the block in a vise. Then use two hands on the tool.

PLANOGRAPHY

There are several types of planographic, or lithographic, printing; but all are possible because of the fact that grease attracts grease and repels water. The illustration and the copy are drawn or photographed on a plate of thin metal or heavy paper. The areas that print are greasy. These areas pick up ink from a roller; but the other areas, being wet, do not. The image is then printed on the paper under pressure.

Crayon-Stone Lithography. Stone lithography was the first of these processes. It was invented in Germany by Alois Senefelder in 1798. A slab of limestone is ground perfectly flat with another stone and some abrasive. The design is then drawn on the surface with a grease pencil or

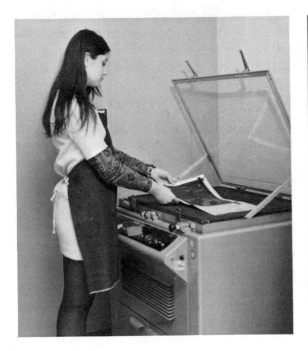

Fig. 6–28. Platemaking is a key step in offset printing. The original copy is photographed to produce a full-size negative. The negative is placed over a sensitized plate by this student. After exposure to light the plate is developed and is ready for the press.

Fig. 6–29. (top) An artist touches up the negatives made in the plate-making camera.
Fig. 6–30. (bottom) A modern four-color printing press.

crayon. The background is "etched," that is, desensitized chemically, with a nitric acid and gum arabic solution. The surface, moistened with water, resists the ink on the roller while the lines drawn with the grease pencil attract it. A dampened paper is laid over it and the impression is printed in a press. The stone is reground each time a new design is needed. Beautiful tones are obtained by this method.

Offset Lithography. On a flexible metal or paper plate, the design is drawn with special crayons or is typed with a special typewriter ribbon. More often it is photographed.

The photographic process is used in commercial printing. This begins with photographing the page. The negative is placed in contact with the *litho plate*. It is exposed and developed. The areas that are to print become slightly greasy in the development. The rest of the plate is smooth and shiny.

The plate is fastened on the plate cylinder. Two sets of rollers contact the plate: the wetting rollers first, and then the ink rollers. As the cylinder revolves, the image is printed on a rubber blanket covering the offset cylinder. The paper is then printed from the blanket.

Offset lithography is adaptable to any printing job. It can be used to print on metal toys, nameplates, milk bottles, and the like. It has replaced the letterpress for many kinds of printed matter. This book was printed by photo offset.

Photo Lithography. This process is used to reproduce illustrations by lithographic plate. The layout is photographed on a sensitized metal plate and a halftone screen is used, just as is done in letterpress platemaking.

Photo Gelatin Process. The design is photographed on sensitized gelatin without the use of a screen. The areas of gelatin exposed to the most light become hardened and quite waterproof. Those with little or no exposure remain absorptive of water. In printing, the most absorptive ink areas pick up the most water and the least ink. The hard areas take the ink. This process reproduces photographs most beautifully because little detail is lost.

INTAGLIO PRINTING

Intaglio printing is done from a surface below the face of the plate. The image is cut into the plate with a tool or is etched in with acid. When a tool is used, that is *engraving*. The acid process is *etching*. The plates can be made by hand or by photographic methods. You can easily make some of the types of intaglio plates and print from them.

For printing, the ink is down in the hollows and not on the top of the plate. Plates are made flat for platen presses and curved for rotary presses.

LINE METHODS FOR PLATEMAKING

Certain methods are used to make plates when the illustrations are drawn in lines, with no halftones as there are in photographs or paintings. Shades and shadows are made by varying the width, depth, and spacing of lines.

Line Engravings. The design is drawn in reverse on a metal plate and lines cut with *gravers,* which vary in shape and make varying lines. This process is similar to cutting a linoleum block.

Zinc, copper, and steel are used for plates. Steel, being the hardest, reproduces the finest detail and lasts the longest. Our paper money is printed from steel engravings. Magnifying lenses are used by the engraver as he works on fine detail. Engraved plates are also made by machine. An original is placed in the machine and is used as a guide for the cutting of others.

Line Etchings. This method depends on acid biting down into the plate to make the lines. A plate of copper is cleaned and polished on one side. This is coated with an acid-resisting *ground*. It is a mixture of asphalt paint, beeswax, and gum mastic. The design is transferred in reverse on this ground and is cut through it with a hard metal point called a *stylus*. The point cuts through to the bare metal.

The back and edges of the plate are also coated with the ground. The acid eats into the copper wherever it is not covered by the ground. When the bite is deep enough, the plate is cleaned, trimmed, and tacked to a block of wood, making it type-high.

Drypoint. This is an interesting process for platemaking, and you can use it for making greeting cards and sketches. Finest results are obtained with a plate of copper, although you can use aluminum or celluloid. Transfer the design in reverse with carbon paper. With a hard, sharp stylus slowly pull the point over the marks, pressing firmly. This procedure raises a burr along the marks. Ink is applied with a dauber and wiped off the surface of the plate. Use a soft paper and make the print in a hand press. The lines will be fuzzy wherever the burrs print. This fuzziness identifies a drypoint print. The plate lasts for only a few points as the burrs break down.

TONE METHODS FOR PLATEMAKING

These methods reproduce the shades and shadows in an illustration by means of dots. Those described here are *soft ground, aquatint,* and *mezzotint.*

Soft-Ground Etching. This technique can be easily done in the school shop. A polished plate of copper is coated with an etching ground. The drawing is made on thin rough paper, dampened and laid over the plate. As the drawing is traced with a pencil, some of the ground sticks to the back of the paper. The pressure used and the kind of pencil point determine how much ground is picked up. When the paper is removed, bare copper shows here and there. The metal is then etched in the nitric acid.

Aquatint. This is a process for reproducing paintings. Rosin powder is dusted on the copper plate for the ground. This is heated to melt it onto the plate. The etcher works with different areas, cutting them as deeply as desired. The deeper the etching, the darker the printing.

Mezzotint. The mezzotint process is used for portraits. The copper plate is coated with an etching ground and then, with a special tool that pricks holes through the ground, the plate is roughened. The picture is made by scraping away the rough burrs.

PHOTOGRAPHIC METHODS

Photogravure and *rotogravure* are commercial methods for producing tones on plates.

The flat photogravure plate is dusted with rosin and heated to melt the particles. These tiny bubbles serve as the photographic screen. A photographic positive is printed and developed on a sensitized glass plate. From this a print is made on a sensitized sheet of special gelatin tissue. This is pressed onto the plate, which is then etched. This is essentially a very slow hand process, not suited for ordinary commercial use. It is used for fine printing of photographs and works of art.

In rotogravure work, the illustration is photographed through a very fine screen with as many as 20,000 dots to the square inch. The rest of the process is the same as for photogravure, except that the plates are curved to fit rotary presses for fast production. Wallpaper, oilcloth, cloth, and books are printed by this process.

HOW MAGAZINES ARE PUBLISHED

The next time you pass a newsstand try to count all the different magazines. Some are weekly, some monthly, others quarterly, and a few are annuals.

Fig. 6–31. A small offset press suitable for graphic arts in industrial arts classes.

Fig. 6-31a. Punching holes in printed sheets for inserting binders.

Fig. 6-32. (top) The exposed printing plate is developed. Note the printed covers on the left.

Fig. 6-33. (bottom) Today's newspapers get much of their news over leased wire services from news-gathering agencies such as the Associated Press (AP) and United Press International (UPI). You can find these initials on the first line of a story after the name of the city from which it was filed. This teletype machine types news stories automatically as they come in by telegraph.

They all go through the same processes in publication. After the advertising space is sold and the articles or stories are on hand, the editorial and art staffs go to work. The main production steps are:

1. *The dummy.* Rough page layouts are made so that the available space can be best used.

2. *Illustrations.* The various drawings, sketches, and so on, for the ads and the articles are made. The cover is designed and the photographs are selected.

3. *The page layouts.* All type is set and prints are made. These and the illustrations are pasted on sheets of cardboard to make the page layouts as they are to appear in the magazine.

4. *The platemaking.* Magazines may be printed from stereotype, electrotype, or offset plates. (See page 256.)

5. *The printing.*

6. *The binding and trimming.* Magazines are either stapled or stitched with wire.

HOW NEWSPAPERS ARE PUBLISHED

The printing of a newspaper is a race against time. Just imagine how well planned the entire opera-

tion must be at the New York *Daily News* plant when over two million copies are printed daily and more than three million on Sunday. Before one edition is ready for the street, the next edition

Fig. 6–34. News stories are edited and head-lines written in the newsroom of the newspaper plant.

Fig. 34a. (top) The Linotype machine sets individual letters and numerals into lines one column wide. It then casts these in type metal. Each line of type is called a slug. Fig. 6–35. (bottom) The composing room. The slugs from the linotype are set up into pages.

is ready to run. Such terrific production shows what industry in our country is capable of. No matter where you live in the United States you can get a newspaper.

The first newspaper in the American colonies, *Publick Occurrences,* appeared on Sept. 25, 1690. Today there are some 1750 daily newspapers in the U.S. selling about 60 million copies. The world's largest newspaper is the *Asahi Shimbun* in Tokyo, with a daily circulation of 9 million.

Modern newspapers go through the following steps before they are ready for circulation:

1. *The news is gathered.* Teletype systems bring in national and world news from such sources as A.P. and U.P.I. (Associated Press and

Fig. 6–36. (top) A page of type is composed. This is locked up and used to make the matrix. Fig. 6–37. (top right) The matrix is the paper mold from which the stereotype is cast in hot metal type. Fig. 6–38. (bottom right) Metal stereotypes ready for mounting on the press.

United Press International). Reporters cover local news. Advertising is sold.

2. *The pages are laid out.* Advertising art work is done. The compositors set the type and the engravers make the cuts.

3. *Full-page mats are made.* (See page 256). From them, the plates are cut.

4. *The paper is printed.* The plates are installed in the presses. The paper is fed from rolls. The pages are cut, assembled, and folded automatically, and the paper is ready for delivery.

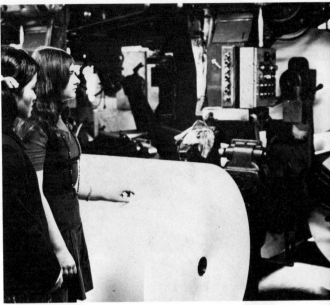

Fig. 6–39. (top left) The pressman places the stereotype on the press. This is a cylinder press. The cylinder, with its plates in place, revolves against inking rollers and then against the moving paper.

Fig. 6–40. (top right) Each roll of newsprint contains approximately five miles of paper. This paper will travel through a press at 20 to 25 miles per hour.

Computerized Printing

Computers have been added to printing in recent years. The copy is typed, producing a perforated tape. This is fed into a computer, which produces a second tape with lines justified. This tape operates an automatic typesetting machine. The plates are then made and the printing done on offset presses.

HOW BOOKS ARE PRODUCED

The publishing and printing of books today is a huge business. With most of the processes done by machine, there is little resemblance to those processes used centuries ago. Books were once copied letter by letter with a pen on handmade paper or vellum.

A publisher undertakes a book only if he feels that there is a good market for it. Let us assume that an author sends him a *manuscript,* a typewritten copy, for his examination. It is turned over to an editor who sends it out to several critics who pass judgment on it. When the decision is made to proceed with the publishing, the book is given a thorough editing. This includes checking for misspelled words, poor sentence structure, errors in punctuation, and the like.

Book designers plan page layouts, select type faces and paper, and design the cover. Then type is set, *galley proofs* are made, and errors are corrected. *Page proofs* are then printed and are checked carefully by the editor and the author.

Fig. 6-41. The printed and folded newspapers are sped up a conveyor to the mail room for distribution. Modern newspaper printing plants can print hundreds of thousands of complete papers each hour.

Soon plates are made and the book is printed. After the pages are bound, the cover is attached and the book is ready for the market. Publishing a book takes from six to twelve months.

Types of Book Binding

Bookbinding is the process of making a book from the loose sheets. Sheets are printed usually in multiples of sixteen book pages and are folded in groups called *signatures*. These are bound together by any one of several means.

Padding. This is the method for making tablets and scratch pads. The individual pages are clamped securely to a cardboard back and a glue, the padding compound, is applied to one edge. The hinge of a cover is pressed into the glue, which dries in a few minutes. Trimming is done after the pads are dry. A similar type of binding (called *perfect binding*) is done with books—for example, paperbacks and your telephone directory. A lining cloth is glued to the padded edge and a paper cover attached to this.

Wire Stitching. Fastening the pages with wire is a fast, economical method. *Side wire stapling* is used when the book is thick. *Saddle wire stitching* is done through the fold when the book is thin.

Mechanical Binders. Your loose-leaf notebook cover is a type of binder. So is the wire or plastic spiral on a pocket notebook. There are

Fig. 6-42. Books are made up of signatures. The individual sections are assembled in the correct order and are then sewn together. Completed "folded, gathered, and sewn" books are being removed at left.

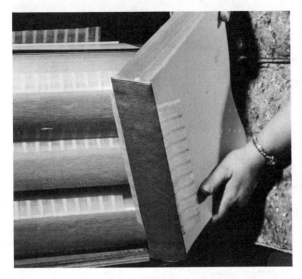

Fig. 6–43. (left) The spine of the book is reinforced with paper and cloth, and decorative headbands are attached. The book is now ready for the covers that have been made and imprinted separately. **Fig. 6–44.** (right) With covers glued to the heavy-stock endpapers, the binding is now complete.

many of such devices. All require that holes be punched in the pages. Special paper drills are available for use in a drill press.

Case Binding.   The book you are now reading was bound by this method, as were most of the books in your library. Many variations are used by different binderies. The actual sewing together of the signatures with thread is the usual basic method. *Endpapers* are attached to the outside pages, cloth applied to the spine, and the cover, made of covered heavy cardboard, is glued to the whole. Case binding, sometimes termed *edition binding,* gives the most permanence and makes the finest type of book.

PAPER AND PAPERMAKING

Fifteenth-century paper measured about 18 by 24 inches. This size was limited by the width of the papermaker's reach and the maximum size mold

he could handle. It happened to be also about the same size as the sheepskins from which vellum was made.

Most of today's printing is done on paper. Without paper there would be no graphic arts industries. We know that in the early days paper was made by hand. Today it is made by machine. Old paper, wood pulp, and rags are the raw materials. The highest quality paper is made from rags or has a lot of rag in it to give it strength.

How Paper Is Made Commercially

The first paper mill in America was built near Philadelphia by William Rittenhouse in 1690. Today there are some 800 mills in 44 states. More than 800 other mills convert paper into products.

Most of today's paper is made from wood pulp. The raw material is cooked in great vats. It is then put into huge beaters, which stir it into a

Fig. 6–45. A pulp mill. Most of our paper is made from wood. The wood is literally taken apart fiber by fiber in the mill.

thick, heavy mass. Bleaching and dyeing follow. The material is now pulp, a mixture of tiny fibers and water. It is thinned with water and flowed out over wide copper screens. The screen is an endless belt that constantly shakes. This causes the pulp fibers to settle down. They arrange themselves lengthwise as logs do when floating downstream. This provides a grain direction that is very important in book printing, binding, and in the way a book lies flat when opened.

The water drains through the screen, leaving a rough mat of pulp fibers. This mat is conveyed between rollers. They squeeze out more water and press the mat thin. The surface finish is pressed on by further rolling. After this the paper is dried and wound on rolls or cut into sheets.

You Can Make Paper
You can make some very attractive paper from old paper or white rags. Make several sheets and print on them. They will look very rich.

Making the Pulp. Cut the paper or rags, or both, into small narrow strips. Soak them in water and cook until they become a thick, heavy mass. Beat this smooth in a kitchen mixer. Then pour it into a pan of clean cold water and stir until the pulp is well thinned.

Forming a Sheet. A wooden frame covered with a piece of copper fly screen is the *mold*. It is placed inside another frame, the *deckle*. Together these are used to dip the pulp. Be sure the mixture is well stirred.

The deposit of pulp held on the screen will be the sheet of paper. Remove the deckle and lay a sheet of heavy felt over the pulp and press out the water. The pulp now clings to the felt and can be lifted off. Stack several of these paper and felt sheets and press until almost dry. They can be sandwiched between pieces of cardboard and run through a clothes wringer to press out water.

Sizing. Sizing is applied to the sheets when they are intended for writing with ink. Otherwise

Fig. 6–46. Manufacture of paper pulp is begun by reducing the logs to chips.

Fig. 6–47. (top) The pulp has been cooked and is now being beaten to separate the fibers. This machine is a beater. Fig. 6–48. (bottom) The pulp is washed to remove impurities and is bleached if other than kraft paper is to be made. Kraft paper is brown wrapping paper.

they will behave like blotters. Get a box of unflavored gelatin at the grocery store and dissolve it in hot water according to the instructions. To a pint of it add a half ounce of alum as a hardener. Now spray or dip the sheets.

Finishing. Different surfaces for the sheets are possible. Lay one between coarsely woven cloths, press it dry with a hot iron and you will get a coarsely textured paper. Pressing directly with the iron gives a smooth surface.

The edge of a sheet when left rough and ragged as it comes from the deckle is called *deckle edge*. It is very attractive for some uses.

Kinds of Paper

Papers are made for many purposes. You find them in paper money, laminated plastics, insulation for homes, picnic plates, as well as for hundreds of other uses.

Fig. 6–49. (left) The Fourdrinier machine. In it the pulp, mixed with approximately 98 percent water, is flowed over a vibrating screen. This is seen in the foreground. As the water drips through, the fibers mat together. This mat is carried through the machine on a succession of rollers where it is dried and finished to become paper. **Fig. 6–50.** (right) This is the dry end of the Fourdrinier machine where the emerging paper is wound into huge rolls.

Newsprint. More of this is made than any other paper. It tears easily but serves its purpose well. It is made from wood pulp. (See Fig. 6–46 to 50.)

Kraft Paper. Made from wood and paper pulp, wrapping paper is commonly brown in color. Paper bags are made of this.

Writing Paper. Most writing paper is made of wood pulp and rags. Better grades are *watermarked*. The manufacturer's name or mark is in the paper, not on it. Hold a piece to the light to read it. The more rag in stationery, the stronger it is.

The standard size of writing paper supplied by the mill is 17 by 22 inches. A package contains a *ream* (500 sheets) and weighs from 13 to 40 lb. This size is cut into four reams 8½ by 11 inches, which is the standard size for writing and typing paper. If you buy a ream of this in the 20-lb

weight, it means that four reams of it weigh 20 lb. The heavier the paper, the more costly it is.

Book Papers. These are usually of wood pulp. Many different finishes are used, depending on the nature of the book and the methods of printing. A finely grained, smooth paper is used when the greatest detail in the illustrations is wanted.

The basic sheet size for book paper is 25 by 38 inches. Books today come in various sizes. These are determined primarily by the size of presses and binding machinery available to the publisher.

Cardboard Stock. Such stock comes in many grades, weights, and finishes. *Bristol* board is a light-weight board of good quality used for such things as tickets, cards, and announcements. *Railroad board* is a heavier card for printed posters and signs. *Showcard board* is a very heavy card with a dull surface, or *tooth,* used especially for hand-lettered posters.

The standard sizes for cardboard vary, but most of them are 22 by 28 inches, or close to it.

DUPLICATING PROCESSES

Duplicators are office-type machines that can print copies of typewritten material and drawings. You will see one or more of such machines in your school.

Stencils are used by one group of such processes. One type of stencil is a thin, waxy, fibrous sheet on which the original is typed or drawn, making actual cuts through the stencil. It is placed on an inked cylinder that rotates. Ink squeezed through the cuts in the stencil produces multiple copies, one at a time.

Automatic copiers are fast replacing earlier types of duplicators. They use photographic or heat and chemical processes. Specially sensitized copy paper is required but no form of stencil needs to be prepared.

BLUEPRINTS

Originally this was a process for making blue copies of drawings. Today the term includes several processes for making prints of drawings. The prints may be blue with white lines, white with black, white with brown, or other color combinations. Blueprinting is a contact printing process. (See page 270.)

The sensitized paper and drawing are fed together into the printer, where electric lights make the exposure. A developer in liquid or fumes brings out the image. Washing and drying follow, with the finished print ready for use in a few minutes. Look for such machines when you visit industries. They are usually found in the drafting departments.

Making a Blueprint. A sun frame is the simplest device for making blueprints. It is made like a picture frame with glass and an easily removable back. A layer of felt lines the back. Cut a piece of blueprint paper a little larger than the tracing of the drawing. Lay the tracing in the frame, face down on the glass. Place the printing paper on the tracing, sensitized side down. Clamp the back in place and expose the drawing to the sun for two or three minutes. Remove the printing paper and quickly wet it all over in water. Sponge with a dilute solution of potassium bichromate dissolved in water. This brings out the image. Wash it well and hang it up to dry. You may have to try different exposures to get the best print. Exposure to bright sunlight requires less time than dim light.

STENCILING

Stenciling is a simple production process for making many identical prints. It can be used on flat or curved surfaces of paper, cloth, wood, metal, leather, clay, and other materials. Tempera colors are used on paper and sometimes wood. Special textile paints for cloth and leather are available. Enamels are used on metals and woods. Glazes, underglaze colors, or *engobes* can be stenciled on unfired clay pieces.

Cutting a Stencil. Lay a piece of transparent stencil paper over your design and with a sharp knife cut out the openings where you want them. It is a good idea to spread the parts out over the sheet so that there are no narrow sections between openings. These usually tear out.

Printing. Lay your material on a drawing board when possible. Stretch it if necessary and hold it taut with thumb tacks. Put the stencil in the correct position. Use thick, heavy paint with a special stencil brush. Dip it in and scrape off any excess. Swipe it across some waste paper to catch any clots of paint. Stroke from the edge to the center of the opening. Use a very dry brush and let the brush marks show for interesting results. Or fill the openings solidly with color.

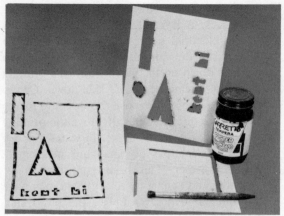

Fig. 6–51. (top left) A hand-cut stencil and a print. Fig. 6–52. (top right) Cutting a silk-screen stencil in lacquer film. Fig. 6–53. (bottom right) Adhering the film to the screen with adhering liquid. The wax paper backing is then carefully pulled from the film.

SILK-SCREEN STENCIL PRINTING

Silk-screen printing is a fast form of stenciling that can produce identical prints with considerable detail. It has many commercial uses, from billboard advertisements to labels on milk bottles. Several variations in the method are possible. The lacquer-film method is the hand process and is described here. Use it for signs, posters, Christmas cards, T-shirt emblems, and so on.

Equipment Needed. Make the frame of soft wood as shown in Fig. 6–54, with mitered corners and a chamfer for tacking the silk. It should be not less than four inches larger each way than the design to be printed.

Use silk that is made for screen printing. Organdy can be used, but it doesn't print as sharply. Moisten the silk with water before tacking it on the frame. Drive a tack or staple at the center of each side. Tack toward the corners,

stretching the silk as you go. It must have no wrinkles.

Printing is done with a *squeegee,* which should be slightly longer than the width of the design. Take good care of the sharp rubber edges. When nicked, they make streaks on the print.

The lacquer film is sold under several brand names. Do not keep a large supply on hand. It deteriorates with age.

Get the adhering liquid recommended for the film you will use. This is very volatile and inflam-

Fig. 6–54. (left) To print, ink is forced through the openings in the stencil by the squeegee.
Fig. 6–55. (right) The finished print.

mable, so keep the lid on the container when not in use.

If you want water-soluble paints, use a good grade of tempera colors. For signs and posters to be hung out-of-doors, use the regular oil-base silk-screen printing colors. Textile paints should be used on cloth. In all cases the paints should be thick, like gravy, so that they do not seep under the edges of the stencil.

How to Print with Silk Screens

1. To cut the stencil, thumbtack the lacquer film over the drawing with the wax paper side down. With a sharp knife cut through the film, not through the wax paper, and carefully peel off the film from the areas that are to print.

2. Lay this film at the center of the frame—under the silk. Fold a soft absorbent cloth into a pad about two inches square. Dip this into the adhering liquid and rub it over the silk, pressing against the film. Cover only a small area at a time and immediately rub it dry with another cloth. Continue until all the white patches between the silk and the film have disappeared. If you use too much liquid and the film dissolves away (this happens so easily), coat the area with some thin lacquer. After a few minutes for drying, lift the frame and carefully peel the wax paper off the back.

3. With masking tape and some wrapping paper, seal off the screen around the film so that the ink cannot leak out.

4. Position a sheet of printing paper under the screen and mark the upper and left edges.

Stick two strips of light cardboard along these marks with masking tape. These are guides to hold the paper in place.

5. Spread a spoonful of paint along the upper edges above the design. Hold the squeegee in both hands and pull it through the paint over the stencil. Try to cover it all in one stroke. Now lift the screen and remove the print.

6. Wash out the paint when you are finished, otherwise the screen will be clogged. To remove the film so that the screen can be used again, dissolve the film in the adhering liquid and soak it up with cloth.

7. For each additional color in the design a separate stencil is cut and printed.

PHOTOGRAPHY

The simplest camera is a light-tight box in which a light-sensitive film is held at one end and a lens at the other, with a shutter for admitting light. This is a box camera. Louis Daguerre made the first such camera in 1839, but used a heavy, clumsy, expensive plate instead of film. Since then the camera has been so improved that a photographic industry has grown up around it and its uses. It was not until 1885 that paper roll film became available, followed in 1889 by film on a transparent base. The invention of roll film, by George Eastman, made it possible for anyone to take pictures. It promoted the discovery of dozens of uses for photography which had not existed up to that time. For example, through the use of the roll film, Thomas Edison was able to make a motion picture camera. From this has grown the movie industry. Color photography was invented by Gabriel Lippmann in 1891 and perfected after World War I. With the giant mirror of Mount Palomar, photographs are made of planets and

Fig. 6–56. Camera parts are being stamped out automatically.

stars millions of miles away. With the electron microscope, scientists can make photos 100,000 times the size of matter so tiny that it is invisible under a regular microscope. Astronauts are making pictures of the earth from miles out in space, and cameras taken to the moon have recorded much scientific knowledge for us.

New uses are constantly being found for photography: the process of recording lifelike images of objects on film and paper.

ABOUT CAMERAS

A camera is designed according to its intended use. Both a miniature camera and an aerial camera take pictures, but they have little in common

Fig. 6–57. The skilled craftsman is indispensable to the creation of a lens. Here, the final assembly adjustment is made on a precision lens.

except the principle by which they work. The automatic miniature camera is intended for the person who wants to take snapshots without having to study the camera. No adjustment and no figuring are needed. Just aim and shoot.

The more advanced photographer wants to take pictures with a camera on which he can operate all controls. He needs one that can be *focused* and that has variable *shutter speeds* and *diaphragm openings*. He can choose from among several sizes and types. There are 35 mm, half-frame (one-half 35 mm), 2¼-inch square, single lens reflex, double lens reflex, and rangefinder types of cameras.

The news photographer needs a *press camera*. This takes a large-size picture but can adjust for distortion and needs little enlarging. He also uses an assortment of lenses.

The commercial photographer uses a *studio camera* with a large negative and *ground glass* focusing.

Movie cameras are in a class by themselves. The home-movie size usually takes 8 mm (the width of the film) or Super 8 film. The professional photographer uses 35-mm cameras. 16 mm is in between. With a "still" camera you take one picture at a time, then move the film. The movie camera also takes one at a time, but at the usual rate of 16, 24, or 32 pictures per second. When the film has been developed, it is projected on a screen at the same rate.

Camera Features

The camera is a machine that must be operated skillfully to get good pictures. You need to know the purpose of the following features:

View Finder. This is a device in which you see the picture as it will appear when taken.

Focusing. The distance from the lens to the film (called *focal length*) is adjusted until the image in the view finder or on the ground glass is sharpest. Sometimes this focusing is done by setting the distance in feet. On some cameras, focusing is done through the finder. On others it is done on a ground-glass back.

Lenses. The camera lens condenses and admits the light reflected from the object being photographed. The light rays pass through the lens and strike the film when the shutter is opened. Lenses have different focal lengths. A standard 35 mm lens usually has a focal length of 50–55 mm (about 2 inches). Wide angle lenses have short focal lengths but they include a wider field of view.

Telephoto lenses have long focal lengths, for example, 135–200 mm. They make closeups of distant objects. Zoom lenses have an adjustable focal length.

Lens ratings are given in "f" numbers. This is the ratio of the focal length to the lens diameter. F/1.4, f/2.0, and f/2.8 are called "fast" lenses.

This simply means that they can admit more light than slower lenses at the same shutter speeds. An f/1.4 lens can take a picture in half as much light as an f/2.0.

The glass in the lens is very soft and is easily scratched. Never touch it with your fingers. Wipe it when dusty with a lens brush or a soft, clean tissue.

Diaphragm. The amount of light entering the camera is controlled by the size of the diaphragm opening. The *stops* or "sizes" are found on the lens mount. For a camera with an f/4.5 lens, the stops will usually be 4.5; 5.6; 8; 11; 16; and 22. The 4.5 opening admits twice as much light as does the 5.6, the 5.6 twice as much as the 8, and so on. The smaller the opening, the more of the picture will be in sharp focus. The lower the f number, the faster the lens.

Shutter. This is the device that determines how long the light is permitted to enter the camera. Common shutter speeds are 1/25, 1/50, 1/100 second. More expensive cameras have both lower and higher speeds. With poor light, use a slower shutter speed. Use a faster speed in bright light or if the subject is moving.

Exposure. Exposure is a combination of diaphragm opening and shutter speed. For example, using these given above, f/4.5 at $\frac{1}{100}$ second gives the same exposure as 5.6 at $\frac{1}{50}$ second, and 8 at $\frac{1}{25}$ second. The first setting would stop fast action in the picture; the last would not. The first would have little depth of focus; the latter, much more.

There is a best exposure for each picture. Some cameras have built-in light meters. Some have a device that prevents tripping the shutter when the light is inadequate. When you use a separate meter, first set it for the film speed. Then set the shutter speed. $\frac{1}{50}$th second is about right for most shots. Take the light reading from the subject. The "f" stop is read opposite the shutter speed. Keep a record of all your exposures in a notebook. This can help you to use the meter more accurately.

Film

Many different films are available for the same camera. Most of the black and white film used is *panchromatic*. It is sensitive to all colors of light.

Fig. 6–58. How a camera takes a picture.

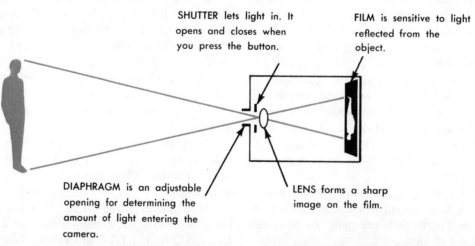

SHUTTER lets light in. It opens and closes when you press the button.

FILM is sensitive to light reflected from the object.

DIAPHRAGM is an adjustable opening for determining the amount of light entering the camera.

LENS forms a sharp image on the film.

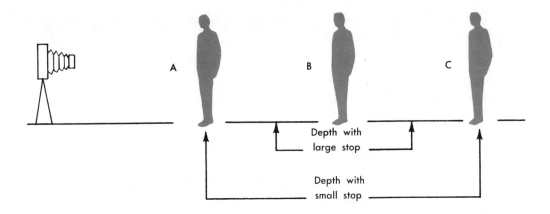

Fig. 6–59. The diaphragm opening determines the depth of focus in a picture. The depth of focus is the distance from the nearest to the farthest objects that are in focus.

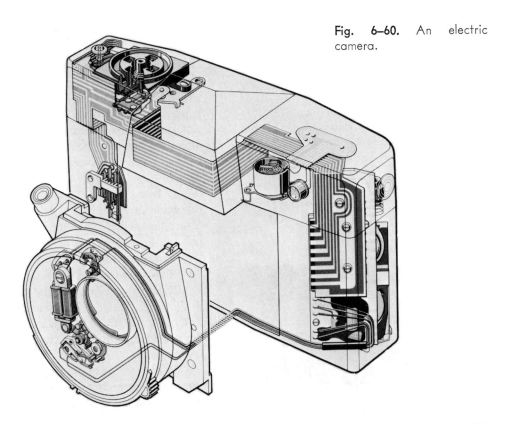

Fig. 6–60. An electric camera.

It must be developed in total darkness. There are different color films as well as films for special purposes. Films are given ASA (American Standards Association) ratings for speed. A rating of ASA 100 is twice as fast as one of 50. Be sure to read the instructions that come with the film.

An Image Is Formed on the Film. The side of the film that is exposed to the light is coated with a light-sensitive *emulsion* which contains a compound of silver. When the light strikes the emulsion, the silver is affected. The image formed is not visible to the eye until the film has been treated with chemicals, a process called *developing*. This process changes the silver compound to metallic silver. Then the film is fixed in a solution to make it insensitive to light. The result is a negative that has its light and dark areas in reverse of the object photographed.

How to Develop Roll Film

Today's black and white films are so light sensitive that they should be developed in a tank. Use the

Fig. 6–61. The three developing solutions for black and white film. Development must take place in total darkness with most film. Trays can be used for sheet film and rolls when tanks are not available.

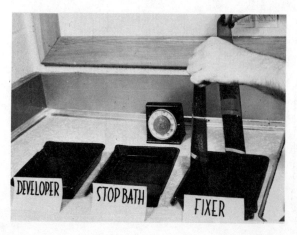

developing solutions recommended for the film. Temperature is very critical. Load the tank in total darkness. Specific loading instructions are necessary for each different brand of tank.

Pour in the developer as quickly as possible. The correct time, temperature, and amount of agitation are given for the developer used. When the time is up, pour out the solution. Pour in the stop bath (a weak solution of acetic acid). Agitate for two minutes and pour out. Replace with the fixer solution for the specified time. Pour out the fixer and wash the film in running water. The reel may be removed from the tank. After thorough washing remove the film from the reel. Hang it up and wipe gently with a moist soft sponge. It should dry before it is used.

How to Make a Contact Print

A contact print is a picture printed directly from and in contact with the negative. It is the same size as the negative. If your negatives are bright and clear, use a No. 2, or normal, paper. Single weight, glossy paper is preferred. When the negatives are thin and pale, use a No. 3 or contrast paper. When real black and thick, use a No. 1 or soft paper. Mix the developer according to the directions and arrange the trays as shown in Fig. 6–61. Use a yellow-orange or yellow-green safelight.

1. Insert the negative, shiny side down, into the guide on the printer. Adjust the mask to fit. Insert the paper, close the lid, and turn on the printing light. Time the first exposure by counting "one-thousand-one," "one-thousand-two," to "one-thousand-five." Then turn off the light.

2. Slip the paper into the developer, wetting it all over quickly. Use developing tongs to keep it moving for the recommended time. Lift out, drain, and rinse it in the tray of stop bath for a few seconds.

3. Immerse it in the fixer and keep it moving for a few seconds. After 15 minutes of fixing, prints should be washed in running water for 30

Fig. 6–62. The development process for black and white film.

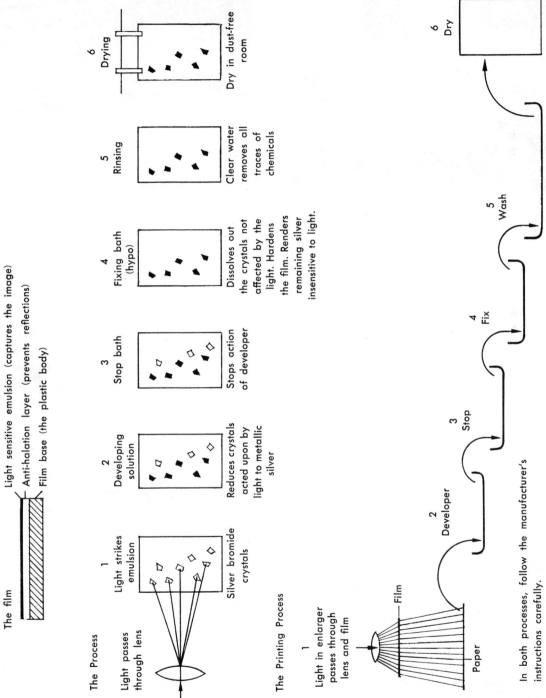

The film

Light sensitive emulsion (captures the image)
Anti-halation layer (prevents reflections)
Film base (the plastic body)

The Process

1
Light strikes emulsion

Light passes through lens

Silver bromide crystals

2
Developing solution

Reduces crystals acted upon by light to metallic silver

3
Stop bath

Stops action of developer

4
Fixing bath (hypo)

Dissolves out the crystals not affected by the light. Hardens the film. Renders remaining silver insensitive to light.

5
Rinsing

Clear water removes all traces of chemicals

6
Drying

Dry in dust-free room

The Printing Process

1
Light in enlarger passes through lens and film

Film

Paper

2
Developer

3
Stop

4
Fix

5
Wash

6
Dry

In both processes, follow the manufacturer's instructions carefully.

Fig. 6–63. Place contact paper glossy side down on the negative in the printer.

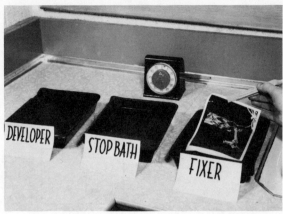

Fig. 6–64. (top) The print is made permanent in the fixing solution. Fig. 6–65. (bottom) A print made on glossy paper is washed and is then rolled face down onto a wet ferrotype plate. When it is dry, the print lifts off and has a glossy surface.

minutes. To dry a print, place it face down on a wet ferrotype plate, cover with a newspaper, and squeeze the water out with a *print* roller. When the print is dry, it comes loose by itself.

4. If, however, after the first few seconds in the fixer this particular print is too light, take another piece of paper and expose it twice as long. Develop this and then you can tell whether to change the time again. The longer the exposure the darker the print. If the first one was too dark, cut the time in half for the next print. Developing time is kept constant. The printing time is varied to produce the right contrast.

How to Make an Enlargement

Prints can be made larger than the negative by a process called *enlarging*. For your first attempts, use a single-weight, glossy paper, 5 by 7 inches and of normal contrast. Pick out a negative that is neither too light nor too black. Mix the developer for enlarging according to directions. Use trays just large enough to take the paper. A yellow-orange or yellow-green safelight is needed.

1. Place the negative, shiny side up, in the carrier in the enlarger. Adjust the easel to fit the

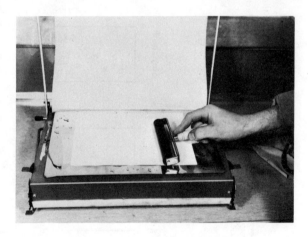

paper, allowing ¼-inch margins. Insert a piece of plain white paper for focusing. In the dark darkroom turn on the enlarger light and bring the image into focus on the paper in the easel. Adjust the image for size and position and refocus it until it is as sharp as possible.

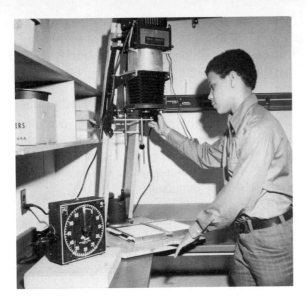

Fig. 6–66. A photo enlarger "blows up" the negative to the desired print size. The exposure is timed by the electric timer on the left.

2. Cut some test strips of enlarging paper about an inch wide. Place one of these diagonally across the opening in the easel and put the rest back in the envelope. Turn on the enlarger light and count to 10 as before, quickly placing a coin on the strip. Count to 20 and place under another coin. Continue counting to 50 and turn off the light. This gives a series of exposures. Greater accuracy is obtained with an enlargement exposure meter.

3. Develop this strip for the specified time. Then rinse, and fix as for contact prints. Examine the spots; the lightest is for an exposure of 10. Pick the spot that shows the most detail as the correct exposure time. Insert the sheet of enlarging paper and make that exposure. Develop, fix, wash, and dry as before.

GETTING STARTED IN PHOTOGRAPHY

A good way to get started in photography is through your industrial arts program. If this is not possible, try to find a camera club and attend meetings. If neither is possible, get a developing and printing kit from a camera store. The instructions are easy to follow. You can build much of your equipment. Any small room that you can completely black out will serve as a darkroom. See the Project Section for details on a printer and a safelight. Much information can be had from good photography books and magazines.

CHOOSING A CAMERA

Select a camera according to the use you will make of it. Many photographers consider the 2¼-inch square reflex the most versatile. It uses 120 or 220 film.

Camera Type	Uses and Characteristics
View	Uses 2¼ x 3¼″ film or larger, cut film or plate. Requires tripod. Focusing very precise. For professional use.
Press	Uses 2¼ x 3¼″ or 4 x 5″ cut film. Ground glass or rangefinder focus. Heavy but portable and versatile.
35 mm	Uses 35 millimeter film. Small negative. Excellent for color transparencies. Original "candid" camera. Reflex or rangefinder focus. Very portable. Wide price range. Manual, semiautomatic, or automatic.
Roll film	Uses roll film, 127, 120 or larger. Reflex or rangefinder focus. Gives larger negative than 35 mm camera. Better for contact prints. Portable and versatile.
Automatic	Electric eye determines exposure and controls diaphragm. User sets shutter speed. Shutter will not trip if light is insufficient. May use film cartridge with no threading.

WHAT IS A GOOD PHOTO?

Consider these measures of a good photograph each time you make one:

1. The print has good quality. It has a range of tones from white to black. It is not muddy. There are no stains from improper handling.

2. The subject has been well selected so that the picture includes only what is needed to tell the story. Make every picture tell a story. The background should not detract from it. One object or group should be dominant.

3. Those objects in the picture that should be sharp, are sharp. The others are soft and fuzzy.

4. The arrangement of details in the photo is known as *composition*. This should be so handled that the photo story is most effective.

GRAPHIC ARTS AND THE ENVIRONMENT

The need for paper in the field of graphic arts makes heavy demands for wood pulp. Paper is made from this. Thousands of acres of pulp trees are being grown in the southeastern states. Pulp wood is a major crop here. As we have learned in the chapter on Wood Technology trees are necessary for man. They manufacture oxygen and filter dust out of the air. It appears then, on the surface that the pulp industry is working for clean air.

One of the major sources of river pollution in certain states is paper making. Much water is required for this. Waste from the process includes pulp fibers and chemicals that spoil the water. Improved filtering is being used at present in the effort to minimize pollution.

Photographic chemicals are a water pollution problem. The free silver left in fixing solutions is being filtered out and reused.

Used newspapers and magazines are recycled into wrapping paper and packing cardboard. Paper is *biodegradable*. This means that it will decompose if buried and return to the soil.

Make a survey of the amount of waste paper produced in your school. Could you figure out a way to cut this production? How about a campaign? Any progress will be significant. It may also cut down on littering. How about it?

GRAPHIC ARTS AND RECREATION

The field of graphic arts is full of recreational and hobby possibilities. Photography is the nation's most popular hobby. Most camera fans do not process their own films. Those known as amateur photographers usually do. The many different printing methods offer invitations to hobbies, too. Check over the following possibilities. You can undertake any one by yourself. There are club possibilities, also. Perhaps you and your industrial arts teacher can get one organized.

Field	Hobby Possibilities
Photography	*Specialize in:*
	Wildflowers, animals, bugs
	Landscapes
	Candid
	Pets
	Babies, children
	Portraits
	Weddings
	Architectural
	Environmental
	Developing, printing, enlarging
	Old camera collecting
	Camera repair
	Construct cameras and darkroom equipment
Printing	Silkscreen: greeting cards, bookplates, designs on textiles, T-shirts
	Hand letterpress: stationery, business cards, invitations
Paper	Paper making, papier mâché
	Photo sensitizing on handmade paper
	Paper decoration, sculpture
	Origami

GRAPHIC ARTS AND OCCUPATIONS

The technical nature of the graphic arts field has gone through something of a revolution in the last two decades. Electronics, automatics, and automation have replaced hand methods rather generally. Even cameras have become so automatic that one doesn't need to understand the camera to take pictures. Special training or apprenticeship is a necessary qualification for technical jobs. Photography not only requires special training but the talent for innovative ideas. There are many jobs that can be learned while one works. But the whole field of graphic arts is becoming highly computerized. Computer controlled printing presses, for example, print newspapers and magazines. College or technical institute preparation is necessary for an increasing number of positions. New machines and processes continue to make jobs obsolete. Study the following chart.

Occupational Field	Typical Occupations		
Photography		*Mass Media*	
Commercial	Cameramen	Newspapers	Reporters
Industrial	Lighting experts	Magazines	Writers
Portrait	Color experts	Books	Editors
Scientific	Film processors		Typesetters
News	Printers		Compositors
Movie			Platemakers
			Press operators
Advertising			Binders
Newspaper	Commercial artists		Engravers
Magazines	Photographers		Photographers
Billboards	Printers		Computer operators
Direct mail	Draftsmen		Technicians
			Servicemen
Illustration			
Cartoons	Commercial artists	*Job Printing*	
Comics	Book designers	Forms	Designers
Books	Photographers	Maps	Platemakers
Magazines	Letterers	Charts	Typesetters
Newspapers		Posters	Press operators
		Signs	Engravers
Paper		Advertising	Photographers
Manufacturing	Machine operators	Book binding	Servicemen
Paper products	Chemists		
	Designers	*Your Own Business*	
		Photography	
Materials and		Printing	
Equipment Manufacture		Silk Screen	
Cameras	Researchers	Book binding	
Lenses	Technicians		
Presses	Chemists		
Films	Physicists		
Paper	Machinists		
Chemicals	Assemblers		
	Service		

TO EXPLORE AND DISCOVER

1. The early history of your local newspaper. Talk with some old-timers. Look up some old issues in your library or in the newspaper "morgue."
2. Why antimony is used in type metal.
3. Why the letters on typewriter keys are arranged as they are.
4. How a Linotype machine works.
5. How automobile license plates are printed.
6. How long you would have to serve as an apprentice before you could become a printer in your town.
7. How roll films are developed and printed in commercial processing plants.
8. How to read a photoelectric exposure meter.
9. How an aerial camera works.
10. What makes slow-motion movies.
11. Why does a lens magnify? Make a large diagram.
12. The history of cameras. Illustrate this with photos and sketches. Visit a museum of technology.

FOR GROUP ACTION

1. Design and print a souvenir school calendar.
2. Print a small handbook to be given to each new student in your school. The writing can be done by your student government officers or faculty sponsors.
3. As a civic service project, print a pamphlet about the library, about a historic site, or some other feature of your town.
4. Make an exhibit telling the story of the development of printing. Include some old books and newspapers. Make a model of an old-time print shop. Get some photographs of modern high-speed presses. Prepare cards to explain your story. Your local printers' union and local publishers might be pleased to help you, or write to newspapers in your capital city.
5. Can you imagine a box camera big enough for several of you to get inside? You could build a wood frame and cover it light-tight with boxboard. Try these dimensions: 4 feet wide x 5 feet long x 6 feet high. Make a light-tight door. Install a lens from an old camera at the center of one end. Make a holder for cut film and install on the inside back wall. You can be inside while the exposure is being made.

FOR RESEARCH AND DEVELOPMENT

1. Photosensitize some paper. Make your own solution. Consult a book on formulas. Make some prints and arrange an exhibit. This would be terrific if done on your handmade paper.

2. Make a pinhole box camera of cardboard. Design it to use cut film. For one test use a pinhole in place of a lens. For comparison add a simple convex lens. What difference does the lens make? Good sharp pictures are possible with the pinhole if it is very tiny.
3. Make a comparison of types of camera shutters. Use pictures and drawings to illustrate. Find out the advantages of each type.
4. Can you photograph the flash of an arc welder? Try different exposures. What causes the phantom circles on the prints?
5. Can air pollution be photographed with black and white or color film? Make a series of comparisons. Use different lens filters and exposures. Prepare an exhibit.
6. Design and print a three-color silk screen poster. What differences can be obtained with opaque and transparent inks?
7. How much load can a piece of paper carry? Build several structures using only one sheet of typing paper for each, plus the necessary model airplane cement. How much weight is necessary to crush them? What are the characteristics of the strongest?
8. How can photography be adapted to silk screen printing? Read up on the photostencil process. Set up an experiment to demonstrate to your class.
9. How can one get the most camera per dollar? Make a buyer checklist of all the items one should consider. Which are most important? Consult some expert photographers.
10. Make an occupational survey of the graphic arts in your community. List the businesses and industries. What are the job classifications, qualifications, salaries, fringe benefits? Which jobs are most likely to become obsolete? Consult personnel managers in the companies.

FOR MORE IDEAS AND INFORMATION

Books

1. Carlsen, Darvey E. *Graphic Arts*. Peoria, Ill.: Chas. A. Bennett Co., Inc., 1965.
2. Cleeton, G. U., C. W. Pitkin, and R. L. Cornwell. *General Printing*. Bloomington, Ill.: McKnight & McKnight Publishing Co., 1963.
3. Kagy, Frederick D. *Graphic Arts*. S. Holland, Ill.: Goodheart-Willcox Co., Inc., 1965.
4. Schlemmer, Richard M. *Handbook of Advertising Art Production*. Englewood Cliffs, N. J.: Prentice-Hall, Inc., 1966.

Booklets

1. *Bibliography of Pamphlets and Books on Photography*. Eastman Kodak Co., Sales Service Division, 343 State St., Rochester, N. Y. 14604

2. *From Forest Tree to Fine Papers.* Hammermill Paper Co., Erie, Pa. 16512
3. *How to Choose the Right Ink.* IPI Division, Interchemical Corp., 67 W. 44th St., New York, N. Y. 10036
4. *List of Aids for Teachers of Photography.* Consumer Markets Division, Eastman Kodak Co., 343 State St., Rochester, N. Y. 14604
5. *Pocket Pal for Printers, Estimators, and Advertising Production Managers.* International Paper Co., 220 E. 42nd St., New York, N. Y. 10017
6. *Should You Go Into the Printing Industry?* Career Information Services, New York Life Insurance Co., Box 51, Madison Square Station, New York, N. Y. 10010

Charts

1. *Diagram of Paper Cutter. Diagram of Platen Press. Oiling and Instruction Chart for Platen Presses.* Chandler & Price Co., 6000 Carnegie Ave., Cleveland, Ohio 44103
2. *Printing Processes.* International Paper Co., 220 E. 42nd St., New York, N. Y. 10017
3. *Pulp and Paper Making Flow Diagram.* Public Relations Dept., West Virginia Pulp and Paper Co., 299 Park Ave., New York, N. Y. 10017

Films

1. *Educational Films.* (catalog of 16 mm sound films on free loan) Classroom Service Dept., Modern Talking Pictures Services, Inc., 1212 Avenue of the Americas, New York, N. Y. 10036
2. Free Film News. (catalog) Sterling Movies, Inc., 43 W. 61st St., New York, N. Y. 10023
3. *Great White Trackway.* Hammermill Paper Co., Erie, Pa. 16512
4. *History of Ink.* Higgins Ink Co., Inc., 271 Ninth St., Brooklyn, N. Y. 11215
5. *The Power of Paper.* P. H. Glatfelter Co., Advertising Manager, Spring Grove, Pa. 17362

GLOSSARY—GRAPHIC ARTS

Aquatint printing process for reproducing paintings.

Blueprint a print of a drawing, originally white lines on blue paper.

Contact print a photographic print made in direct contact with the negative.

Daguerre, Louis Frenchman who made the first camera—1839.

Deckle wooden frame that holds the mold for handmade paper.

Deckle edge the rough natural edge on paper.

Diaphragm adjustable opening between camera lenses for controlling the amount of light.

Dragon's blood powdered rosin used in etching zinc printing plates.

Drypoint a type of print made from a plate on

which the image was cut by a metal stylus.

Dummy rough page layout or paste-up.

Duplicator machine for making office copies from stencils.

Eastman, George American inventor of roll film.

Electrotype a printing plate made by electroplating.

Engraving process of plate making in which the image is cut into the metal plate.

Fixer final solution in the film and photo development process.

Focus to bring the image of the object being photographed into sharp view in the camera finder.

Fourdrinier an automatic paper making machine named for its inventor.

Font complete assortment of type in one face and size.

Galley a flat tray for holding type after it has been set.

Galley proof the first printing proof for a book.

Gauge pin devices for holding the paper in a letterpress.

Grippers the fingers of a gauge pin.

Ground acid-resisting coating applied to a metal plate when etching.

Halftones middle tones between black and white.

Intaglio method of printing from a surface below the face of a plate.

Italic style of type with slanted letters.

Justify to make lines of type equal in length.

Kraft paper heavy brown wrapping paper.

Leads (leds) spacers in strips used between lines of type, usually two point width.

Lens the glass objective that brings light rays to focus on the film in a camera.

Letterpress original type of printing press.

Lino block type-high block of linoleum-faced wood for making cuts.

Linotype a machine operated from a keyboard that casts lines of type.

Lithography printing process in which printing is done from a flat surface.

Manuscript the original copy for a book produced by the authors.

Matrix papier mâché form from which metal printing plates are cast.

Mat short for matrix. Also a paper plate used in offset printing.

Mezzotint a hand method for making printing plates used for portraits.

Mold wooden frame with a screen bottom used in handmade papermaking.

Movable type individual type characters.

Newsprint the paper used for newspapers.

Padding binding method for making tablets and pads.

Page proof proof copies of book pages.

Papyrus writing material, before paper, made of reed originally grown along the Nile river.

Parchment originally the prepared skin of sheep or goats used as a writing surface.

Photogravure a hand method for plate making, characterized by delicate tones.

Plane to level the surfaces of docked-up type with a block of wood and mallet.

Planography process of printing from a flat, or plane surface, as lithography.

Plates metal or plastic sheets generally for printing illustrations.

Pulp solution of wood fibers, water, and chemicals used in making paper.

Quads type spacers made of metal, of same point size as the type.

Reglet thin strip of wood spacing for type.

Relief printing printing from a raised surface.

Rotogravure process for plate making using very fine halftone screens.

Serif fine line cross stroke at the tops and bottoms of letters.

Shutter movable gate that admits light to the film in a camera.

Silk screen a fine grade of silk stretched over a frame for printing.

Slugs strip spaces for use between lines of type, usually 6-points thick.

Stencil sheet of card or metal through which a design is cut. Ink is forced or brushed through the openings to produce prints.

Stereotype in newspaper printing, the metal plate for printing an entire page.

Stop bath acid bath for stopping development of film and paper.

Type individual characters, letters, figures for composing lines.

Type face a particular style of type.

Tympan paper heavy paper used in a letterpress to hold gauge pins.

Watermark manufacturer's trademark stamped in paper.

Wood block carved wood block for printing.

Ceramics Technology **7**

Ceramic materials are nonmetallic minerals. They become hard and permanent with high temperature heat treatment. The field of ceramics includes clay, glass, enamel, gypsum products, and portland cement. You can learn about all of these in this chapter.

Clay was the first of the ceramic materials used by man. It is one of the most abundant materials in the earth's crust. The Chinese were using it for pottery 7,500 to 10,000 years ago. They were the first to find a clay that fired white, and from it they made the first *porcelain* (*por*-se-lin). They discovered that this clay, called *kaolin* (*kay*-o-lin), could be fired high enough to make it translucent and glasslike.

The *potter's wheel* was invented by the early Egyptians. Until the turn of this century it was a basic production process. It has been displaced by the *jigger*. This machine automatically forms plates, platters, cups, and saucers.

Early man fired his clay pots in an open fire. Today ware is fired automatically in *kilns* (pronounced either "kils" or "kilns"), which move it continuously from cold, to hot, to cool.

CLAY PRODUCTS

Begun as pottery making, the industry today includes the manufacturing of products from inorganic (nonvegetable), nonmetallic minerals, using high temperature heat, called *firing*. These products include those of clay, glass, enameled metals, cements, and plasters.

PRODUCTS OF THE CLAY INDUSTRIES

Products used in houses, buildings, bridges, highways	Brick—common, face, fire, insulation
	Block—hollow building tile
	Conduits—large tile for carrying water, sewage, etc.
	Drain tile—porous tile for draining swamps, lowlands, etc.
	Flue liners—for chimneys
	Tile—for floors, walls, roofs
	Terra cotta—architectural decoration
Products used by industry	Abrasives—for grinding, cutting, polishing, i.e., grinding wheels, abrasive cloth, and paper
	Filters—for liquids
	Insulation—against heat and cold, and electrical insulators
	Laboratory equipment—chemical vats, apparatus
	Refractories—heat resistant linings for furnaces, kilns, crucibles
Products used in the home	Art pottery
	Dinnerware
	Kitchenware
	Sanitary ware
	Stoneware
Personal products	Jewelry—man-made gems, ornaments
	Pencil lead—made of clay and graphite
	Medicines, dentures

Centers of Production Brick, tile, and cement plants are found in every state. Potteries and glass plants are not so widely scattered. Most American pottery is being made in Ohio, New York, New Jersey, and California. Glass production is largely centered in Ohio, New York, and Pennsylvania.

New uses for ceramic materials are changing the nature of the industry. They are being used in chemical combination with other materials to pro-

Fig. 7–1. (top) A student at the U. S. Department of the Interior Institute of American Indian Arts in Santa Fe, New Mexico, using the old hand-method of pottery making by the coil technique. **Fig. 7–2.** (bottom) Another student demonstrating the Pueblo Indian technique of smoothing a coil-built pot with a gourd shell.

duce materials previously unknown. For example, *cermets* are hybrid materials made of ceramics and metals. Glass fiber is being combined with plastic, textiles, and rubber.

A Guide for the Study of Clay Products Manufacturing

The following steps represent basic processes used in the manufacture of clay products. They are listed in the usual order of application.

Basic Steps in the Preparation of Clay

1. Mining—digging clay from the earth.
2. Crushing—breaking up lumps.
3. Grinding—breaking clay down into fine particles.
4. Blending—measuring and adding various ingredients to obtain the desired qualities of the clay.
5. Blunging—mixing with water in a large tank called a blunger (*blun*-jer).
6. Filtering—sieving to remove foreign matter. Iron is caught by magnets; other matter, by lawns (cloth sieves).
7. Pressing—squeezing out excess water, in a filter press.
8. Pugging—kneading to make the clay plastic, in a pug mill.
9. De-airing—removing air to make the clay dense. This is usually combined with pugging.

Basic Methods for Forming Clay Into Products

10. Hand building—this is used only in small potteries.
11. Throwing—the potter's wheel is used when quantity production is not essential.
12. Jiggering—clay is shaped on a revolving mold by a stationary template (*tem*-plat) or pattern.

Fig. 7–3. Pugged and de-aired clay emerges from a pugmill. In this machine, clay taken from the filter press is mechanically kneaded in a vacuum chamber to make it plastic and to remove air. It is now ready for jiggering.

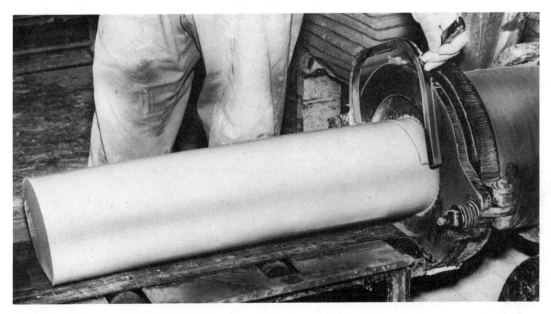

Fig. 7–4. (a) (top left) Make the clay round and true by pressing down and in as it turns. Use plenty of water for lubrication. (b) (top right) Start the hollowing with the thumbs pressing down at the center. (c) (bottom right) Stretch the inside to open it, then lift the clay between inside and outside fingers as it turns.

Fig. 7–5. A jiggerman hand jiggers a plate over a plaster mold.

13. Dry pressing—clay powder is formed into objects in steel dies under tremendous pressure. Floor tile is an example of dry pressing.

14. Ramming—moist clay is pressed into shape between the two halves of a plaster mold, to form such products as plates and saucers.

15. Extruding—moist clay is forced through a die. It emerges in the desired shape; for example, brick and tile.

16. Slip casting—liquid clay is poured into plaster molds and a thin layer adheres to the cavity. The excess is poured out, leaving vases, lamp bases, and the like.

17. Turning—semidry clay is turned to shape on a lathe. This forms electrical insulators, precision parts, and the like.

Fig. 7–6. (top) Filling plaster molds with liquid clay, called *slip*. After a few minutes the molds are drained, leaving a clay shell lining the cavity in the mold to form cups. In approximately a half hour, the molds are taken apart to remove the clay pieces. Fig. 7–7. (bottom) Cup handles are poured in plaster molds. They are cut from the stem and are stuck to cups with slip.

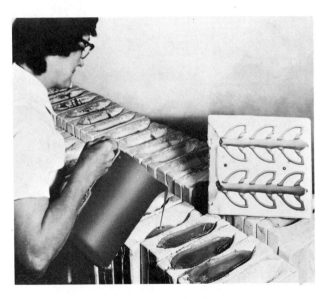

18. Molding—soft clay is pressed into molds, just as is done for brick.

Basic Processes Used on the Formed Ware

19. Decorating—decals (decalcomanias), are designs printed on paper in ceramic colors. They are applied over or under the glaze.

Lining, or banding—this is the application of border lines, done by hand and machine.

Hand painting—this is done with underglaze and overglaze colors.

Rubber stamping of designs is done by hand and machine.

Silk screen printing is done by machine.

20. Glazing—the ware is covered with a coating which, when fired, becomes glasslike.

21. Firing—this is the key process common to all clay products. They are heated until the particles are fused together. Tunnel kilns are commonly used. The ware moves continuously, day and night. The first fire is the bisque fire. The glost or glaze fire is the second. When a third is needed it is usually a decorating fire. Overglaze decorating colors are adhered to the glaze.

How Dinnerware Is Made

The clay usually goes through the processes 1–9 inclusive, and 12, 14, 16, 19, 20, and 21.

How Brick Is Made

Brick, an entirely different product from dinnerware, goes through processes 1, 2, 3, 8, 9, 15 or 18, 20 (for some types), and 21.

What Clay Is

Clay is rock. It was formed millions of years ago in volcanic action. Since then it has decomposed into clay. Pure clay is $Al_2O_3 \cdot 2SiO_2 \cdot 2H_2O$. It is a mixture of aluminum oxide, silica, and chemical water. Under a microscope the particles of clay are flat, like leaves. When wetted they slide over one another. This explains the plasticity of clay.

Fig. 7–8. (top left) Glazing produces the lustrous finish and a hard wearing surface in fine china. Cups are quickly dipped here in a tub of glaze, which is a kind of liquid refractory glass. After 36 hours in the glost kiln, the lustre appears. Fig. 7–9. (bottom right) Ware is stacked in open racks on a kiln car for the first or bisque fire. This kiln is 200 ft. long. The ware moves slowly through it. It is gradually heated until near the center of the kiln the temperature reaches 2300° F. Continuing to move, it slowly cools and emerges from the opposite end after 70 to 90 hours total time. Fig. 7–10. (top right) To paint lines on the glazed plate is a delicate skill. Plate is attached to a small turntable which the artist rotates with the left hand while painting a gold line with the right. This is called decorating. These colors are fired on at low temperatures, 1400–1600° F.

The water you add is called the *water of plasticity.* It evaporates as the clay dries.

Types of Clay. There are many types of clays. Natural clays as found in Nature can be used. Brick and tile are made from these. Prepared clays are blends of kaolin, ball clay, and other ingredients. These are used in dinnerware and art pottery.

These are the common types:

Kaolin—the purest of clays, least plastic. Fires white, not used alone.

Ball clay—very plastic, not used alone, mixed with kaolin.

Earthenware clay—the common clay found near the surface of the earth. Fires red, very plastic, low firing, cones 06-02.

Stoneware clay—plastic, good for throwing. Fires gray to brown, cones 1-8.

Porcelains—mixtures of kaolin, ball clay, and other materials. High fire, cones 6-12.

In a pottery you will see, for example, large bins of kaolin, probably from Georgia or Florida; *ball clay* from Tennessee; *feldspar* from South Dakota; and *flint* from Michigan.

Selecting a Clay. The clay for pottery, such as you will be using, should be suited to modeling. It should not fire higher than your kiln allows. An underfired clay has little strength and takes glazes poorly. A clay that fires between *cones* 06-02 is recommended. (See page 295.)

Keeping Clay Moist. The easiest way to keep clay moist is to store it in a plastic bag. To

Fig. 7–11. Some basic shapes in clay. (See Fig. 2–26 for ideas of other shapes.)

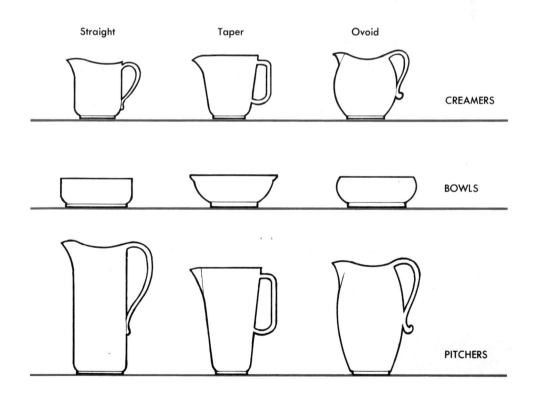

Straight   Taper   Ovoid

CREAMERS

BOWLS

PITCHERS

keep your project moist from one period to the next, slip it into the plastic bag.

Preparing Clay for Use. If your clay is a powder, mix it with water according to the recommendations of the manufacturer. If it is moist, it may be ready for use. Cut a chunk in two on the *wedging wire*. If there are no holes, the clay has probably been *pugged* or *wedged* and is ready to use. If there are holes or lumps, you should wedge the clay. As you cut the chunk in two, slam one piece down hard on the other. Repeat this a dozen times or so to drive out the air. This makes the clay dense and plastic. In industry this is done to clay in a pugmill and de-airing machine.

Welding Clay. To stick two pieces of clay together permanently, they must be equally moist. When they are soft and plastic, press them together as you gently twist them back and forth. If they are stiff, put a layer of *slip* (liquid clay) between them and press them together. To add a handle, use slip in the joint and twist and press at the same time so the joining is snug.

Smoothing Clay Surfaces. Use wet fingers or a moist sponge on clay that is moist; use a barely moist sponge on clay that is dry. Too much water cracks dry clay.

Drying Your Ware. Some clays can be left to dry out in the room without cracking; others must be dried slowly. If your clay cracks, use a plastic bag to retard the drying. Ware that is fully dried in the room is *bone dry*. Ware that is half dry, moist but not plastic, is *leather-hard*. Pottery and ware that is dry and ready for firing is *greenware*.

Fig. 7–12. Tools used in making pottery.

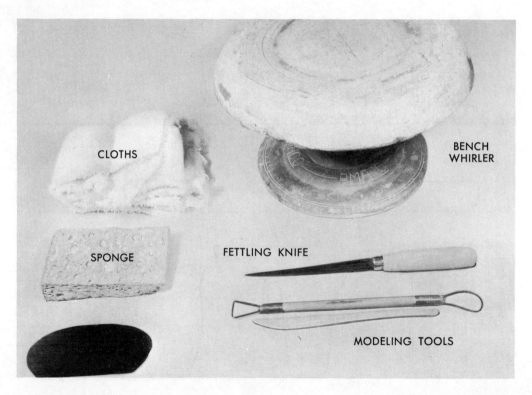

CLOTHS

BENCH WHIRLER

SPONGE

FETTLING KNIFE

MODELING TOOLS

Fig. 7–13. (top left) Clay pots can be kept from one class period to the next in plastic bags. When you plan to add moist clay to a pot at a later time, be sure to keep it moist. Fig. 7–14. (top right) A wedging bench on which clay is made more plastic and from which air is driven.

Fig. 7–15. (bottom right) Coils can be used for a variety of constructions. The body of the figure on the left was made as a small bowl. The sculpture was made from a tangle of coils welded into a single solid.

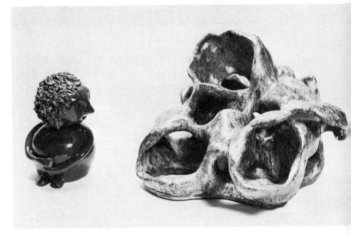

How to Make a Coiled Pot

This is the best hand process for such pieces as pitchers, bowls, and lamp bases.

1. Wedge your clay and roll a soft plastic piece into a ball about 1½ inches in diameter. Place this at the center of a *bat* (plaster slab) on a bench whirler and press it flat with a palm to about ⅜ inch thickness.

2. Cut this clay bottom to shape with a sharp pencil or fettling knife.

3. Take another piece of clay and roll it out into a rope on the table. It should be about as thick as your thumb. Coil this rope around, and on top of, the base and cut it off to the right length.

4. With a forefinger or modeling tool weld the rope to the base. Clay must be moved from the rope into the base.

5. Add one coil at a time, welding as you go. Vary the diameter of the coils to fit the shape of the pot. When the pot is finished, let it dry leather-hard, so that you can do the smoothing.

6. Then turn the pot upside down and with a *wire loop tool* scoop out the underside about ⅛ inch deep, leaving a foot, or ridge, around the outside. Sponge the pot smooth.

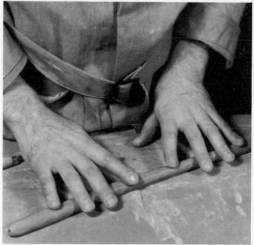

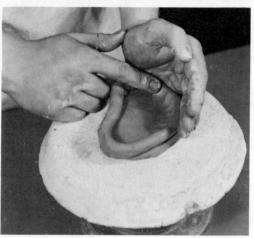

Fig. 7–16. (top left) Cutting the bottom of a pot to shape. Fig. 7–17. (top right) Rolling out a clay rope for coil building. Fig. 7–18. (bottom right) Welding the first coil to the base.

7. With a sharp pencil, mark your name and the date on the underside, then set the pot aside to dry.

How to Make a Low Dish with a Slab of Clay

1. Cut out a full-size paper pattern; trace it on a piece of cardboard and cut it out.

2. Shape a piece of clay on the bench to fit this outline and to fit the form of the inside of the dish. This is the mold, bottom side up. Stretch a damp cloth over it. (See Fig. 7–21.)

3. Wedge a chunk of clay so that it is soft and plastic, but not sticky. On a damp cloth press this clay out into a flat sheet, about ½ inch thick. Use the heel of your hand. Lay a damp cloth over the clay and with a rolling pin roll it to thickness.

4. Peel the clay from the cloth and lay it over the mold. Smooth it to the shape of the mold and trim the edge.

5. Roll out a rope for the foot and weld it in place.

6. Let the dish dry just enough that it will hold its shape. Lift it off and smooth the edge. Add your name and date. Let it dry to leather-hard and smooth it with a moist sponge.

How to Model a Figure

1. Cut and roughly shape the parts.

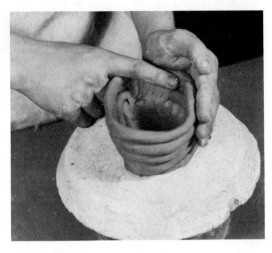 

Fig. 7–19. (top left) Adding the final coil. Fig. 7–20. (top right) When the pot is leather-hard, hollow out the underside, leaving a foot or rim for a base.

2. Weld them together with slip.

3. Bend and shape the figure to give the expression you want. Let it stiffen until it will just hold its shape.

4. Model in the details or add them as separate pieces of clay.

5. Any part of the figure that is thicker than one inch should be hollowed out to prevent cracking in drying and firing.

Slip-Casting Pottery

You can make several pieces just alike by means of slip-casting in *plaster of paris* molds. This method is used in industry to produce quantities of identical ware; for example, art pottery or tea pots. Slip is liquid clay that has been sieved. When it is poured into the plaster mold, water is absorbed by the mold. A clay shell is formed in the cavity. After a few minutes the slip is poured out. The cast is allowed to stiffen until the mold can be taken apart.

Make the Model

1. Make the full-size model of solid clay.

Build it bottom side up on a piece of board, which is about one inch larger on each side than the model. Waterproof the clay side of the wood with paste wax before you add the clay.

2. Tack a strip of cardboard around the block. It should be high enough to allow about an

Fig. 7–21. Clay slab is formed over a plaster mold. This is called a drape mold.

inch of plaster above the model. Be sure there are no leaks in this box and handle it carefully.

Determine the Amount of Plaster Needed

1. One pint of water mixed with 1½ pounds of pottery plaster makes 35 cubic inches of plaster.

2. Find the volume of the box, subtract the volume of the model, and divide the remainder by 35 to get the number of pints of cold water.

Mix the Plaster

1. Measure the water into a clean mixing bowl. Shake the plaster into the water and let it *slake* (soak) for three or four minutes. Use either the Regular grade or the No. 1 grade of pottery plaster. The latter sets slower.

2. Slip a hand into the bowl and, keeping it below the surface, stir until the plaster is ready to pour. In three to five minutes, when it is as thick as a heavy malted milk, pour it quickly over the mold. Wipe the bowl clean immediately with some crumpled paper, then rinse with water. Never pour plaster down a drain. It hardens under water.

3. After about thirty minutes, the box can be taken off and the clay model removed. Rinse out the cavity and set the mold aside to dry.

Make the Slip

A ready-made slip is preferred. Follow the manufacturer's instructions for mixing and aging. If this is not available, soak some of your dry clay scrap in water until it can be stirred into a heavy, smooth paste. A gallon of this will make a great many casts. Add some liquid sodium silicate to the paste. A teaspoonful per pint of paste is usually enough. This thins the clay to a liquid. If this does not seem to work, add a small amount of soda ash dissolved in water. A teaspoonful per pint should do it. Slip should be aged for several days.

Pour the Slip

1. When the mold is dry, it is ready to use. If it feels warm it is dry; if cool, it is wet. Sieve the slip just before using.

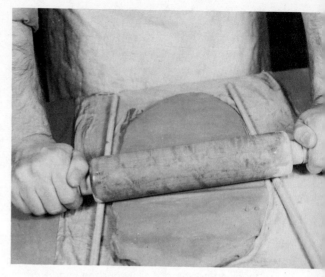

Fig. 7–22. A slab of clay is rolled to a uniform thickness between two sticks.

Fig. 7–23. Lay the slab over the mold which has been covered with a moist cloth; then cut it to shape.

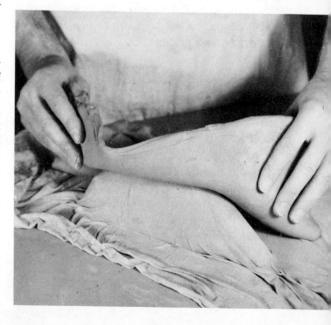

Fig. 7–24. Smooth the dish with a moist sponge.

Fig. 7–25. Figures can be easily made from pieces of clay, as shown at the left. Weld them together. Then shape the figure to express an action. Prop it, if necessary, until it is leather-hard. Add details as in the figure of the "hog caller."

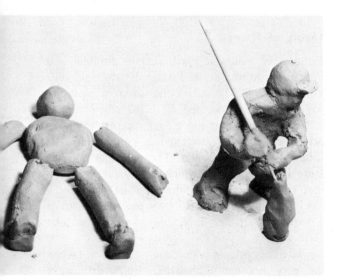

2. Pour the cavity full, noting the time on your watch. As the level drops, add slip. After five minutes, scrape the edge of the cavity to see how thick the wall is. The longer the slip is left in, the thicker the wall.

3. When the wall is about $\frac{3}{16}$ inch thick, pour out the slip. Let the mold drain, upside down. In about an hour, the clay shell will have shrunk free. When it is stiff enough to hold its shape, remove and trim. The mold should dry again before using.

Clay Prospecting

Try clay prospecting for an interesting experience. You can usually find it in a building excavation, river bank, strip mine, or other such place. Test it for sand content. If a small dry piece breaks

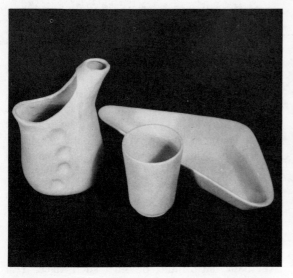

Fig. 7–26. (top left) These various-shaped pieces were made by slip casting. Fig. 7–27. (top right) Place the clay model upside down on a board around which cardboard is tacked. Fig. 7–28. (bottom) Pottery plaster is poured over the mold.

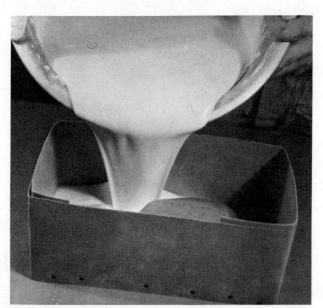

easily, it has too much sand. If it breaks with difficulty, it is probably usable. Let the lumps dry. Then crush them and soak in water. When the clay is soft make a thick slurry (unsieved liquid) with water. Brush this through a 30-mesh screen. Dry the slip on plaster bats. Work it into chunks and age it in plastic bags. If the clay is not plastic enough, add some very plastic clay to it.

Theory of Firing

Firing clay makes it hard, strong, durable, and impervious to weather and water. By the time your kiln reaches 600°F, all of the water of plasticity will have been evaporated. By 1200°, the chemical water (the water in the clay molecules) will be driven out. Clay heated to this point cannot be softened again with water. The *maturing* temperature is that point at which a clay takes on the best combination of qualities for its use. There has been only enough fusion of particles to make it hard and strong. It is porous enough to take glaze well.

High temperatures produce *vitrification* (when the clay becomes glasslike). Full vitrification makes the clay watertight and it will fracture like glass. Further heating will cause slumping and melting.

If you watch the firing through a peephole, you will notice a deep red glow at about 1000 degrees. Increased heating changes the color to a bright red, orange, yellow, and eventually to white, if the kiln is capable. Note the color at the maturing temperature of your clay. When watching this, keep at a cool distance so that the heat does not singe your hair.

Temperature Indicators. You need some means for telling when the kiln reaches the right temperature. If the kiln is equipped with an accurate *pyrometer,* it is just a matter of reading the dial.

Pyrometric cones are also used as temperature guides. In fact, in ceramics, temperatures are referred to by cone number. A cone is a slender clay pyramid, which softens and bends with heat. The correct cone or cones are selected to match the maturing temperature of the clay or glaze. The following information is for large Orton pyrometric cones:

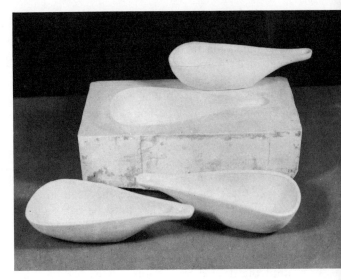

Fig. 7–29. (top) Dishes and the mold from which they were slip cast. Fig. 7–30. (bottom) Here is a kiln loaded for a bisque fire. Note the cone placed on the top shelf where it can be observed through the spyhole when the door is closed.

Cone	Temperature Equivalent	Cone	Temperature Equivalent
020	1202	07	1814
019	1220	06	1859
018	1328	05	1904
017	1418	04	1940
016	1463	03	2039
014	1526	01	2093
012	1607	1	2120
010	1643	10	2381
08	1742	20	2786

The manufacturer of your clay recommends the firing range. If, for example, it is cone 06–02, you should not fire it higher than cone 02 nor lower than 06. Cone 05 may be just right for the bisque.

Set the cone in a plaque of clay. If your clay matures at cone 05, use cone 05. Cone bases are cut at the correct angle for setting so that the cone bends against the flat side. Cones are read by the

position of the tip as at 3 o'clock, 4 o'clock, and 6 o'clock on the face of a clock. The maximum heat indication for a cone is at 6 o'clock. Added heat melts it and makes it useless. Shut off your kiln when the cone tip is at 6 o'clock or before.

The Bisque Fire. The *bisque fire* is the first fire. The ware must be bone dry before firing or it may explode. In loading, do as follows:

1. Arrange the pieces to conserve space. Small pieces may be placed inside large ones. Put those of uniform height on the same shelf.

2. Place the cones where they can be seen through the spy hole. They must be free to bend without touching the ware.

3. Close the kiln and turn on the low heat for an hour. Use medium for an hour, and then high until the desired temperature is reached. At this point shut off the heat and let the kiln cool to room temperature before opening. Different kilns require somewhat different operation. Follow the manufacturer's recommendations.

The Glost Fire. This is the glaze fire that fuses the glaze to the ware. It is preferably higher than the bisque fire. Loading requires much care so that pieces neither fuse together nor fuse to the shelves. Each piece must be placed on a *setter,* which for your ware will be a *stilt.* A stilt is a double-pointed, three-pronged setter of fired clay. Use as large stilts as possible. Place the pieces not closer than ¼ inch. Select the cones to suit the glazes, place them, and fire as for the bisque. In any one fire, all the glazes should mature at the same cone.

Glazes

A glaze is a glasslike coating fused to the clay to add usefulness and beauty. It makes ware waterproof and in addition makes it easier to clean.

Selecting a Glaze. The best glaze for you to use is one that flows in melting. It may be glossy or dull, transparent or opaque. It should melt within the firing range of your clay.

Fig. 7–31. (left) The upright cone is set at the proper angle. The kiln should be shut off when the cone tip is at 6 o'clock, or before. **Fig. 7–32.** (right) For the glaze fire, the pieces are set on stilts.

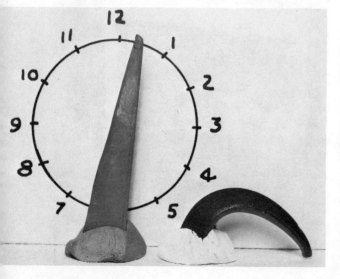
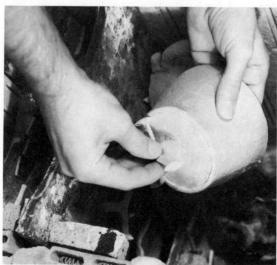

Fig. 7–33. Some unglazed bisque figures. These are sand castings made by pouring liquid clay into sand molds.

To find the best color for a particular pot make some tests. Take some pieces of clay about half the size of golf balls and press them into little bowls. Scratch numbers on the underside for identification, then dry, and fire them. Try your glazes first on these. Make a set of samples as follows, applying each to a different test piece.

1. Take a pair of glazes and mix together a teaspoon of each.

2. Apply a coat of one of these glazes, then lay a coat of the other on top.

3. Reverse the above order.

Try several pairs of glazes in these combinations, then add a third color and repeat the trials.

What Makes the Color? Ceramic colors are usually the oxides of certain metals. Rust is an oxide of iron. Browns, reds, tans, and blacks can be made from iron oxide. Copper oxide gives blues, greens, blacks, or reds; cobalt gives blue. If you have time to experiment with such oxides you are in for some surprises.

Applying Glazes. Ready-made glazes are usually sold in powder form. Add water a little at a time and stir until the mixture is thick but fluid. Brush it through an 80- or 100-mesh sieve to break up lumps. Add water until it is like coffee cream. Give the bisque pot a quick rinse in running water to remove dust and finger marks. With a varnish brush lay on a heavy coating, about $\frac{1}{32}$ of an inch. Do not brush it like paint. Cover the inside first. Let the glaze dry before firing. Clean the foot so it won't stick to the stilt in firing.

Fig. 7–34. Unloading a glost fire (glazed ware). Note the pyrometer on top of the kiln. This measures kiln temperatures and is used as a guide to increasing the heat when firing.

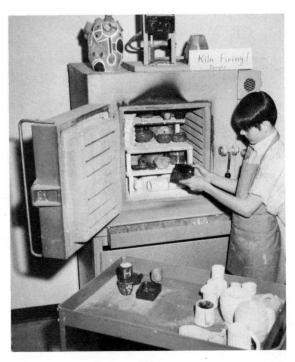

Glaze Defects. Once in a while a glaze behaves badly. Rather than feel discouraged, try to find the cause and the cure. Here are two defects and their remedies.

1. Sandpaper surface—the roughness means that the glaze is too thin. Next time add a second coat.

2. Crazing—the glaze has a fine network of cracks. It was either underfired or on too thick to melt.

A Simple Glaze. The early Egyptian potters used glazes made of clay and borax. For a test sample of a similar glaze mix 2 tablespoons of dry clay powder with 3 of common powdered borax (obtainable at the grocery). Add water and stir until smooth. It should be about as thick as paint. Brush it through a fine sieve to break up the clots. Apply a very heavy coating. It must be used immediately because it hardens quickly. Fire at cone 05. The glaze should be clear, transparent, and glossy. To reduce the firing temperature, add more borax.

How to Decorate Pottery

There are so many ways to decorate pottery that we cannot describe them all. The following are recommended because they are always appropriate and are easily done.

1. *Finger marks* and *tool marks* lend interest to handmade pottery. They tell how the piece was made. They also add textures that make unusual patterns under a glaze.

2. *Incising* is an ancient method for applying designs. They are scratched into the leather-hard surface. Try a sharp pencil for fine lines and a modeling tool for broad. The grooves may be filled in with a contrasting color of clay or with *engobes*, which are colored clays.

3. *Sgraffito* is a technique developed long ago in Italy. Paint a coat of an engobe on leather-hard clay. Use color contrasting to that of the fired clay. With a modeling tool, nail, toothpick, or other such tool, scratch through the engobe to expose the clay. After bisque firing, apply a transparent glaze.

Fig. 7–35. A glaze preparation center in an industrial arts laboratory.

4. *Designs* can be painted on leather-hard clay with engobes or *underglaze* colors. After bisque firing, apply a transparent glaze.

5. *Carving* is easily done in clay. Outline the design on the leather-hard clay and then cut away the unwanted clay. Fine detail will likely be hidden by the glaze.

6. Invent some decorating methods of your own. Try them on samples first.

The Potter's Wheel

The potter's wheel is sometimes thought to be the original use of the wheel. The first one was a flat stone turned on another stone. The kick wheel, foot-powered, is centuries old. It is still preferred by many professional potters. Industry added a mold and a template to the wheel to make the *jigger*. This is the mass production machine for plates and cups.

The process of throwing is a simple one. Moist clay is lifted upward and out while it is turning. Clay is the only material that can be formed in this manner. But to master the wheel takes years of experience.

1. Wedge a chunk of clay thoroughly to remove any air pockets. It should be moist and plastic as for coil building. Press it firmly onto the wheel.

2. Center the clay. Hold your hands around the clay while it turns. Press down and in hard enough so that the clay moves into round. Use water for lubrication.

3. Hollow the clay. Press the thumbs gently down into the center of the clay as it turns.

4. Lift the clay. Squeeze it gently between the finger tips and lift up slowly. Lift clear to the top, and then take another lift.

5. Shape the pot. Stretch it slowly outward, with the inside finger tips, into a bowl. Choking it in at the top reduces the diameter of the opening.

6. Our best wishes go with you. Throwing is great fun. If at first you don't succeed, you may be glad to know that we didn't either.

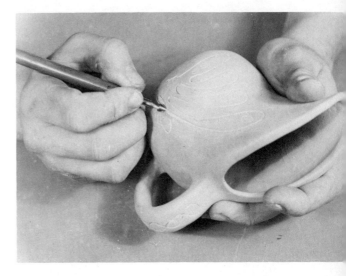

Fig. 7–36. Incising (cutting in) a design on leather-hard clay.

Fig. 7–37. Scratching a design through a layer of a contrasting color of clay or glaze shows the clay body underneath. This is called sgraffito.

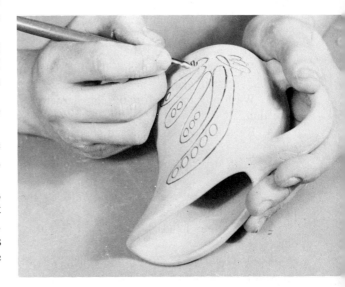

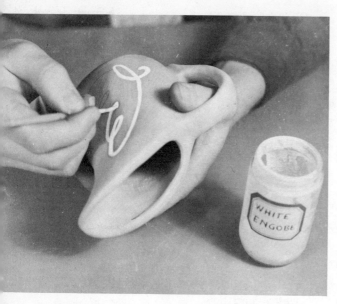

Fig. 7–38. Painting on a design with engobes (colored clays). Use these like tempera colors. For best results, put them on greenware.

Good Housekeeping Makes Good Sense

The clay you use is clean even though it may look like mud, and it must be kept clean if you want good results. Remember:

1. Before rolling your clay on a bench, make sure there is no dust or dried clay on it.

2. You will spoil your clay if you wedge it on a clay-crusted wedging bench.

3. Wash your tools before the clay hardens, to save time.

4. Rinse out bowls, pans, and sponges before using.

5. Keep dry or liquid glazes covered when not in use.

6. Take pity on the sink. Clay will clog the drain unless there is a clay trap.

7. Rinse your hands frequently so that they don't get crusty.

8. Don't blow clay dust off the bench; wipe it up with a damp cloth.

Fig. 7–39. A print from the Diderot Encyclopedia published in 1751–1772 shows a French glass factory of the period. The "gatherers" (m) collect molten glass through openings (h) in the sides of the wood-burning glass furnace. The "servitor" (n) takes the blowiron to form the rough shape. Next,

9. After washing your hands at the end of the period, "excavate" beneath your fingernails.

10. If you get clay on your clothes, let it dry, then brush it off.

1. Never open the kiln until the ware is cool enough to handle. This protects the ware, the kiln, and your fingers.
2. Watch for razor-sharp burrs on the bottoms of pieces being unloaded from a glaze fire. These can be chipped off with an old file, but watch for flying chips.
3. Such burrs can be ground smooth on a grinder. Use a face shield. Hold the pot lightly against the wheel, but do not use the tool rest.
4. Spilled slip is slippery.

GLASS

Glass is as old as the earth. Nature made the first glass through the intense heat of volcanic action. Back in the Stone Age, primitive man made arrowheads, knives, mirrors, and jewelry from obsidian, a volcanic glass. If you have a collection of rocks, you probably have some samples of obsidian. It is hard, smooth, glossy, and usually black, although it may be brown, red, or green.

The first method for making hollow ware of glass involved the dipping of a sand core packed around a metal rod into molten glass and then winding the trailing tail around the core. Several dippings were necessary.

The use of the blowpipe was discovered by Phoenician glass workers about two or three hundred years B.C. The worker dipped the end of a hollow metal tube into molten glass and gathered a ball of it. Then by blowing, turning, and swinging the glass bulb he shaped it into the desired form. This invention is considered one of the great discoveries in history. Until then glass products had been rare luxuries. With the invention of

the "gaffer" or master craftsman (o) seated at the bench finishes the item. His apprentice (p) sometimes must reheat the object. The annealing arch (a) for gradually cooling the glassware is at the top of the furnace. Many of the tools and processes are essentially the same today.

Fig. 7–40. (top left) The technique of blowing into a mass of molten glass, as you would blow into a toy balloon, is still used today both for hand-blown and machine-blown ware. Here is the modern "gatherer." Fig. 7–41. (top right) A shape is blown and the glass is being prepared for transfer from the blow-iron to a pontil rod by the servitor. It is then held on this rod until all forming is complete. Fig. 7–42. (bottom right) The hot, plastic glass is cut, pressed, and stretched into the desired shape. Using simple tools, many of which are wooden, the gaffer at the bench completes the object.

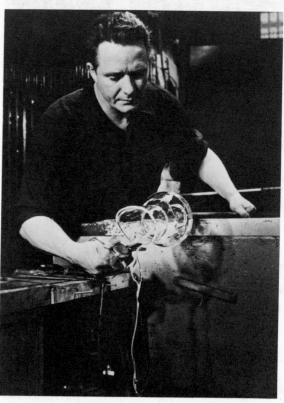

the blowpipe, they became common. The use of molds for blowing and pressing followed. This further simplified the production of hollow ware. Rolling and drawing processes made sheet and tube glass possible.

Fig. 7-43. In this photo, the piece is supported by an asbestos-covered fork while the base is cracked off the pontil rod.

TYPICAL GLASS PRODUCTS

Products used in buildings	Windows Block Plate glass	Wall panels Insulation— foam glass Glass fiber
Products used in the home	Lighting fixtures Mirrors Tables	Dishes Art glassware Book ends Stemware Fiber glass cloth
Lighting products	Lamps Signs Television Photography	Signals Projectors Insulation

Products used by industry	Glass piping
	Flasks, pots
	Bottles, retorts
	Jewel bearings
	Automobile
	  safety glass
	Sports
	  equipment
		Chemical
		  apparatus
		Optical instru-
		  ments, lenses
		Filters
		Insulation,
		  insulators

Other products	Safety glass Bullet-proof glass	

Fig. 7-44. Giant glass jugs are blown in a mold. This mold has been opened to show you the jug. Note the blower's cheeks.

The Making of Glass

Glass is a close relative of clay chemically, but in other ways it is quite different. It is not simply dug out of the earth and used. It is shaped while hot and soft and hardens on cooling.

Batch Mixing. The raw materials are silica sand, lime, potash, soda, lead oxide, cullet (crushed glass), and metallic oxides for color as in ceramic glazes—for example, cobalt for blues, copper for greens, and gold for reds. These are combined in the correct proportions for the type of glass desired.

Melting. The batch is melted in a fire-brick-lined furnace, usually gas fired to temperatures

Fig. 7–45. (top) Glass products must be cooled slowly to prevent internal strains and cracking. Here some pieces can be seen in the lehr or annealing oven for controlled cooling.

from 1800 to 2800°F. It becomes soft and viscous like molasses.

Forming and Shaping. Glass is shaped while in the molten state. *Blown* by mouth or machine into cast-iron molds, it produces hollow ware.

Molten glass is dropped into the cavity of a mold and a plunger *presses* it into shape.

Glass is formed into sheets by *rolling* between large steel rolls like paper through the rollers on a typewriter.

Glass tubing is formed by *drawing*. It is pulled through a die, which determines the size and shape.

Annealing and Tempering. As soon as the glass has been formed, it is reheated, in a furnace called a *lehr*. The temperature is somewhat less than the softening point. This annealing relieves strains and prevents cracks that would otherwise result.

Annealing followed by quick chilling increases the strength of glass and is called *tempering*.

Grinding and Polishing. The surfaces of sheet glass are ground and polished to make it smooth and easy to see through.

Decorating. The techniques used for decorating glass are different from those used for clay. Decoration is cut, ground, etched, sand blasted, painted, gilded, silvered, and stained.

How to Cut Sheet Glass

With a little practice you can cut window glass easily and quickly. The trick is to score an unbroken line on one surface from edge to edge. The glass breaks evenly on this line. If it breaks elsewhere, it is because the line was not evenly and completely scored.

Tools and Equipment. You need a flat level table and a piece of soft wallboard larger than the

Fig. 7–46. The largest piece of glass ever cast —200 in. wide, 26 in. thick, and weighing 20 tons. This is the telescope lens for the Hale Obervatory on Mt. Palomar.

Fig. 7–46a. A drawing of a glass furnace. The batch ingredients are fed into the hopper at the rear. The gas furnaces on the sides melt the batch into a liquid. Here the molten glass is drawn off and formed.

glass. Add a sharp glass cutter, a pair of canvas gloves, and some scrap pieces of window glass.

The Cutting. The wallboard provides a flat surface for supporting glass. Make your practice cuts on the scrap glass.

1. Start with the cutter wheel at the far edge. Press firmly, and with a smooth, uniform stroke draw the cutter toward you across the glass. You must make the score in *one* stroke. Do not retrace the score. The cutter will be dulled and made useless.

2. Practice until you can make the complete cut in one stroke, before you try it on good glass.

3. Place the glass score over the edge of the wallboard and table. Hold the one side down

firmly. Grasp the other and press down sharply. The glass will snap in two.

4. Now make some practice cuts against a straight edge before you cut your glass to size.

Smoothing Glass Edges

With Abrasive Powder. This hand method is used for smoothing edges of flat glass as well as those of cut bottles, jugs, and the like. Sprinkle some coarse (No. 150) abrasive powder on a piece of plate glass or a flat cast-iron block. Add enough water to eliminate any dust. Hold the glass in both hands. Slowly rub it back and forth edgewise over the abrasive. When the rough edges are cut away, use a finer abrasive.

Fig. 7–47. The plunger at left presses molten glass to form a headlight lens. The automatic machine operating as a merry-go-round is called a rotary press.

With Power Sander. A wet sander will speed up the grinding. Be sure the belt has silicon carbide abrasive. Do not use a power sander unless water can be used with it for cooling. Hold the glass edgewise, firmly but lightly against the belt.

Cutting Off a Bottle

Cutting a bottle in two is not difficult with the tool described here. You can easily make it. The current should be only enough to heat the wire. An auto storage battery with a rheostat, or a toy transformer, will provide sufficient current for this purpose. It is not advisable to connect it directly to 110/115 volts.

To use the tool, clamp the resistance wire loop around the bottle. Adjust the length with the screws so that it fits snugly with about a $\frac{1}{16}$ inch gap between the terminals to prevent a short circuit.

Turn on the current for about twenty seconds, shut it off, then quickly remove the clamp and immerse the bottle in a pail of cold water. This sudden chilling should crack the bottle cleanly. If not, increase the current for the next try.

Drilling a Hole Through Glass

Holes can be cut through glass with an abrasive but not with an ordinary drill. Use a short length of brass or copper tubing of the same outside diameter as the desired hole. Slot one end with a hacksaw about $\frac{1}{16}$ inch deep. This keeps the abrasive cutting.

Insert the drill in the chuck of a drill press that is set for its slowest speed. Place a soft wallboard pad under the glass. Build a dike of clay on the glass around the hole location as in Fig. 7–56. In this put about a half teaspoon of No. 120 silicon carbide grit and about a teaspoon of water. Use

Fig. 7–48. Glass engraving is one of the most highly skilled arts and crafts. This example is entitled "The Performing Arts." It was done by Don Weir at the Steuben Glass Plant, Steubenville, Ohio. The cutting tool for engraving is either an abrasive wheel or a copper wheel charged with abrasive.

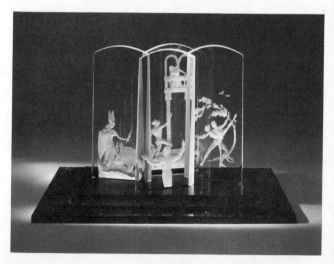

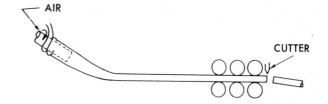

Fig. 7–49. The diagrams on this page show the main methods for machine forming of glass: In making glass tube, molten glass is poured onto a revolving mandrel and slides off as a tube. Air is blown through it to keep it from collapsing until it hardens.

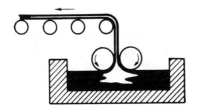

Fig. 7–50. To make glass sheet, molten glass is drawn through rollers. As the glass travels over rollers, it cools into a sheet.

Fig. 7–51. Pressing glass. A "gob" of molten glass is placed in the mold. The plunger forces the glass to fill the mold.

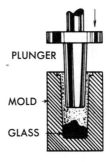

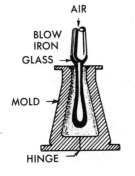

Fig. 7–52. Blowing glass in a mold. Molten glass is gathered on the end of the blowiron. A bubble is blown and inserted in the open mold. The mold is closed and the bubble is blown to fill the mold. When cool, the glass holds its shape.

only enough pressure to keep the drill cutting. Too much pressure will crack the glass. Keep the abrasive fed into the drill with a small paintbrush. Along with this you will need a good supply of patience because glass cuts slowly.

Glass Blowing

The glass blower that you may have seen at work at a fair or exposition is known as a *lamp worker* among glass craftsmen. From soft glass tubes and rods he fashions vases, ships, birds, horses, jewelry, and the like. As for most skilled craftsmen, his techniques look much simpler than they are. He heats the glass over gas burners until it is plastic but not melted. Then he can draw it out into hairlike threads or blow it into beautiful forms.

To experiment with glass blowing try Pyrex glass rod and tube. For heat use a Bunsen burner or a propane torch. Heat the end of a 12-inch length of tube until it melts together. Then heat a

section until soft. Blow gently into the tube while it rotates in the flame.

Glass Slumping. Bottles and flat glass can be shaped by heating. Lay a bottle in a kiln. Heat it slowly until it sags into a tray. Let it cool down slowly in the kiln. This is an annealing process and prevents cracking.

Lay a piece of flat glass over a clay bisque mold and place both in the kiln. Heat until the glass slumps. This will be at approximately 1400 to 1500 degrees. A mold can also be made from a ceramic insulation brick. Shape it with a knife and coarse file. The colored glass as used in stained glass windows forms easily at low temperatures.

Fig. 7–53. Sheet glass being scored with a cutter.

Fig. 7–54. When the glass has been evenly scored, it snaps in two easily.

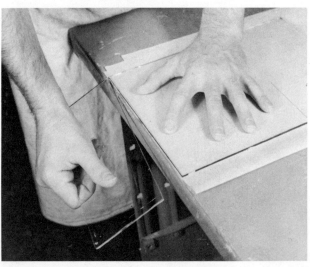

Fig. 7–55. You can cut glass bottles and jugs with this simple tool. It is shown in detail in the Projects Section.

Fig. 7–56. Glass can be drilled almost as easily as metal. A piece of aluminum tubing is being used here.

PORCELAIN ENAMEL

Porcelain enamel is a vitreous (glasslike) coating fused to a metal base at high temperatures of 1000 to 1600°F. It is not porcelain as in clay products. The industry that has developed around this process got its start in this country just before the turn of the past century. The first porcelain enameling on ferrous metals here was done on the insides of cast-iron kettles and pots. By 1900 enameled bathtubs, sinks, and lavatories were being produced in quantity. Since 1925 porcelain enameling on sheet steel has become the major part of the industry. At present, enameling on aluminum with low-temperature enamel is possible.

The early Egyptians successfully fused vitreous enamel to bronze and copper. To decorate their armor, they made shallow hollows in the metal. They filled them with the enamel, and fused it with heat.

In foreign countries, enameling for art purposes reached perfection long before industrial enameling. However, in the United States industrial enameling got a head start. Copper and silver enameling are the common art forms.

PORCELAIN ENAMEL PRODUCTS

Products used in structural work	Prefabricated building Wall panels Theater and store fronts Outdoor signs Wall tile
Products used by industry	Chemical tanks, piping Hospital equipment Plumbing ware Scientific equipment Jet engine lining Flue linings
Products used in the home	Kitchenware, pots and pans Appliances, refrigerators Stoves, sinks, washing machines, table tops Plumbing fixtures
Decorative products	Architectural murals Artware Jewelry

Composition of Porcelain Enamel. The enamel is composed of borax, silica sand, soda ash, feldspar, clay, zinc oxide, and other chemicals. They are melted together into molten glass. Then this is plunged into cold water, producing shattered bits called *frit*. This is finely ground and applied to the metal.

Why Metal Is Enameled. Porcelain enamel increases the usefulness of many metals. It has all the qualities of glass. It resists acids, corrosion, weather, heat, abrasion, and stains. It is easy to keep clean and sanitary. The colors never fade and decorating is easily done. Metal enameled with porcelain is as permanent as the porcelain itself.

How Porcelain Enameling Is Done in Industry

The metal pieces are formed before the porcelain enamel is applied. Here is the typical processing:

1. *Cleaning.* All dirt, grease, and oil is cleaned from the metal. Cast iron is sand-blasted.

2. *Pickling.* Immersed in an acid bath, the metal is chemically cleaned. The surface is roughened by acid action. This makes a better bond with the enamel.

3. *Ground coating.* The metal is given a first coating of the enamel frit by dipping or spraying.

4. *Firing.* The ground coat is dried. Then the metal is passed through an electric furnace to fuse the enamel.

5. *Cover coating.* If additional enamel is desired, it is sprayed on, dried, and reheated. This coat may be for color or decorative design.

Fig. 7–57. Edward Winter, Cleveland enamelist, truing up design on enlarged sketch from the preliminary full-color small sketch. The enlarged drawings were made with the fiberboard resting on a large table, but the actual application of enamel on the completed 16-gauge steel murals was done on the floor. Then each section was picked up individually and fired for three minutes at 1500° F. Each panel received approximately twelve separate firings.

Fig. 7–58. Architectural panels pass through the enameling furnace on a continuous chain. Furnace temperatures run approximately 1500° F.

Equipment for Enameling

Enameling is a very interesting process and is easily done with a minimum of equipment. You will need the following:

Heavy sheet copper, 24 to 32 ounces.

Enamel for copper. Specify it as vitreous enamel for copper, ground finely. Many colors are available, either transparent or opaque.

Pickle. Add ½ ounce nitric acid to 1 pint cold water in a glass or pottery bowl. Prepared pickle solutions are available.

Adhesive. This is mixed with the enamel to make it stick to the metal until the heat takes over. Mix 1 teaspoon of powdered gum tragacanth in 2 tablespoons of alcohol. Add this to ½ cup of water and it is ready to use. Prepared adhesives are available.

Steel wool. Grades 000 or 00 are used for cleaning and polishing the metal.

Furnace, kiln, or burner. A small electric furnace capable of 1400 to 1500°F, such as a pottery kiln, will do. If this is not available, use a large Bunsen-type gas burner. A blow torch will work for small pieces if it will heat the copper to a bright cherry red.

Tongs. A pair of long-handled tongs or a fork for placing and holding the metal during the heating are needed.

Setters. Stilts used for setting glazed pottery can be used for holding the metal during heating.

Miscellaneous. You will also need some small artist's brushes. Salt shakers are used for sprinkling enamel powder.

How to Enamel Copper

For the first project, let's make an identification tag or pendant with an initial or other simple design on it. Follow these steps:

1. Clean the copper well with steel wool.

2. Immerse the copper in a pickle for 10 to 15 minutes, while you prepare the enamel. Suspend it on a wire to keep your hands out of the acid. If you do get acid on them, rinse well immediately in running water.

3. Rinse the tag in clean water just before you apply the enamel. Blot it dry between paper towels, without touching it with your fingers.

4. Mix the enamel. Put ½ teaspoon of the desired color in a small glass cup and add enough water to cover it. Stir, then let it settle. Pour off the water. Add a few drops of adhesive and stir.

5. Apply the enamel ground coat. With an artist's brush pick up a load of the enamel. Lay it on the face of the copper, patting it on evenly. Add more so that the thickness is between $\frac{1}{32}$ and $\frac{1}{16}$ of an inch. Another method is to paint on a coat of gum. While it is wet, sift on the ground coat.

6. Dry out the moisture by holding the tag over low heat. Do not let it get hot enough to form steam.

7. Place the tag in the hot kiln or the flame and heat until the enamel is melted. When it looks wet and smooth all over, move it to a cooler part of the kiln. Let the enamel harden.

8. When the piece is cool enough to handle, brush a thin coat of gum over the hardened enamel. Apply the colors to make the design. Sift them through a stencil with a salt shaker or apply with a brush. Dry out and refire.

If you wish to make designs with fine lines and detail, use *ceramic decorating colors*. They are applied on the fired enamel and then the piece is refired. Use them like paint, with an artist's fine brush. Follow the manufacturer's instructions for firing. The colors should come out bright and shining. The firing range is usually from cone 017 to 019. (See page 295.)

Good Craftsmanship Needs Good Housekeeping

To enamel copper successfully you must work carefully. Take great pains to keep everything clean. Keep your fingers off the copper after it has been pickled and rinsed. Use only absolutely clean bowls for the pickle and for mixing enamel.

PORTLAND CEMENT

You may have watched concrete being poured for a sidewalk, a basement foundation, or a street. You probably have seen a truck with a built-on mixer churning away as it traveled to the job with a load of concrete. All of us have walked and ridden on a lot of concrete. Today concrete is

Fig. 7–59. Ground enamel is sifted onto a copper pendant.

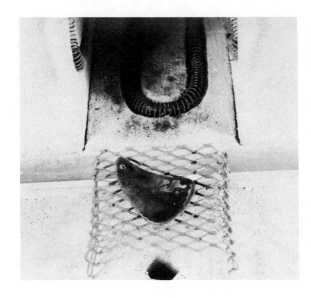

Fig. 7–60. The pendant has been fired and is covered with a glassy coating. This is the furnace described in Fig. 5–8. Metal lathe makes a convenient setter.

indispensable in the construction of highways and bridges, docks and dams, houses and skyscrapers.

The Romans used cement. Some of the roads they made with it still stand. After Rome fell in A.D. 400, however, the art was lost. In America the first large use of cement was in the construction of the Erie Canal in the early 1800's. Reinforced concrete was developed about 1850. Prestressed concrete appeared in 1927 and air-entrained concrete in the 1940's.

Two types of cement are of concern to us here, *portland* and *gypsum*. They are close relatives, yet products made from them are very different. Both are ceramic products from which other products are made. They are called cements because, mixed with water, they harden to rocklike masses.

An English bricklayer, Joseph Aspdin, discovered in 1824 a cement that became hard and strong under water. He called it "portland cement." Its color resembled the natural stone

being quarried on the Isle of Portland off the British coast. In 1872 the first portland cement was produced in the United States at Coplay, Pennsylvania. Today the United States produces some 65 million tons a year. This is one-fifth of the total world production. We use more than two and a half times as much as any other country.

Portland cement is the chief ingredient in concrete, in fact most of the portland cement produced is used in concrete. It is also used in mortar, stucco, masonry paints, asbestos shingles, and the like.

How Portland Cement Is Made

Portland cement is made of limestone, shale, clay, slate or blast furnace slag, shells, and chalk or marl. Gypsum is added to control the setting time of the cement. The ingredients go through the following processing:

1. *Quarrying.* The materials are dug from the earth.

2. *Crushing.* The large rocks and chunks are broken.

3. *Hammer milling.* The pieces are crushed still smaller.

4. *Ball milling.* The pieces are finely ground.

5. *Blending.* The ingredients are proportioned and mixed.

6. *Drying.* Moisture is driven off.

7. *Burning.* The ingredients are fused into masses, called *clinkers,* in a kiln at 2600 to 2800°F.

8. *Crushing.* The clinkers are crushed.

9. *Grinding.* The particles are ground into a dust.

10. *Sacking.* The cement is blown into a closed sack through a small flap. The standard weight is 94 pounds per sack.

Concrete

Portland cement concrete is composed of sand and gravel (crushed rock, slag, or other coarse material) bonded together with a paste made of

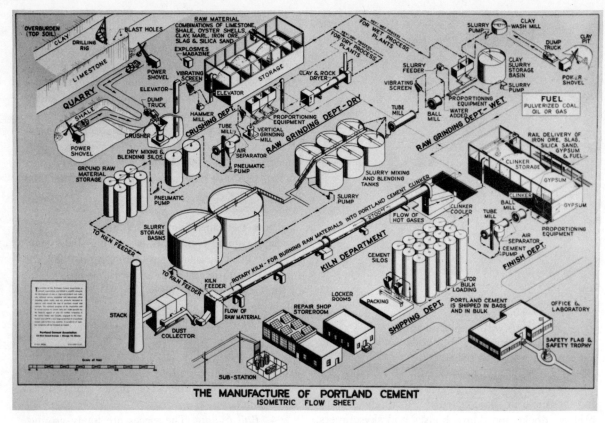

THE MANUFACTURE OF PORTLAND CEMENT
ISOMETRIC FLOW SHEET

Fig. 7–61. The manufacture of Portland Cement. Follow the flow from the upper left corner to the lower right.

cement and water. When cured or completely set, it is a hard, rocklike mass. It is resistant to fire, water, weather, and abrasion.

The qualities of concrete are varied to suit the job. It can be made so heavy and hard that it weighs 250 pounds per cubic foot. It can be so light and soft that it weighs as little as 30 pounds. This can be sawed and nailed like a board. Air-entrained concrete contains tiny cells that help prevent cracking caused by weather extremes.

The Making of Concrete. For large construction jobs, the concrete is made at central "ready mix" plants and is delivered by truck. Trucks, which are portable mixers, carry the dry

materials and water. The driver starts the mixing at the right time. For small jobs, portable mixers make the concrete right where it is to be used. In all cases the correct proportion of the ingredients is determined in advance.

When mixing is complete, usually in two or three minutes, the batch is poured into the forms. These have been built to the desired shape. It is immediately tamped so that it settles to all parts of the form. The concrete gradually hardens or cures. After a few days the forms can be removed. During this aging, the concrete is kept damp so that it will be as strong as possible when completely cured.

As soon as walks, drives, floors, and the like, are poured, the concrete is leveled off. It is allowed to set for a few hours. Then it is finished, or given a surface treatment to make it smooth or rough as required.

An All-Purpose Concrete Mix. The following mix (called 1:2½:3) makes a good concrete for general use. It is suitable for floors, walks, walls, driveways, fish ponds, and the like. Use clean water, washed and screened sand, and gravel. One and one-half cubic feet of the dry mix will make one cubic foot of concrete. When concrete is thoroughly mixed, it is of a uniform, gray color because each grain of sand and gravel is coated with the cement paste. Mix the water and cement thoroughly, then add to the aggregate.

Gallons of water per sack of cement	Portland cement	Aggregate	
		Sand	Gravel
5½, if sand is damp	1 sack	2½ cu ft	3 cu ft
4¾ if wet	(1 sack cement = 1 cu ft)		

Proportions for convenient measuring are:

1 measure (gallon, quart, bucket)	1¼ measure	3 measures	3¾ measures

Fig. 7–62. (top) In kilns like this, more than 500 ft. long and 12 ft. in inside diameter, raw materials are burned at 2700° F. to produce Portland Cement clinker. The kilns rotate once each minute. Fig. 7–63. (bottom) For major paving projects, cement, sand, stone, and water are mixed at a central mixing plant. The resulting concrete is then hauled to the paving site in regular dump trucks.

Some Things to Make of Concrete

To make such projects as described here, these items are needed: a mixing pail, a wooden mixing paddle, a finishing trowel, and a five-gallon can. The can holds water for rinsing cement from the tools. Do not pour this down the drain. You can buy the portland cement and the aggregate locally. Concrete mix is available in bags. Just add water.

A BOAT ANCHOR

The Mix
¾ pint water
1 pint portland cement
3 pints dry sand
4 pints gravel, ½ to ¾ inch size

Fig. 7–64. (top left) An outstanding example of the architectural use of concrete in the Bahai Temple, Wilmette, Illinois. Note the forms in the foreground. These are used in casting the concrete in the unusual shapes.

The Mixing

1. Add the cement to the water in a mixing pail and stir until smooth.

2. Mix in one-half of the aggregate, then mix in one-half of the remainder. Add as much of what is left as is needed to make a thick, heavy, not watery concrete.

3. Pour it into the form, insert the hook, and tamp by bouncing the can a few times. The concrete should cure for a period of three days before the can is cut away.

4. Wash your tools thoroughly in the rinse can.

A BASEBALL HOME PLATE

The Form

Make the form of scrap lumber. Coat the inside with old crankcase oil to seal it. Be sure to use the official dimensions of the plate. Cut the bottom board to the size of the plate and nail the sides around it.

The Mix

Use the 1:2½:3 mix starting with 3½ quarts of cement. Figure the amounts of sand and gravel needed. Use only enough water to wet the particles. The concrete should be very stiff, not runny. Check your answers with your teacher.

The Mixing

Mix as for the boat anchor and then pour the form full and bounce it a few times. After it has set for an hour or two, smooth the top with a finishing trowel. Don't trowel too long because you will bring water to the surface. Let the plate cure under damp cloths for at least three days.

Fig. 7–65. (bottom right) Wooden forms are needed in constructing concrete walls and foundations for a house, such as is shown here, or for large buildings or bridge piers. After the concrete has hardened, the forms are removed.

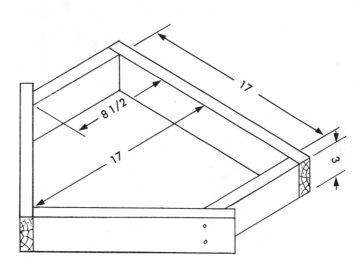

Fig. 7–66. (left) An anchor for a small boat. Bend an "eye" at one end of a piece of hot rolled mild steel rod. At the other end make an "L" bend. The form is a discarded can or plastic jug. **Fig. 7–67.** (right) Wooden form for home plate.

A FEEDER FOR
DOGS, CATS OR RABBITS

The Form

Shape a piece of moist clay to the form of the inside of the feeder. Lay this upside down on a board that is about 1½ inches wider than the clay on each side and end. Keep the clay moist until the cement is poured. Nail sides on this base, which are just high enough to allow a 1½ inch thickness for the bottom of the feeder. Waterproof the inside of the form with a coat of old motor oil.

The Mix

¾ to 1 quart water
1½ quarts cement
3 quarts sand

The Mixing

No coarse aggregate is used because the walls are thin and the surfaces should be very smooth. Mix the water and cement into a paste. Add most of the sand and mix well. The final mixture should be thick, heavy, and plastic. Add the rest of the sand if needed. Pour the form full and bounce it a few times to settle the mix. Then strike it off level with a straight stick. Let it cure for two days before removing the form, then soak the feeder in a pail of water for three days. After it is dry, coat it with a suitable color of enamel. Get a nontoxic paint that can be used on cement.

GYPSUM CEMENT

Gypsum is a soft rock, which chemists know as *calcium sulphate*. In this country it is found chiefly from Texas north through Kansas into Michigan, and from California to Utah.

The early Greeks and Egyptians found that this rock could be ground up and burned. When the residue was mixed with water, it hardened into

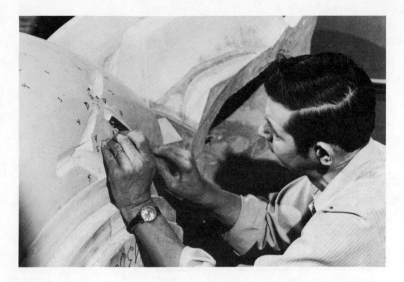

a rocklike mass. They used it for wall plaster and for mortar in their buildings. The Egyptians used it in the great pyramids. The first gypsum plaster plants were near Paris, France; thus the name "plaster of paris."

Most of the mined gypsum rock is converted to cements and plasters. It is ground finely and then heated to drive out most of the water. This is called *calcining*. Then it is ground again and is ready for use as plaster of paris. Gypsum plasters have an unusual quality. When mixed with water they harden and become like the rock from which they are made.

COMMON USES FOR GYPSUM

In paints—for filler and pigment.

In rubber—to make it hard and strong.

In matches—to form the heads.

In concrete—to control the setting.

In toothpaste—to polish.

In movies—for stage sets and snow.

In wallboard—for the core of plaster board and rock lath.

In roofing—for insulation, light weight, and resistance to fire.

In farming—for conditioning soils.

In dentistry—for molds to cast crowns.

In jewelry—for molds in casting precious metals.

In orthopedics—for casts to hold broken bones.

In potteries—for molds in mass-producing ware.

In art—for sculpture.

In building—for plastering walls.

In police work—for making impressions of footprints, tire tracks, and so on.

Type of Gypsum Cements

There are many different gypsum cements, each suited for different uses. Following are the common ones for your use. They deteriorate with age, so keep only a sack or two on hand. Only clean, cold water is used for mixing.

Pottery Plaster. This is very porous, to permit absorption of water from the clay slip. It is available in three grades—Regular, No. 1, and No. 4. Regular is fast-setting, No. 1 is medium fast, and No. 4, slow. The recommended proportions are 1 pint of water to 1½ pounds of plaster. Setting begins in 3 to 5 minutes.

Casting Plaster. This is the plaster for general use. It is less porous than pottery plaster. It becomes hard and strong, and takes paint well. Use 1½ to 2 pounds to a pint of water. The less water, the stronger and the less porous the plaster.

Keene's Cement. Much of the marblelike wall covering in public buildings is Keene's cement. Centuries ago Italian craftsmen perfected the process for coloring gypsum plaster to resemble marble. Threads imbedded in the wet plaster, when pulled along, left the colors in streaks.

Keene's cement is a slow-setting plaster that becomes very hard and strong. After it is set, it rings like clay bisque when struck.

How to Make a Plaster Mold

The method for making plaster molds for pottery has been described (see page 291). You may make a mold from a metal, china, or glass object by this method, too. Select an object, which is not ornate, and imbed it in clay up to the parting line, then follow the procedure as given. There should be not less than one inch of plaster around the object. Instead of slip being used to make the cast pieces, use plaster.

Plaster Separator. To keep plaster from sticking to plaster or other porous material, a coating of a separator is necessary. Potters' mold soap is used in slip molds. For casting plaster in plaster or wood molds, use a mixture of two parts kerosene and one part petroleum jelly. Brush this into the cavity before the plaster is poured in.

Painting Plaster Objects. Plaster objects take paint well if they are thoroughly dry. First seal the surface with two coats of thin shellac. When dry, this forms the base for enamel or tempera colors.

Using Keene's Cement

Keene's cement sets more slowly than do the other plasters. It remains plastic for about an hour. Use it for book ends, lamp bases, desk sets, and the like. Here is a list of the materials you will need.

Keene's cement. Buy it in 100-pound bags from a builder's supply store. Three grades, Regular, Fine, and Superfine, are made. Regular is less costly and quite adequate.

Colors. Get lime-proof plaster colors, red, yellow, brown, and green.

Waterproofing wax. Mix one pint of kerosene in a pound of melted paraffin. Do not heat the kerosene. This wax is for waterproofing the insides of a mold.

Abrasives. 2-0, 4-0 and 8-0 waterproof abrasive papers are used for shaping and polishing the hardened cement.

Sealer coat. Mix four parts of boiled, refined linseed oil with one part of commercial paint thinner or turpentine.

Polishing wax. Use paste polishing wax for a final finish.

Molds for casting. A mold is needed to hold the cement securely while it hardens into the desired shape.

The surfaces of wooden molds that will be exposed to the cement must be sealed well with the waterproofing wax. Metal molds should be coated also to prevent rust from staining the cement. Dishes and bottles, when used as molds, should be given a very thin coating.

Mixing. The first step in mixing is to calculate the volume of the mold in cubic inches. Then find the number of pints of water and the pounds of cement from this proportion. One pint water and 3 pounds cement makes 35 cubic inches. Now follow these steps:

1. Shake the powder into the water and let it slake for 3 to 4 minutes.

2. Mix with a strong spoon until it is free of lumps.

3. Shake in some of the colors, but use them sparingly. Too much weakens the casting. With the edge of the spoon cut the color into the cement. Do not stir them in. Crisscross the cutting to distribute the colors.

4. Dump the batch into the waxed mold and bounce it several times. You can now restreak the colors if you wish. Let the cast harden overnight before removing it from the molds.

Drying. From three days to a week are required for drying. Shaping and polishing can be done as soon as the casting is hard enough to ring when tapped.

Shaping. When the cast is hard it can be sawed, drilled, planed, and filed as easily as wood.

The cement is corrosive to tools. Be sure to clean them well after using.

Polishing. Polishing is done under water. Use a dishpan in which the cast can be immersed. Start with the coarse abrasive and follow with the fine in succession.

Finishing. The cement must now be bone dry. Sand it very lightly with some worn 8-0 paper. Apply a heavy coat of the sealer and let it soak in. Rub with a clean, soft cloth and set it away to dry overnight. The next day rub the surface to a gloss and apply paste polishing wax. Several coats of the sealer can be used to give a greater gloss.

LAPIDARY

The cutting and polishing of gem stones is called *lapidary*. The person who does it is also called a lapidary. Primitive man eventually found that all of his time wasn't needed to provide food. He

Fig. 7–69. Colors are added to the Keene's cement and then are streaked through to produce the marble-like effect.

Fig. 7–70. Pouring the cement into the form to make a flat tile.

began to make ornaments of stones and shells, which he wore. These were eventually used as money. With the development of methods for cutting and polishing, certain stones became more valuable than gold.

The first examples of cut and polished stones were the official seals of the governments of Egypt, Syria, and Babylonia about 5000 years ago. The early lapidary used sapphire points to cut other stones and sapphire dust for polishing. Many of the gems we see today are synthetic or man-made. Synthetic gems can be so perfect that it is impossible for anyone but an expert to tell them from natural stones. Synthetic rubies are made by melting oxides of aluminum with chromic oxides for color.

Hardness of Stones

Some stones are harder than others. This quality is used in identifying them. The standard for comparing hardness is the Moh scale.

Diamond	10	Apatite	5
Corundum		Fluor	4
(ruby, sapphire)	9	Calcite	3
Topaz	8	Gypsum	2
Quartz	7	Talc	1
Feldspar	6		

This is a rating based on the diamond, which is the hardest of stones. It does not mean that the diamond is twice as hard as apatite nor ten times as hard as talc. A file has a hardness of 6 or 7 on this scale.

The Lapidary Process

The lapidary follows a step-by-step process. The stone, if large, is split along a natural line with a hammer and chisel, as one would split wood. This is called *cleaving.* The object is to divide the stone so that several smaller and more valuable ones can be made. These pieces are then cut to the approximate desired size on a diamond saw. This is a metal disc with diamond dust imbedded in its rim. The dust cuts the stone.

The smaller pieces are now cemented to *dop* sticks for easy handling. They are ground to the rough shape on silicon carbide abrasive wheels. Stones that are transparent are sometimes *faceted,* or cut with many sides. Holes are drilled with a small brass tube and diamond dust. (See page 306). Polishing is done on a revolving horizontal disc called a *lap.* This is coated with abrasive and water.

Lapidary Equipment. Complete outfits for home and school use are available at reasonable cost; however, you can assemble your own to include the following:

¼ hp electric motor and a grinding head.
120 and 220 grit.
1″ by 6″ silicon carbide grinding wheels to fit the head.
1″ by 8″ hard felt buff.
Several wooden sanding discs with waterproof silicon carbide abrasive paper to fit, in grits 280, 320, 360, and 400.
An assortment of polishing compounds: tripoli, rouge, and tin oxide.

Several dop sticks made of 8″ lengths of ⅜″ dowel rod and a stick of dop cement (stick shellac).

Since grinding is done on wet wheels, a hood over the wheel and a pan under it is necessary.

How to Polish a Stone

Some of the stones that you may find need only be polished to bring out their beauty. Select one that has a smooth, even surface, then follow these steps:

1. If the stone is small, cement it to a dop stick. First warm the stone in hot water. Melt some cement onto an end of the stick, using a hot piece of iron. While the cement is warm, press it onto the stone.

2. Use the felt buff at a speed of 400 to 500 rpm (revolutions per minute). Hold the tripoli against the face of the revolving buff to charge it with the abrasive. Hold the stone lightly against the wheel. Rotate it and move it back and forth across the face.

3. After the stone has been brought to a gloss, proceed with the rouge. Use the tin oxide for final polish. It is best to use a different buff for each different abrasive. Polishing is a slow process. The harder the stone, the slower it polishes. When it no longer gets smoother, it is time to stop. Pry the stone loose with a knife.

How to Grind and Polish a Cabochon

A *cabochon* (*kab*-a-shon) is a stone that has been cut and polished to smooth, rounded surfaces. It may be of any shape. Start with a stone that is not much larger than you want it to be. Make sure it is free from cracks, then:

1. Cement it to a dop stick so that you can grind the bottom first.

2. Use the coarse grinding wheel with water dripping on it. Hold the stone lightly against the wheel, moving it back and forth over the surface. No tool rest is used. Grind until the surface is flat.

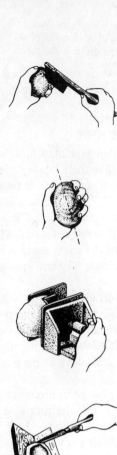 Brush rock briskly with stiff wet brush to clean.

Examine rock specimen carefully for best grain selection.

Place rock firmly in vise for first slab cut, blocking out with wedges.

If rock is too irregular, cast in plaster of paris form.

 After rock is slabbed, examine carefully and lay out for trimming, choosing best patterns.

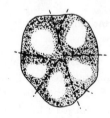 Check speed of diamond saw carefully. Always make sure that saw reservoir is filled with water-soluble coolant.

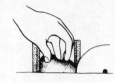 Position rock against diamond saw before starting.

 Hold slab firmly during trimming operation.

 After blank is trimmed, use template for drawing final shape directly on the stone.

 Rough grind or sand blank close to final desired shape.

 Mount blank stone on dop stick by means of an alcohol lamp and small amount of dopping wax. Don't let stone get too hot.

 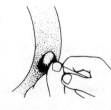 First sand or grind base of stone, using 100 grit. When saw marks are gone, finish with 200 grit.

Use felt wheel with grit to polish stone, washing off excess grit carefully.

Buff on wheels, always removing previous grit to prevent scratching.

Stone is now ready for top side polishing and final forming. Heat wax, remove stone with knife, and remount base down.

Repeat buffing and sanding operations, mixing and polishing compounds such as tripoli, rouge, cerium oxide, tin oxide, or chrome oxide.

Apply compounds to buffs with a small brush. Do not mix compounds on the same wheel. Rubber polishing wheels can be used for sanding and grinding operations both vertically and horizontally.

Drilling the stones can be done with diamond drills or with small diameter thin wall tubing with 220 grit silicon carbide and light machine oil mixture.

Fig. 7–71. (left) The basic process of gem cutting and grinding.

✚ SAFETY SENSE

1. Keep the dust to a minimum when working with cement and plaster.
2. Keep a firm grip on the dop stick when buffing stones.

Fig. 7–72. (below) This stone is being ground on a lap, which is an iron disc. Oil and abrasive powder are used on the lap.

3. Crack off the cement and reset the stone. Grind it roughly to shape on the coarse abrasive wheel. Follow with the fine abrasive wheel, 220 grit.

4. Polishing is started on the sanding discs. First use the coarse, and then the finer. Dip the stone in water frequently. Do not hold it in one place on the disc. Keep it moving. The final polishing is done on the buff, as has been described on page 323.

You Can Make Synthetic Gems. From the glazes you use on pottery you can make jewels. Put a thin layer of ground flint on a piece of broken bisque, and on the flint drop some globs of glaze. When the glaze is dry, put it in the kiln and fire it. The glaze draws up into balls and when cool can be tapped loose.

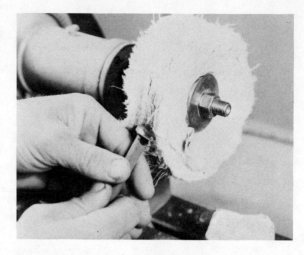

Fig. 7–73. The final polishing is done on a felt wheel using rouge (iron oxide) or tin oxide abrasive.

OCCUPATIONS IN CERAMICS

Occupational opportunities have changed considerably in the past two decades. During this time many dinnerware and art pottery industries have gone out of business. Competition with imported clay products and with glass and plastics has been strong. Numerous small independent pottery shops have opened. They specialize in handmade items of aesthetic quality. The glass industries have been expanding with new types of glass and new glass products. Occupations in the use of concrete are increasing as construction of buildings, highways, airports, and dams increases. Industrial positions in enameling, lapidary, cements, and plasters are relatively few. The following chart lists general occupational classifications. Apprenticeships are the usual route to jobs in the industrial manufacturing areas as well as in the building trades. Make a survey in your community of the types of occupations in the ceramic fields. Include data on wages, salaries, fringe benefits, unions, qualifications, hazards, and opportunities.

Industrial Manufacturing Occupations

Clay Products	*Glass Products*	*Enamel Products*
Clay mixers	Batch mixers	Batch mixers
Kiln operators	Furnace operators	Furnace operators
Jiggermen	Glass blowers	Spray men
Mold makers	Polishers	Decorators
Glazers	Cutters	
Decorators	Engravers	
	Decorators	

Lapidary
  Gem cutters
  Gem polishers
  Monument makers

Cements, Plasters
  Mixers
  Kiln operators

Concrete Products
  Form builders
  Finishers

Occupations Common to All Ceramics Industries

Unskilled laborers
Quarry men
Truck drivers
Heavy equipment operators
Machinists
Welders
Electricians
Mechanics
Production machine operators

Business management
Industrial engineers
Sales, marketing
Safety engineers
Inspectors

Occupations Related to Ceramics

Building contractors
Brick masons
Stone masons
Cement finishers
Plasterers
Rough carpenters
Tile layers
Terrazo layers
Architects
Interior designers
Civil engineers
Artists
Environmental engineers
Structural designers
Industrial arts teaching
Art teaching

Professional, Private Occupations

Potters
Sculptors
Glass blowers
Enamelers
Ceramics hobby shops
Recreation equipment
  construction

CERAMICS AND THE ENVIRONMENT

It can be said that ceramics is our environment. Clay is nearly everywhere on the earth's crust. It is commonly found just below the topsoil. Clay with rock and minerals make up much of our natural environment and resources. They are what the planet is composed of. The early permanent buildings and construction in this country were of brick or stone. Today concrete and steel have been added as building materials.

The quarrying of clay and rock has left large cavities in the surface of the earth. These are sometimes convertible into lakes for recreation and for depositing land-fill. Brick from rased buildings can be reused. Secondhand brick usually costs more than new. Why? Broken brick and concrete are used as rip rap and ballast along water fronts to control erosion. They make good fill for roads. Old glass is reusable as cullet. This is mixed with new glass batches in the recycling process. It is also usable in concrete for roads. All ceramic products are of permanent materials. This gives them an advantage for some types of reuse.

Factories producing ceramic products tend to give off dust, smoke, and gases. There is little solid waste to be discharged into rivers, however. The use of gas to replace coal for factory fuel has reduced air pollution significantly.

CERAMICS AND RECREATION

Having been through this chapter you will have some ideas on its recreational possibilities. Can you add to the following?

Clay

Modeling: pots, figures, lamp bases, sculpture
Slab work: tiles, plaques, boxes, draped ware
Casting: mold making, slipcasting
Decorating: painted decoration on bisque and glazed ware
Throwing: pots, bottles, dishes, trays, lamp bases
Glazes: experimenting with glaze bases and colorants
Clay: clay prospecting, refining, testing

Glass

Making: from raw materials
Slumping: bottles, flat glass
Decorating: with ceramic decorating colors
Stained glass: windows, decorative pieces, lamps

Lapidary

Rockhounding: prospecting, collecting, identifying
Stone cutting: splitting, sawing, polishing
Gem making: grinding, faceting, polishing

Cements, Plasters

Keene's cement: carving, casting
Access to a source of heat is necessary for working with clay, glass, and enamels. Study the catalogs for descriptions and prices. Perhaps you could have your clay pieces fired in a local pottery or ceramic shop. Electric kilns and furnaces can be made in industrial arts. A potter's wheel can also be constructed. Other items can usually be made or adapted from common utensils.

TO EXPLORE AND DISCOVER

1. How did the American Indians make pottery? Did they use glazes? Kilns? Molds? Could you make some pottery in this way?
2. What is adobe? How did the Southwest Indians make adobe brick?
3. What ceramics industries are there in your state?
4. Why does ceramic insulation, as in your kiln, insulate? How is it made? Could you make some from your clay?
5. What discoveries and developments did the early Chinese and Egyptian potters make?
6. Which of our good design qualities or principles apply to handmade pottery?
7. What is glass cloth? Fiber glass? How are glass fibers made?
8. What is different about the nature of glass that is heat or oven proof?
9. What type of glass is safest for glass doors?
10. How are eyeglass lenses shaped? Of what is this glass made?
11. Why should one not touch a camera lens with fingers?
12. What is the nature of bullet-proof glass? How is it made?
13. Why is safety glass safe? How is it made?
14. What is the difference between window glass and plate glass?
15. What does curing mean when applied to concrete? How is it done?
16. How are abrasive materials made? How do they cut? Could you make an abrasive experimentally?
17. How are sandpaper types of abrasives made? What systems of grading are used?
18. How is pencil lead made? What are the ingredients?
19. Make a list of the types of ceramics products in your home? Include the construction, furnishings, appliances, art objects?
20. What would happen if suddenly all ceramics products disappeared? Give possible examples.

FOR GROUP ACTION

1. For a store window in your town, or a display at your local library, make an exhibit showing how pottery is made. Detail the steps from the raw clay to the finished ware. Include examples of pottery made in your class.
2. Set up an exhibit of porcelain-enameled ware and explain how porcelain enameling is done. Conduct demonstrations to which you can invite other students and your parents.
3. Organize your class into a pottery industry to produce a quantity of a single item by slip casting. This might be a school souvenir.
4. Set up a production line to make copper enameled pins or charms.
5. For neighborhood service projects, design and build concrete park benches, flower planters, bird baths.

6. Replace a broken section of sidewalk on the school grounds.
7. If there is a children's home in your vicinity, three or four of you might arrange to teach the youngsters to make things of clay. Perhaps you could do this one evening per week. A Cub Scout pack would also appreciate this.

FOR RESEARCH AND DEVELOPMENT

1. Make a comparative analysis of native clay taken from a river bank, stream bed, or excavation. Dry it, break it up, soak it into a slurry (unsieved liquid clay). Test batch No. 1: sieve through window screen. Test batch No. 2: sieve through 20- to 30-mesh sieve. Test batch No. 3: sieve through 40- to 50-mesh screen. Compare the handling characteristics, plasticity, rate of drying, shrinkage, maturing temperatures.
2. How many different colors can be obtained from two prepared color glazes? Make 1 x 1 inch test tiles.
3. Can you make five warp-free tiles 4 x 4 x ½ inches in size? Here is a clue to warpage: the uneven strains set up in the moist clay during the working into tiles. Tiles warp for several reasons—strains are set up in the clay when it is rolled out, uneven drying on top and bottom, uneven heating during firing. How can you eliminate these problems?
4. Develop a color temperature chart using 100-degree intervals beginning at 1000 degrees. Try to match the colors with tempera paints.
5. Find out how glass is blown into molds. Design and make a two-piece clay bisque mold for a small glass bottle. For the glass, melt some glass chemistry tubing. A bisque mold is made of clay and fired.
6. Ceramic colors are obtained from the oxides of metals. Make some very fine filings of copper, iron, aluminum, bronze, zinc. Add small amounts of these to measured quantities of a clear transparent and an opaque glaze. Make sample test tiles. Do all of these metals add color? Now try cobalt, manganese, and red iron oxides.
7. Make some glass. Get a batch recipe for a low temperature glass. You can make a crucible from stoneware clay.

FOR MORE IDEAS AND INFORMATION

Books

1. Bates, Kenneth F. *Enameling: Principles and Practice*. New York: The World Publishing Co., 1951.
2. Brennan, Thomas J. *Ceramics*. S. Holland, Ill.: Goodheart-Willcox Co., 1964.

3. Kenny, John B. *Ceramic Design*. Philadelphia: Chilton Book Co., Publishers, 1963.
4. ———. *Complete Book of Pottery Making*. Philadelphia: Chilton Book Co., Publishers, 1959.
5. Kinney, Kay. *Glass Craft*. Philadelphia: Chilton Book Co., Publishers, 1961.
6. Nelson, Glenn C. *Ceramics*. New York: Holt, Rinehart and Winston, 1971.
7. Sanders, Herbert H. *Pottery and Ceramic Sculpture*. Menlo Park, Calif.: Lane Books, 1964.

Booklets

1. *Clay Modeling Methods*. American Art Clay Co., 4717 W. 16th St., Indianapolis, Ind. 46222
2. *For Career Opportunities Explore the Wonder World of Ceramics*. The American Ceramic Society, 4055 N. High St., Columbus, Ohio 43214
3. *The History of American Glass*. Glass Crafts of America, 816 Empire Bldg., Pittsburgh, Pa. 15222
4. *Laboratory Glass Blowing*. Corning Glass Works, Laboratory Glass Sales Dept., Corning, N. Y. 14830
5. *The Making of Portland Cement*. Portland Cement Association, Public Relations Bureau, 33 W. Grand Ave., Chicago, Ill. 60610
6. *Fiberglas Is the New Basic Material*. I. A. Meeks, Owen-Corning Fiberglas Corp., Fiberglas Tower, Toledo, Ohio 43659
7. *This Is Glass*. Corning Glass Works, Corning, N. Y. 14830

Chart

1. *How Fiberglas Is Made*. (chart) Owens-Corning Fiberglas Corp., Fiberglas Tower, Toledo, Ohio 43659

Magazines

1. *Ceramic Industry*. Industrial Publications, Inc., 5 S. Wabash Ave., Chicago, Ill. 60603
2. *Ceramics Monthly*. Professional Publications, Inc., 4175 N. High St., Columbus, Ohio 43214

Films

1. *Building a Highway*. 16 mm, sound, color, 16 min. Portland Cement Association, Public Relations Bureau, 33 W. Grand Ave., Chicago, Ill. 60610
2. *Free Educational Films*. (catalog) Modern Talking Pictures Service, Inc., 1212 Avenue of the Americas, New York, N. Y. 10036

3. *The Nature of Glass.* 16 mm, sound, color, 23 min. Corning Glass Works, Corning, N. Y. 14830
4. *Old as the Hills.* 30 mm, abrasives. Modern Talking Picture Services, Inc., 1212 Avenue of the Americas, New York, N. Y. 10036

GLOSSARY—CERAMICS

Abrasive a material used for grinding or polishing.

Aggregate the sand and rock used in concrete.

Ball Clay a natural, very plastic clay.

Ball Mill a machine for grinding materials into very small particles. Porcelain balls tumble about pounding the material to bits.

Bat a slab of plaster of paris.

Bisque clay products that have been fired the first time.

Bisque fire the first fire.

Blowing the process of forming glass in molds by air pressure; the process of forming glass holloware by blowing into it with a tube held in the mouth.

Blowpipe the tube used in blowing glass.

Blunger the machine for mixing clay and slurry.

Bone dry clay that has dried fully in the atmosphere.

Cabochon a gemstone cut and polished with round surfaces.

Calcine to heat a ceramic material to drive out the chemical water.

Ceramics the field of study and work in nonmetallic, inorganic materials for which high temperature heat treatment is required.

Cermets man-made material that combines ceramics materials and metals.

Cleaving the process of splitting stone.

Coil a rope of plastic clay.

Concrete a hard dense substance made of sand, rock, cement, or water.

Crazing a network of fine cracks in a glaze.

De-air to remove the air from, as to de-air moist clay.

Decal decalcomania decorative transfers adhered to pottery or glass.

Dopstick the stick to which a gem stone is adhered for cutting and polishing.

Drawing the process of pulling molten glass into tubing.

Earthenware pottery made from surface, low-temperature clays.

Enamel the glass fired onto metal surfaces.

Extrude to force a material out of a hole or a die to give it form, such as in the forming of bricks.

Facet one of the small polished faces on a gem stone.

Feldspar a mineral found in igneous rocks, essential in glazes.

Fire the heat treating process for hardening clay.

Flint the potter's term for silica (sand).

Glass the hard transparent or translucent material made by melting together silica, soda, potash, lime, and metal oxides.

Glaze the glasslike coating fused to the outside of pottery.

Glost fire the glaze fire.

Greenware unfired pottery.

Ground coat the first or undercoat in enameling.

Gypsum the natural rock from which plaster of paris is made.

Hammer mill a grinding machine in which materials are hammered into particles.

Incising the process of cutting designs into unfired pottery.

Jiggering the process of forming clay over or in a revolving plaster mold using a template to form one side.

Kaolin the purest of clays, main ingredient in porcelain.

Keene's cement a gypsum plaster used for imitating marble.

Kiln the furnace or oven in which ceramic heat treatment or firing is done.

Lap a metal disc on which an abrasive is used for grinding gem stones.

Lapidary one who is an expert at cutting and polishing gems, also the field of stone cutting and polishing.

Leather-hard the state at which clay products are half dried.

Lehr the heat treating furnace for annealing glassware.

Maturing temperature the temperature at which a clay is fired, short of vitrification or melting.

Moh Scale scale for ranking the hardness of gem stones.

Mold a plaster or bisque form in or over which clay is formed.

Obsidian natural glass formed in volcanic heat.

Overglaze color decorating color applied over a fired glaze.

Oxide a compound of oxygen with some other element.

Pickle the acid solution for cleaning metals.

Plaster of Paris a gypsum product used for pottery molds.

Plaster separator a soapy solution for keeping layers of plaster from sticking together.

Porcelain clay products of highest quality. It is translucent in thin sections.

Potter's wheel the wheel on which clay is formed by the hands.

Pressing the process of forming plastic clay by forcing it into molds.

Pug mill the machine for kneading clay.

Pyrometer the temperature indicating device on a kiln.

Pyrometric cone a triangular pyramid of clay used to indicate kiln temperature.

Quarry to mine rock and clay.

Ramming the process of pressure forming plastic clay over plaster molds.

Refractory the capability of resisting heat.

Rolling the process of forming molten glass into sheets.

Sgraffito the decorating process in which a design is scratched through a coat of glaze or clay to reveal the clay body underneath.

Slip liquid clay that has been sieved.

Slip casting the process of forming clay products by pouring slip into plaster molds.

Slurry unsieved liquid clay.

Soda ash carbonate of soda used to deflocculate (thin) slip.

Sodium silicate commonly known as water glass; used as a deflocculant for slip.

Stilt a three-legged clay support for ware in a glaze fire.

Stoneware a high firing clay; clay ware made from this.

Tempering the heat treatment given glass to increase its strength.

Template a pattern or guide used to shape clay.

Throw the process of shaping clay on a potter's wheel.

Translucent semitransparent.

Tripoli a polishing compound used for lapidary work.

Tunnel kiln a continuous firing kiln in which ware enters at one end and comes out at the other end.

Turning the process of shaping clay by cutting as it rotates.

Underglaze color ceramic colors applied to the ware over which the glaze is applied.

Vitrification the state at which clay when fired becomes glasslike.

Wedging the process of kneading clay by hand to increase plasticity and drive out air.

Welding the process of adhering two pieces of clay.

Plastics **8**
Technology

Plastics are man-made. They do not exist in Nature in a raw state as do clay, metallic ores, and woods. Plastics are combinations of chemical substances. They are compounded by chemists to have certain specific qualities for special needs. Consequently, there is an ever-increasing number of plastics materials. Originally plastics were considered as substitutes for other substances. Today plastics have properties that make them superior to natural materials for many uses.

The plastics industry began a century ago. A company manufacturing billiard balls of ivory offered a $10,000 award for a substitute material. The new material, called *celluloid* and invented by the Hyatt brothers, was made from wood pulp and nitric acid. It was found suitable for other uses, too. At the turn of the century, collars and cuffs on men's dress shirts were made of celluloid. With roll film made from this plastic, Eastman was able to produce the first hand-held cameras. Eventually, combs, eyeglass frames, lacquer, safety glass, and many other products were made from *nitro-cellulose*. This first plastic was highly flammable and consequently dangerous. Cellulose made from wood pulp proved to be a suitable base for an entire group of plastics called cellulosics. In 1900 the E. I. duPont de Nemours

Fig. 8–1. Building a water-tight pond with a sheet vinyl lining as the bottom.

is the smallest particle of a material having all the qualities of that material.) Plastics molecules are called *polymers*. They are made up of smaller molecules called monomers. The latter are joined into chains and nets under the chemist's control. This process is called *polymerization*. Plastics are available in various forms: sheet, film, laminates, rod, tubing, fiber, foam, liquid, powder, flakes, granules, and pellets.

Types of Plastics

Plastics materials can be divided into two basic groups depending on their reaction to heat. One

Fig. 8–2. Parts of this construction toy were molded of colorful styrenes. Such a toy of other materials than plastics would be too costly.

Company produced *cellophane,* the transparent wrap.

In 1909 Dr. Leo Baekeland obtained a patent for a plastic made from phenol resins. It was called *Bakelite.* This type of plastic is still in use in such items as the agitators of washing machines, distributor caps, furniture, TV cabinets, and electrical insulators.

PLASTICS AS MATERIALS

Plastics are materials that can be formed into shape when soft. When they harden or set, they hold this shape. Some natural materials, namely rosin, amber, asphalt, clay, and tar, have these qualities. The man-made plastics, called *synthetics,* have little similarity to substances that are naturally plastic. It is these synthetics that the plastics industry manufactures and uses.

Plastics as materials are a family of chemical compounds made of large molecules. (A molecule

Fig. 8–3. (top) A fiber glass reinforced pool. The liner is built up on the job. Fig. 8–4. (bottom) This huge top-shaped tank is a blender. In it the ingredients for a particular plastic compound are thoroughly blended.

Fig. 8–5. (top) Butyrate plastic pellets are being inspected. Fig. 8–6. (bottom) Fibers of glass are being spun from molten glass. They are then wound on reels like thread.

THERMOPLASTIC PLASTICS

Kind	Origin	Qualities	Typical Uses
Cellulosics **Cellulose nitrate** **Cellulose acetate** **Cellulose acetate** **butyrate** **Ethyl cellulose** **Cellulose propionate**	1869	High tensile and compression strengths Very tough Beautiful color effects Good electrical resistance Low heat resistance	Eyeglass frames Steering wheels Toothbrushes, combs Wrap, displays Hammer heads Tool handles
Acrylics	1930–37	High optical clarity Easily scratched Rubbery when heated Tends to return to original form when heated: "memory"	Aircraft windows and bubbles Taillight lenses Signs, displays Contact lenses
Polyamides **Nylons**	1938 duPont	High tensile strength Tough, wear resistant Heat resistant, do not change shape in boiling water	Clothing fabrics Rugs Machine bearings Gears Fish lines
Polyolefins **Polyethylene** **Polyproplyene**	England 1933 U.S. 1943	Light weight Good electrical resistance Tough and durable Film can "breathe": permit passage of gases	Thin wrap Squeeze bottles Toys Packages
Styrenes	World War II	Hard, brittle Excellent electrical resistance	Toys, models Cases Wall tile Light fixtures
Vinyls	Germany 1925 U.S. 1933	Flexible, tough Good electrical resistance Low temperature resistance Have memory	Adhesives:white glue Floor tile Inflatable toys, pools Woven fabrics

Fluorocarbons
  Tetrafluorethylene
    (TFE) Teflon
  Chlorotrifluorethylene
    (CFE)
	1943	Very heavy
		Excellent electrical insulation
		Excellent heat resistance, not flammable
			Electrical insulation
			Pipe, hose
			Bearings
			Cooking utensils

Acetals	1959 duPont	High resistance to solvents Toughest of the thermoplastics	Bearings Rollers Housing Cabinets

group softens with heat: the *thermoplastics.* The other group, *thermosets,* hardens with applied heat. When once set, these do not soften again with heat. See the accompanying charts for the types and kinds of plastics.

MANUFACTURING PROCESSES

The variety of plastics materials and their qualities require a number of different processes for converting them into usable products. Some processes were borrowed from other materials. Some are new with plastics. New ones are constantly being developed.

Molding Processes

Several different processes are included in the molding family. All form liquid plastic compounds into shape. These processes are: *compression,*

THERMOSETTING PLASTICS

Kind	Origin	Qualities	Typical Uses
Phenolics	1909 Dr. Leo Baekeland	Very hard, little flexibility Good electrical resistance Low heat conductivity	Cabinets Handles Molds and dies
Casein	Developed in Europe Brought to U.S. 1900	Very strong and flexible Absorb water	Buttons Buckles
Aminos **Urea** **Melamine**	1929–39	Very hard, durable Good electrical resistance	Counter tops Dinnerware Adhesives
Epoxy	1947	Superior adhesive quality to any surface Good electrical resistance	Metal to metal Coatings Jigs, fixtures Dies
Polyesters **Alkyds** **Dacron** **Mylar**	World War II	Very strong and durable Cure at room temperature	Resin in fiber glass products Alkyds used in paints Mylar used for photographic film Dacron used for woolen fabrics
Silicones	General Eectric and Corning Glass Works, 1930	Excellent electrical resistance High heat resistance	Polishes Adhesives Electrical insulators
Urethanes	Introduced in U.S. 1955	Good electrical resistance Excellent forming quality	Heat, sound insulation Cushions Clothing lining

transfer, injection, extrusion, blow molding, dip molding, calendering, laminating, and cold molding.

Compression Molding. This process usually applies to thermosetting plastics. The material in the form of powder, granules, or pellets is preheated. Placed in the mold cavity, it is further heated to melting. The two halves of the mold are forced together under high pressure. This squeezes the liquid into all parts of the mold. The heat also

Fig. 8–7. Compression molding.

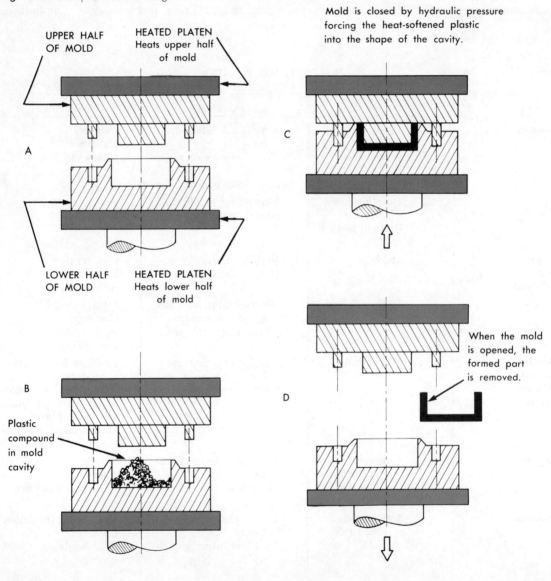

UPPER HALF OF MOLD

HEATED PLATEN Heats upper half of mold

A

LOWER HALF OF MOLD

HEATED PLATEN Heats lower half of mold

B

Plastic compound in mold cavity

Mold is closed by hydraulic pressure forcing the heat-softened plastic into the shape of the cavity.

C

When the mold is opened, the formed part is removed.

D

causes the plastic to cure or harden into shape. For some materials the mold is cooled with water before the piece is ejected.

Transfer Molding. Transfer molding is a variation of the compression process. The liquid is forced into the mold cavity by a *plunger*. This is necessary when the mold includes small deep holes that must be filled.

Injection Molding. Thermoplastics are most commonly formed by injection. The powder or pellets are forced from the open *hopper* into the heating chamber of the injection machine. At temperatures from 300° to 650°F, depending on the plastic, the material liquefies. It is forced or injected into a closed mold. Injection pressures are as high as 40,000 p.s.i (pounds per square inch).

Fig. 8–8. Transfer molding. The ram forces the plastic into the mold. Here it is cured under pressure. The mold is then opened and the part removed.

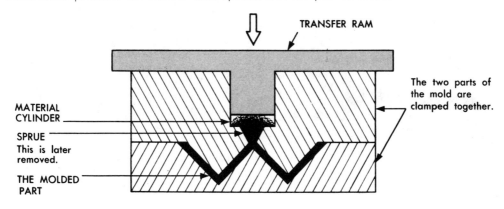

TRANSFER RAM

The two parts of the mold are clamped together.

MATERIAL CYLINDER

SPRUE
This is later removed.

THE MOLDED PART

Fig. 8–9. Injection molding. The plastic is forced through the cylinder into the heated end where it is melted. The spreader holds the plastic in contact with the cylinder wall. Ram pressure injects the melted plastic into the mold where it cools and hardens.

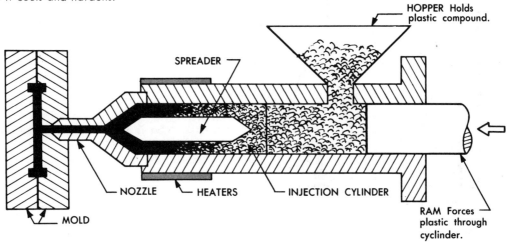

HOPPER Holds plastic compound.

SPREADER

NOZZLE   HEATERS   INJECTION CYLINDER

MOLD

RAM Forces plastic through cyclinder.

339
Plastics Technology

Upon cooling, the mold is opened and the part is removed.

Extrusion Molding. By the extrusion process, liquid plastic is forced through a die in a continuous length, much as toothpaste from a tube. The shape of the die determines the cross-sectional shape of the plastic. The formed plastic is cooled as it moves through a bath of cold water.

Blow Molding. This process is similar to blowing glass in a mold. A heated plastic tube with one end sealed is inserted into a mold cavity. Like a toy balloon, it is inflated with compressed air and takes the shape of the cavity. Plastic bottles are made in this manner, usually of polyethylene.

Dip Molding. A solid mold usually of metal or ceramics is heated to the prescribed tempera-

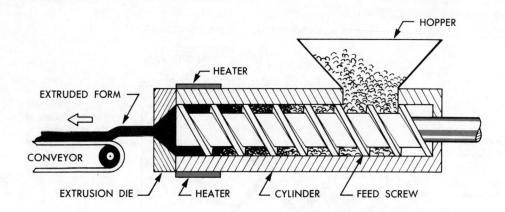

Fig. 8–10. Extrusion molding. The feed screw forces the plastic through the cylinder into the heated end. Continuing pressure causes the melted plastic to be extruded through the die. It takes the cross-sectional shape of the die. The conveyor carries it away in a continuous length.

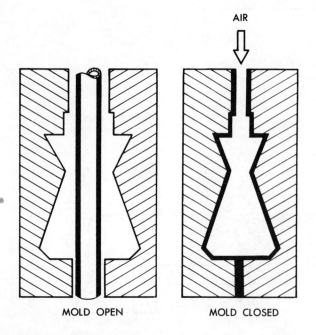

Fig. 8–11. Blow molding. An extruded thermoplastic tube is placed between the mold halves while still hot and soft. The mold clamps shut, sealing the lower end. Air is blown into the tube, which takes the form of the mold.

ture. It is dipped into liquid thermoplastic and is slowly withdrawn. A coating adheres to the mold. When cool, it is peeled off.

Calendering. Sheet plastics are produced by calendering. This process was borrowed from the rubber industry. The thermoplastic material is softened and rolled between a series of heated rollers. When of the desired thickness, it is passed between chilling rolls to cool. Then it is rolled up like paper on take-up rolls.

Laminating. Laminating is a process of bonding two or more layers of sheet materials together, using adhesives, heat, and pressure. The layers are coated or impregnated with a thermosetting resin. The heat used is commonly 270° to 350°F and the pressures, 1000 to 1500 p.s.i. Sheets as well as forms can be made.

There are several variations of the laminating process. They use lower operating pressures. *Bagmolding* uses an inflatable rubber bag to apply pressure against the laminates while curing. *Vacuum molding* uses atmospheric pressure. The

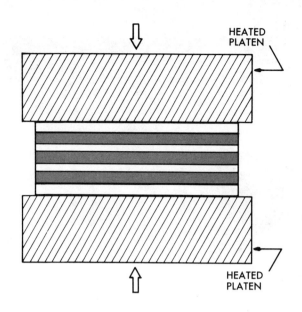

Fig. 8–13. The laminating process. Alternate layers of paper, wood, fiber glass, or other material are bonded together with plastic by means of heat and pressure.

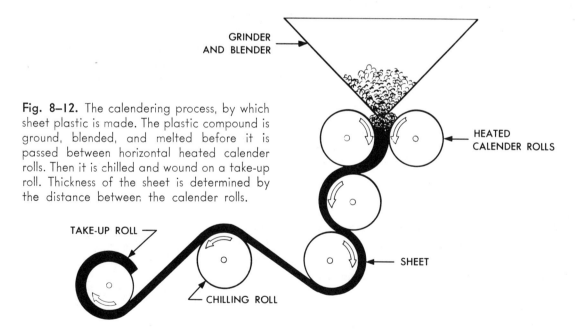

Fig. 8–12. The calendering process, by which sheet plastic is made. The plastic compound is ground, blended, and melted before it is passed between horizontal heated calender rolls. Then it is chilled and wound on a take-up roll. Thickness of the sheet is determined by the distance between the calender rolls.

Fig. 8-14. A mold for an insignia over which a vacuum-molded copy was made.

layers are placed in a mold with the edges clamped airtight. A vacuum pump removes the air from the mold and the pressure of the atmosphere (14.7 p.s.i.) forces the layers into contact.

Casting Processes

Liquid plastic can be poured into molds and cured at room temperature or in an oven. This simple casting requires no pressure or melting heat. It can be done with several different plastics. A liquid plastic is usually mixed with a *catalyst,* an agent that hastens a reaction or process, in this case the curing.

Vinyl resin plastics have the quality of hardening when in contact with heated surfaces. Dip molding makes use of this and is sometimes called *dip* casting. These plastics can be used similarly to clay slip in a plaster mold. Poured into a heated mold, plaster for example, they adhere and build up thickness depending on the time allowed. This is called *slush casting.*

Thermoforming Processes

Thermoplastic sheet gets rubbery when heated. Placed in or over a mold, it can be formed by stretching into shape. This can be done by clamping the sheet over the cavity. A vacuum can be used as already described for laminates. Using air pressure instead of a vacuum gives a variation of thermoforming called *blow forming.* (See Figs. 8–15 to 17.)

Reinforced Plastics Forming Processes

Reinforced plastics are those plastics compounds used in conjunction with fibers of other materials such as glass, threads, and fabrics. Fiber glass is probably the best known. The fibers are usually added at the time that the plastic is being formed. The two common methods are hand building and spray. In the first, a mold either concave or convex is used. The wall is brush coated with a *release agent* to prevent the plastic from sticking. A layer of the plastic is applied. This is usually an epoxy or polyester resin.

Layers of resin-coated fibers or fabrics are laid in place and pressed or rolled into contact with the mold. When cured, the piece is removed. The *spray process* sprays layers of resin as well as layers of fibers from a gun.

Foam Forming Processes

Certain plastics can be caused to foam and to produce a cellular body. The urethanes and styrenes are commonly used for this. A closed mold is filled with beads of polystyrene plastic. When the mold is heated, the beads expand and fuse together, producing a lightweight, porous article. Insulated ice cube buckets are made in this manner.

The resin can be injected into a mold followed by a chemical that also causes a foaming reaction. The foam fills the mold and is then cured with heat. This principle is used in filling flotation tanks

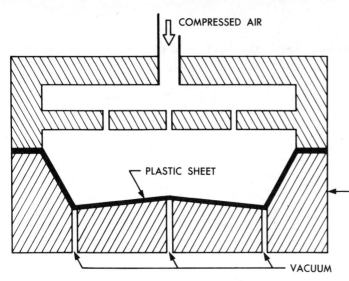

COMPRESSED AIR

PLASTIC SHEET

MOLD

VACUUM

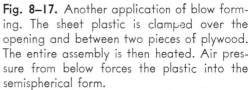

Fig. 8–15. Thermoforming. Heated plastic sheet is formed in a mold, using air pressure on the top side and partial vacuum on the bottom.

Fig. 8–16. Blow-molding bottles. The bottle is partially formed outside and is then enclosed in the mold. The plastic, being hot and viscous, is blown to shape by air pressure.

Fig. 8–17. Another application of blow forming. The sheet plastic is clamped over the opening and between two pieces of plywood. The entire assembly is then heated. Air pressure from below forces the plastic into the semispherical form.

for boats and hollow walls for insulation. Urethane foam is the common plastic foam in cushions.

TOOLS FOR WORKING WITH PLASTICS

Many hand tools for wood and metal can be used with plastics when they are to be worked cold. But some are not suitable, for example: gouges, chisels, knives, and auger bits.

Plastics are sufficiently soft to be easily damaged in a vise. When the plastic has protective paper on, leave it intact until all of the processing is complete. Vise jaws must be smooth.

Saws and Plastics. Plastics can be cut with handsaws. Hard plastics such as Bakelite will dull the teeth quicker than wood. Be sure the sheet plastic is supported close to the saw kerf.

Fine teeth are less likely to chip the edge than coarse. For rough cutting, stick a strip of masking tape over the location of the cut and mark the line. For fine cuts, use a hacksaw, backsaw, or jeweler's saw.

Files and Filing. The common file for wood and metal can be used for smoothing edges. For straight edges hold the file in line with the edge. The sheet should be held between two pieces of wood in a vise. Small pieces of plastic can be held in the hand with the file placed on the bench. Plastic cuttings clog files quickly. Clean them out with a file card.

Planes and Planing. Thin edges held between wood blocks in a vise can be trued with a hand plane. Make a very shallow cut to prevent chipping the outside corners. Obviously, the plane must be very sharp.

Sanding. The surface of sheet plastic is usually not sanded. It is already fully smooth. Rough sanding an edge can be done with 4/0 garnet paper on a block. Use wet-or-dry finishing abrasive and water for the smoothest cut. Sanded under water, the cuttings are flushed away.

Scrapers and Scraping. A cabinet scraper blade can be used for preliminary smoothing of sawed edges and flash, or parting lines. Use a square edge honed sharp.

Drills and Drilling. The hand drill or electric hand drill with twist drill bits is used for making holes. On smooth surfaces, the bit may want to slide when the hole is started. It is not feasible to try to center-punch a starting point. Instead, carefully nick the surface with a sharp scriber or awl. Brittle plastic such as the acrylics should be backed up with a block of wood on the underside to prevent chipping. Either carbon steel or high-speed steel bits are used.

MACHINING PLASTICS

Plastics can be cut with the power saws used for wood. Hard plastics tend to dull the blade, however. Dull powersaw blades will gum the plastic because of the heat of friction. The plastic must be kept firmly in contact with the table. Protective paper coating should be left on acrylic plastics while sawing. Follow all the Safety Sense suggestions for the machines in Chapters 3 and 4.

The Circular Saw. This machine can be used for cutting both thick and thin thermoset plastics. Use a fine tooth blade. With heavy thermoplastics, however, the wide area of the blade can produce enough heat of friction to soften the material. This is not a problem with thin sections.

Band Saw. The band saw is a good general-purpose machine for cutting plastics. The long, thin blade cools well as it travels. Chips are removed quickly by the teeth. Special blades are made for plastics and produce a smoother cut than those for wood. For thin sheet, the blade should have more *pitch,* or more points per inch. For thick material, the skip-tooth type of blade is preferred.

Jigsaw. For rough cutting thin stock to intricate shapes, the jigsaw is suitable. A skip-tooth blade is preferred.

Sander. Machine sanding must be done with care because plastics are poor conductors of heat. The cuttings tend to clog the abrasive. Use light pressure to minimize friction. Wet sanding with the abrasive being rinsed with a flow of water does not heat the plastic. The cutting edges are kept free of particles for a faster cutting.

Drill Press. This is a universal machine suitable for drilling all kinds of materials. The twist drills used for wood are suitable on plastics. The same drill speeds can be used. Should heating be a problem, use cold water as a coolant while drilling. Be sure to place a block of wood or plastic under the piece being drilled. This prevents chipping when the bit breaks through.

Wood Lathe. Plastics may be turned on the wood lathe. The live center at the headstock is not driven into the plastic as for wood. For spindle turning, drill a center hole in each end as for a metal bar. With a backsaw or hacksaw, make a pair of shallow cuts in one end. They should be through the center and 90 degrees apart. The prongs of the live center fit these. Use lubricating oil or wax on the tailstock, or dead center.

A scraping action with the cutting tools is preferred. Lathe speeds can be the same as for wood.

For face plate turning, the work can be held with screws. When this is not desirable, the plastic can be cemented to another plastic block with wrapping paper in between.

Machine Lathe. Plastics can be turned on a machine lathe in the same manner as soft metals. Use a honed cutting tool ground for aluminum.

Buffing. Buffing is the final finishing process by which a high polish is obtained. The first step is that of cutting. The cutting wheel is made of layers of cotton cloth sewed almost to the rim. The buffing compound, an abrasive in wax, is held against the revolving wheel to charge the rim. The wheel turns downward away from you. After a few seconds, the wheel is ready. The work is held firmly in hand and is moved lightly into the rim. It must be kept below the center of the wheel. Move it back and forth across the surface to use all of the abrasive. From time to time recharge the buff with the compound. The second step is polishing. A loose cloth wheel is used without the buffing compound. If the piece has been sufficiently buffed, it will take on a high polish in a few minutes with the second wheel.

Engraving and Carving. Designs can be cut into plastics by engraving and carving. The former is best done with a vibrating electric tool. Carving removes material. This is done with a highspeed rotary electric tool.

FORMING PLASTICS

Several of the plastics manufacturing processes are readily adaptable for projects of your own: *molding, laminating, casting, fabricating,* and *die forming.*

Fig. 8–18. Architecture finds many uses for plastics. Here is a plexiglass dome for a shopping center. Plastics combine well with many other materials.

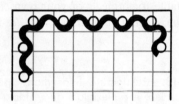

3/4 X 4—4 BLOCK OF WOOD

DOWEL PEGS

PLASTIC

Molding

Molds for Forming Sheet Plastic.　The major problem in molding sheet plastic is the excess material around the edge. The wrinkle-edge trays and bowls commonly seen have the excess taken up in the wrinkles. One-piece molds can be used to form such pieces providing there are no more wrinkles than you can make with your fingers.

To make a mold, drill $\frac{3}{8}$-inch holes $\frac{1}{2}$ inch deep in a 4 x 4 block in the patterns shown. Cut several 4-inch pieces of $\frac{1}{16}$- or $\frac{1}{8}$-inch thermoplastic sheet into squares, circles, and triangles. Make four $\frac{3}{8}$ x 2-inch dowel pegs with rounded, smoothed ends. Place these in the block to form squares and circles. Only three are needed for the triangles.

Fig. 8–19. A simple one-piece mold for use with square plastic sheet. One variation is shown. You can devise others.

Fig. 8–19a. Simple one-piece molds for hand forming thermoplastic sheet into trays. Lay out the design accurately on coordinate paper. Transfer centers of holes for dowels to a piece of $\frac{3}{4}$ in. wood of suitable size.

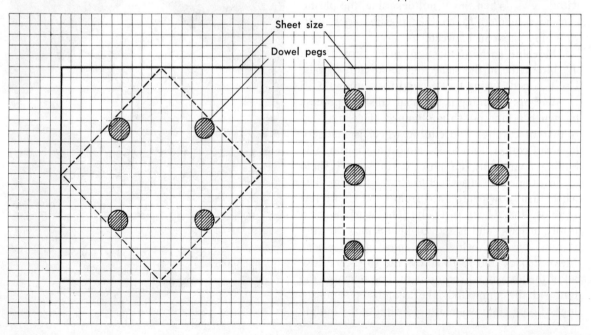

Sheet size

Dowel pegs

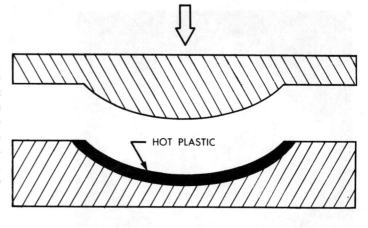

Fig. 8–20. (right) A two-piece tray mold. This illustrates the principle of dies. Use only enough pressure to force plastic into cavity. Line cavity with felt or thin plastic foam to prevent marring the sheet plastic. Fig. 8–21. (bottom) Drape molding for trays, a simplified thermoforming. Heat plastic until very flexible. Lay it over the mold block and let the edges take their own shape.

HOT PLASTIC

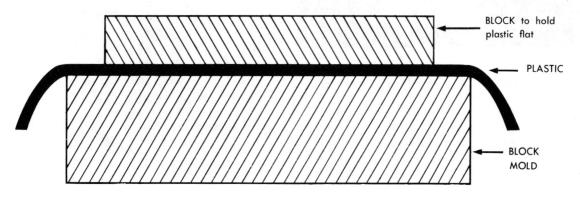

BLOCK to hold plastic flat

PLASTIC

BLOCK MOLD

Heat a piece of plastic in an oven with protective paper removed. When it is rubbery, center it over the pegs and quickly press it into shape. Hold until stiff. This piece may be reheated and reformed several times. Cotton gloves should be worn.

Two-piece Molds. Instead of using fingers to form the plastic, one can make a two-piece wrinkle mold. Make the bottom of the size and shape desired. The closer the pegs, the shallower the wrinkles. Cut a block the shape of the inside of the dish and ½ inch smaller all around than the inside of the pegs. This allows room for the wrinkles. Round the bottom edge of the inner block. Smooth it carefully so that it cannot mar

the soft plastic. For convenience, add a knob for a handle. Heat and insert the plastic as before and press into shape.

Two-piece Turned Molds. A dish-shaped cavity can be cut into a block on the wood lathe. For your first try, keep the cavity shallow and reasonably small, say 1 x 6 inches. Shape the second piece to match the first except make it ¼ inch smaller in diameter. Sand both smooth. Place a blank of soft, heated plastic over the cavity. Center the second part and in the drill press quickly apply the necessary pressure to form the plastic into shape.

Drape Molds. Free-form trays and planters can be formed by placing the soft plastic sheet on

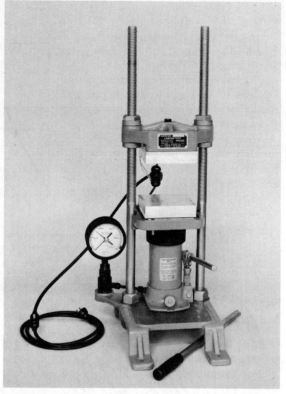

Fig. 8–22. (left) A jig for forming plastics is similar to one for metal forming. Fig. 8–23. (right) A plastic-laminating press. The sheets of plastic are placed between the two heated platens at the top. Hydraulic pressure forces them into permanent contact.

a cloth. Pick up the corners and draw them into any desired position. Hold until cool. The reverse of this method can also be used. Build up a convex mold of wood, clay, or plaster. Lay the plastic over this to cool and harden.

Bending Jigs. Bending jigs for forming parts can be made of wood sawed to shape or with dowel pegs located as desired. Soft sheet may be wrapped around mailing tubes and turned wood cylindrical solids to form cylinders.

Laminating

Special plastics presses are available for laminating flat, or nearly flat objects, within a plastic sandwich. Stamps, photos, coins, leaves, brightly colored feathers and many other items can be permanently sealed in plastic. Both acrylic and vinylite sheet are suitable. The general instructions are:

1. Adjust the thermostatic control on the press to 300–315°F.

2. With two pieces of Plexiglas or Lucite make the sandwich.

3. Place the sandwich on a polished sole plate and lay the other plate on top.

4. Place a blotter on each plate, insert the sandwich in the press and apply 2 to 5 pounds pressure.

5. Heat the sandwich for five minutes. After this, increase the pressure 25 pounds every 2–3 minutes until 150–200 pounds has been reached. The pressure depends on the machine and the thickness of the plastic.

6. Let the sandwich soak in the heat for 5–10 minutes. Then cool the press.

A portable laminating press can be made from a pair of ¼-inch aluminum plates and four bolts with wing nuts. Drill holes in each corner for a bolt. Insert the sandwich and tighten the nuts securely by hand. Insert the press in the oven and draw down the nuts one turn every two minutes until the plastic begins to squeeze down noticeably. Cool the assembly in water.

Casting Liquid Plastics

Castolite is a clear liquid acrylic plastic which can be cast into molds. It is used for imbedding objects such as polished stones, coins, or butterflies. A plaster mold coated with separator can be used with the plastic. For imbedding, a container fashioned from aluminum foil, a milk carton, or a paper cup may do. The resin must be mixed with

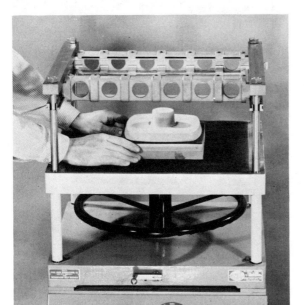

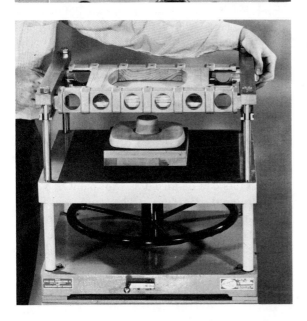

Fig. 8–24. (top) The process of thermoforming begins with the mold being properly positioned. Fig. 8–25. (bottom left) The sheet plastic clamped onto the yoke is positioned and aligned directly over the mold. The yoke is the wood support for the plastic. Fig. 8–26. (bottom right) The plastic is heated to make it stretchable.

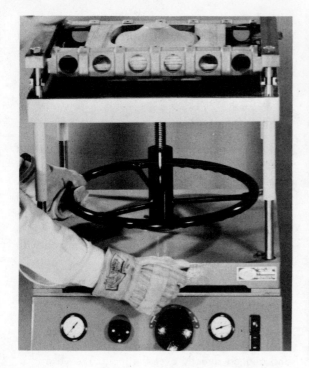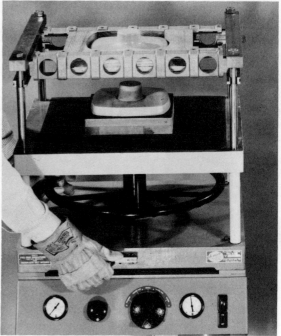

Fig. 8–27. (left) The mold is now pressed into the yoke stretching the plastic. The action alone is sufficient for some objects. Here the vacuum pump is turned on to complete the forming. As the pump reduces atmospheric pressure under the plastic, the pressure of the atmosphere above it forces the plastic into contact with the mold. **Fig. 8–28.** (right) After the plastic has cooled, the mold is removed and the piece is released from the yoke for trimming.

a catalyst according to the manufacturer's instructions. In an oven set for 150°F, it is cured 10 minutes for each ⅛ inch of thickness.

Colored polyester resins can be used for striking cloisonné (kloiz-on-*nay*) *effects*. Original cloisonné was a decorative treatment on bronze. The wall of a vase or a bowl was carved away leaving a network of narrow dikes. The cavities were filled with vitreous enamel which, when fired, fused to the wall. (See pages 309–312.) For plastic cloisonné use 1/16-inch copper, brass, or aluminum brazing rod. This is cut or bent and laid on a tempered hardboard back in the desired design. Col-

ored plastic is poured into the compartments, level with the top of the rod. When cured, the surface is polished with wet-or-dry finishing paper and water. This plastic requires a catalyst and must be mixed according to the manufacturer's instructions. Colors can be mixed as with paint.

Rotational Molding. This process forms plastic in a hollow rotating mold. It is charged with plastic liquid, powder, or granules. The mold is rotated in two directions in an oven. This distributes the plastic evenly over the cavity. It is cured by the heat of the oven. Vinyl plastisol and polyethylene are the commonly used plastics.

Fabricating Plastics

Plastic sheets, rod, and tube can be cut, fitted, and assembled into projects. (See the instructions for cementing, pages 354–355.) This type of assembly is called *fabricating*.

Fig. 8–29. (left) Laying out copper wire forms for a decorative wall hanging. The heavy wire is bent and cemented to hardboard. Different colors of liquid plastic are placed in the different sections. When cured, the surface is leveled and polished with waterproof finishing paper and water. This technique is a variation of the original cloisonné. Fig. 8–30. (right) Examples of decorative pieces using the plastic cloisonné method.

FIBER GLASS

Fiber glass is a material made from plastic and glass fibers. The plastic is usually polyester resin. The glass film may be in the form of woven fabric, called *glass cloth*. It may be in long, continuous strands of spun glass as used in fishing rods. It may be short fibers sprayed onto a resin-coated mold. It is the glass that gives the strength and durability to the material. The more glass, the stronger—so long as there is sufficient resin to bond the fibers together. Cured fiber glass is very hard. It will dull woodworking tools. A hacksaw or metal-cutting band saw is used for cutting. Boats, automobile bodies, airplane cowling, and furniture are made from fiber glass. The possibilities of color and imbedded design are limitless. A mold is necessary to hold the material while curing takes place.

Making Fiber Glass

A flat-bottomed tray or a mat suitable for diffusing the light in a lamp you might design are good first projects to experiment with. Remember to follow the manufacturer's instructions for preparation of the resin. For a mat or tile, cut a pair of blocks from some scrap. Formica-covered particle board such as that used for sink tops is excellent. These should be the size and shape of the mat. Heavy plate glass, too, can be used. These surfaces are smooth and will produce an equally smooth surface on the fiber glass. With these as the mold, follow the steps on page 352:

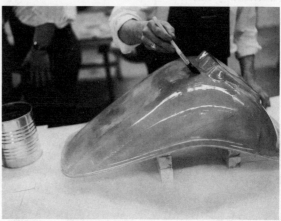

Fig. 8–31. (top) Glass fibers are twisted into yarn. This is wound on bobbins (spools) ready for installation in the looms that weave glass cloth. Fig. 8–32. (bottom) A vacuum-formed plastic hull for a model boat.

Fig. 8–33. Coating a fiber glass chair mold with gel coat preparatory to laying on glass cloth.

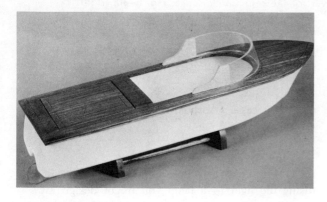

first or *gel* coat for the mold surfaces. Apply the gel coat with a brush to each block. This should be a heavy coat, but uniformly applied. If the coating is not smooth and even, let it dry. Then sand lightly.

4. Mix the laminating resin and catalyst to the manufacturer's recommendations. Apply a thick, even coat. Lay the fiber glass in position on one block. With the brush, work it into the resin until the surface is even. Apply an additional coat of resin to the top surface of the cloth. Lay the upper block in place, carefully pressing and sliding it into good contact.

5. Place a weight on the mold and let the fiber glass cure. This may take two or three hours.

6. When the resin is hard, remove the mat and trim the edges.

Several layers of glass cloth can be built up in this manner for greater strength. Colorful woven or printed cloth can be used in place of the glass for interesting effects.

Dried pressed flowers, and weeds, pine needles, bugs, artificial fishing flies, and other items

1. Cut a piece of glass cloth the same size as the mold.

2. Spray the plastic sides of the mold blocks with silicone spray separator. The polyester does not adhere to this. Or tape a piece of cellophane smooth and wrinkle-free over each mold surface.

3. Mix a small quantity of the resin and catalyst, say two ounces with fifteen drops. This is the

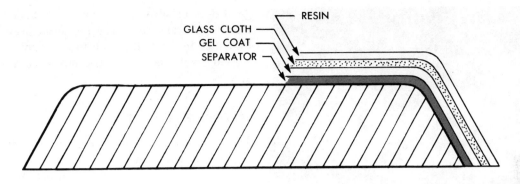

Fig. 8–34. Mold for a flat-bottom tray.

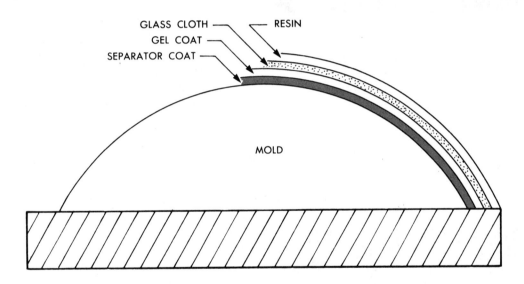

Fig. 8–35. Laminating with glass cloth for a round-bottom bowl. Add as many layers of glass and resin as needed. Mold can be of wood, plaster, glass, metal, fiber glass, or other material.

can be imbedded in mats and trays. Designs can be drawn on glass cloth with wax crayon before laminating.

Tray and dish molds are easily made of moist clay. Shape the tray upside down. Place a cardboard box around it and cover the clay with an inch of plaster. (See Fig. 7–27.) When the plaster is dry remove the clay. Sand the cavity as smooth as possible. Apply several coats of plaster separator (a 50/50 mixture of paraffin and kerosene). Then apply the gel coat as before and proceed to build up the tray.

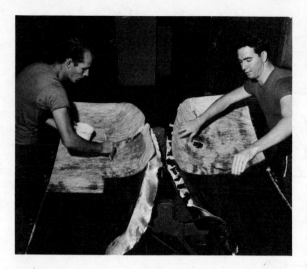

Fig. 8–36. Building fiber glass wing tanks for airplanes. The worker on the right is fitting a layer of glass cloth. The worker on the left is coating the cloth with plastic adhesive.

Fig. 8–37. Mass production of engine cowlings for airplanes. Each is built up by hand with alternate layers of glass cloth and plastic adhesive.

Edges of fiber glass should be sanded with aluminum oxide cloth. The tiny glass fibers tend to irritate the skin. It is well to wear gloves when working with fiber glass. Clean your brushes in acetone.

FASTENING PLASTICS

There are four common means for fastening plastics materials together. These are by *adhesives, cohesives, heat,* and *mechanical means.* Adhesives refer to cements that bond two pieces. There are many of these on the market today and most of these are liquid plastics. Cohesives are solvents that dissolve the surfaces to be bonded. When the pieces are tacky, they are pressed together and held until solid. Different plastics require different solvents. Heat is used when laminating, sealing, and fusing. Mechanical means include bolts, screws, and drive pins as well as laps and folds. The latter are also used with sheet metal. (See Fig. 4–51.)

Adhesives

Many different plastic cements are available. All do not bond all kinds of plastics. That commonly known as model airplane cement will adhere the acrylics and acetates. The fast-setting type has very volatile solvents and must be used with speed. Slower-setting types give more time to position and clamp parts. Epoxy resin cements have made it possible to bond plastics that have resisted all other adhesives. They set slowly and must be mixed according to directions. These also serve to join plastics with other materials.

A messy cement joint can spoil the appearance of your project. It is recommended that you experiment with the adhesives on scrap materials before you use them on your project. The surfaces must be clean and free of oil and wax. The latter may be left on after buffing. Cements adhere better to roughened surfaces than to smooth. They will fill in joints that do not fit perfectly.

Fig. 8–38. Fiber glass wheel covers for light airplane landing gear.

Cohesives

The cohesive action of solvents generally produces a stronger, cleaner joint than do adhesives. The problem is to find the best solvent.

Dissolving some filings of the plastic to be cemented in the solvent makes a solvent cement. This is sometimes easier to handle than is the solvent alone. Being somewhat viscous, it stays in place and also fills tiny cavities. The solvent alone should be used only on snugly fitting joints. Here it quickly runs to all parts and begins its action.

For joining with solvents, soaking is recommended when possible. For a butt joint, for example, the two edges are held in the solvent until they become soft and tacky. Quickly placed together and clamped, they form a strong joint. Pour the solvent into a tray to a depth of no more than $\frac{1}{16}$ inch. Soak the edges until you feel the softness when they are moved along the tray bottom.

For a project of several parts, the soak method may be useful only for a few joints. In this case, the parts are carefully fitted and then are assembled and clamped together. The solvent in a hypodermic needle is introduced at one point in a joint. A drop or two may be adequate. If the joint fits well, the solvent will quickly flow. If not, it will puddle up at a point of contact. Several applications, each a few minutes apart, are made. Let the joint set overnight.

SOLVENTS FOR COMMON PLASTICS

For
Acrylics   Methylene Chloride
Ethylene Dichloride
For maximum strength a mixture of the following is recommended:
Methylene Chloride
    60% (by volume)
Methyl Methacrylate
    40% (by volume)
Hydroquinone (.006% by weight)

For
Acetates   Acetone 70% and
Ethyl Acetate 30% (by volume)

For
Styrenes   Ethylene Dichloride
Toluene

Thermal Welding

Joining by means of heat is called thermal welding. The heat softens and liquefies the surfaces to be joined. By the time they are cool, they have fused. Plastic film wrap is sealed with heat. Your mother may do this with a hot iron when she is filling sacks of fruit or vegetables for freezing. In the supermarket, packages of meat are weighed, priced, wrapped, and sealed automatically.

Friction welding can be done in a drill press with acrylic rod. It is chucked and brought into contact with another piece of acrylic. The speed of rotation combined with slight pressure on the

Fig. 8–39. (top left) A fiber glass tote box being removed from a press. Each half of the die is heated. This speeds up the curing of the plastic. Fig. 8–40. (top right) A fiber glass cowling for a garden tractor is being removed from the die in which it was formed. Note how smooth and shiny the interior of the upper half of the die is. This makes the top side of the cowling smooth. Fig. 8–41. (bottom right) Completed fiber glass parts for the cowling of the garden tractor.

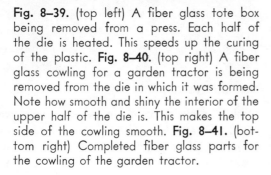

pieces produces enough friction to melt the surfaces. The technique requires practice so that by feel you will know when to stop the machine.

Mechanical Fastening

Plastic screws, and bolts and nuts are used for fastening. They are installed as for metals. Shrink fits can be made when one part is to fit permanently into another. Drill a hole slightly smaller than the diameter of the part to be inserted. Heat the piece having the hole. Press the other into place and cool.

PLASTICS AND OCCUPATIONS

Plastics are a twentieth century material. For this reason the processes used in their manufacture, as well as in the production of plastics products, are generally of a machine type. Most of the plastic products you use and see are made in automatic machines. These processes have been described in the chapter. Fiber glass plastics generally require more handwork. Boats, automobile, and airplane parts of this material are, for the present at least, usually made by hand with molds. The operation of automatic machines and the production of fiber glass products is usually learned on the job. A high school education is adequate. Mold making is a highly skilled field and usually requires apprenticeship. This is especially true when the molds are machined of metal.

Plastics products design, as in all types of product design, requires college preparation. Chemists and chemical engineers also have a college background. This often includes graduate study beyond the Baccalaureate or first degree. Positions in research in plastics chemistry often require the Master's and the Doctorate degrees. The Master's is the first graduate degree and the Doctorate the highest.

Technicians are employed in the maintenance of the automatic production machines. Training for such positions can be had in a vocational trade school or technical institute. Chemical technicians assist the chemists and chemical engineers.

In all plastics industries and businesses, just as in all other types, the management level includes many positions. Among these are clerks, stenographers, secretaries, computer operators, accountants, attorneys, sales personnel, safety engineers, office and business managers. The following is a list of occupations that are typical and related directly to work with plastics.

Plastics Chemistry

Chemists
Chemical engineers
Chemical technicians

Product Design

Designers
Engineers
Artists
Draftsmen
Product specialists: e.g., toys, furniture, fabrics

Factory Technicians

Electricity-electronics
Machining: molds and dies
Mechanics: machine servicing
Hydraulics, pneumatics, fluidics specialists

Production

Machine operators
Assemblers
Packers
Inspectors

PLASTICS AND THE ENVIRONMENT

Plastics materials while man-made are actually made from natural materials. Coal, air, and water, for example, are used to make nylon. Cellulose as obtained from wood and cotton is a common ingredient in plastics. The natural materials are converted to chemicals. These are combined to produce materials that do not exist in Nature. New plastics are continually being developed and new uses for them found. They have come to be indispensible in many ways. This is especially true as chemists are able to develop plastics for particular purposes.

Environmental problems center on discarded plastics products such as bottles, bags, and containers. You have no doubt seen many of these littering the streets and highways. Even when buried, these materials do not decompose in time as do many others. Materials that do break down and decompose are called *biodegradable*. Great quantities of waste plastic are accumulating. With no way to dispose of them they easily become a serious problem. The manufacture of chemicals for use in plastics produces fumes that may be hazardous not only to workers but to the outdoor environment. Discharge from such plants into rivers and streams causes pollution.

The great challenge to the plastics industry is to create plastic containers and other products that are either recyclable or biodegradable. Watch for developments along this line. Meantime, conservationists advise consumers to buy products in returnable or disposable paper containers.

PLASTICS AND RECREATION

For recreational purposes you can think of plastics materials in at least two different ways. First, there is the design and development of ideas using these materials in ways as described in this chapter. There are many ways to process plastics materials without having the specialized equipment illustrated. Second, there is the possibility of the development of useful articles from discarded plastics products. The latter cost little and offer the imagination unlimited opportunity. You have probably seen bird feeders and houses made from plastic jugs and bottles. The problem in reusing such articles is to so design an item that it doesn't look like junk when finished.

Old plastic containers can be used as molds for plaster of paris or Keene's cement casting. See the chapter on Ceramics Technology. They can also be used as molds for handmade candles. Many books are available on this activity.

Christmas ornaments, toys, planters, small parts boxes can be made from used plastic containers. If such ideas turn you on, you will find no end to the possibilities.

The development of molds and plastics forming equipment for the home workshop can be fun. There is the plastic-laminating press as described on p. 348. A strip heater for sheet thermoplastic and a hot wire cutting device for styrofoam suggest other possibilities. There is the problem of odors when heating plastics. This may be objectionable at home. Be sure to investigate this before you proceed.

TO EXPLORE AND DISCOVER

1. Why are plastics materials as used in bottles, bags, and containers not biodegradable? This means that they do not break down or decompose when exposed as do wood, paper, and iron. You may need to consult a chemist.
2. Make a list of consumer products for which plastics are better than other materials. Are there some consumer products for which other materials are better than plastics?

3. Some plastics are inflammable. How can they be identified? What products should not be made from these?
4. Why should there be no sharp corners in a mold for fiber glass?
5. How can one be sure that a mold will work before he tries it?
6. How can one tell whether plastics, wood, or metal is the best material for a design?
7. How does glass fiber contribute to the strength of a laminate?
8. What is a catalyst? Why is it necessary?
9. What happens in the curing process that is necessary for certain plastics? How is this done?
10. Is fiber glass really glass? How is it manufactured? If it is glass what accounts for its bending ability?
11. Using broken pieces of transparent colored sheet plastic, design a panel for a lampshade. How can they be held neatly in place? A stained glass window in a church may offer a clue.
12. Design and develop an appropriate form for imbedding a set of coins in plastics.
13. Make a design of pieces of heavy copper wire suitable for the "cloisonné" process. Keep it simple and try for a good color balance. (See Chapter 2.)
14. Plastic products often appear cheap and gaudy. Must they? See if you can redesign a simple product. Can you make a prototype?
15. Make a laminating press as described in the chapter. Experiment with it until you can produce a set of instructions for its use.

FOR GROUP ACTION

1. As a community service project, mass produce some fiber glass planters. Plant flower seeds or bulbs in them and deliver them to patients at a nursing or rest home. (Sure made you feel good to do this, didn't it?)
2. Mass produce plastic numerals for house numbers. Design an appropriate mounting.
3. Design a neat insignia for your industrial arts department. Make a pattern or a mold. Decide in advance on the material to be used for the copies.
4. Design and mass produce fiber glass notebook or photo album covers. Imbed colorful objects to add to the interest and effectiveness.
5. Design a mold for a bucket seat. Use a slab of moist clay to determine its shape. Cover the clay with thin sheet plastic. Have several of your classmates sit in it. This should give the general shape of the mold. Make a plaster mold over the clay. Design the leg structure to be used. Make the seats of fiber glass.

FOR RESEARCH AND DEVELOPMENT

1. Compare the strengths of fiber glass reinforced plastic with that of aluminum, steel, concrete, or other materials. Make test pieces of the same dimensions. Use

bending and compression tests. For a bending test, hang weights on a 10 inch bar of the material. Suspend this between two points so that 8 inches are left for bending. For a compression test, use a hydraulic press with a pressure reading dial. Both of these tests are called destructive. The samples are destroyed in the process.

2. There is a national problem of the disposal of used plastic products. What is being done to solve this problem? Are biodegradable plastics possible?

3. Could a hollow fishing plug lure be made using a mold and liquid plastic? Outline the steps and develop a prototype.

4. How was the original man-made plastic, celluloid, made? Was it a simple process? Could it be done safely in the industrial arts laboratory? Consult your teacher.

5. Could liquid plastic "cloisonné" be done over a curved or cylindrical surface? This could produce a strikingly beautiful lamp shade or mood lamp.

FOR MORE IDEAS AND INFORMATION
Books

1. Cherry, Raymond. *General Plastics Projects and Procedures.* Bloomington, Ill.: McKnight & McKnight Publishing Co., 1967.

2. Cope, D. W., and J. O. Conaway. *Plastics.* S. Holland, Ill.: Goodheart-Willcox Co., Inc., 1966.

3. Lappin, A. R. *Plastic Projects and Techniques.* Bloomington, Ill.: McKnight & McKnight Publishing Co., 1965.

4. Edwards, Lauton. *Industrial Arts Plastics.* Peoria, Ill.: Chas. A. Bennett Co., Inc., 1964.

5. Steele, Gerald L. *Fiber Glass—Projects and Procedures.* Bloomington, Ill.: McKnight & McKnight Publishing Co., 1962.

6. Swanson, Robert S. *Plastics Technology, Basic Materials and Processes.* Bloomington, Ill.: McKnight & McKnight Publishing Co., 1965.

Booklets

1. *The ABC's of Modern Plastics.* Union Carbide Corp., Plastics Products Division, 270 Park Ave., New York, N. Y. 10017

2. *How to Build a Durable Raft.* Zonolite Division, W. R. Grace and Co., Inquiry Dept., 135 S. LaSalle St., Chicago, Ill. 60603

3. *Rubber* (teacher's manual). Firestone Tire and Rubber Co., Public Relations Dept., 1200 Firestone Parkway, Akron, Ohio 44317

4. *The Story of the Plastics Industry.* The Society of the Plastics Industry, Inc., 250 Park Ave., New York, N. Y. 10017

5. *The Story of the Tire*. Goodyear Tire and Rubber Co., Public Relations Dept., Akron, Ohio 44316
6. *This Is Fiberglas*. Owens-Corning Fiberglas Co., Public Relations Dept., Box 901, Toledo, Ohio 43601
7. *Wonder Book of Rubber*. B. F. Goodrich Co., Public Relations Dept., 500 S. Main St., Akron, Ohio 44316

Chart

1. *How Fiberglas Is Made*. Owens-Corning Fiberglas Co., Public Relations Dept., Box 901, Toledo, Ohio 43601

Film

1. Modern Talking Picture Service, Inc., 3 E. 54th St., New York, N. Y. 10022. Send for list of films.

GLOSSARY—Plastics

Adhesives cements that bond pieces together.

Bagmolding the use of an inflatable rubber bag to apply pressure to the laminates while curing.

Bakelite a plastic made from phenolic resins; developed by Dr. Leo Baekeland in 1909.

Blowmolding the use of air pressure instead of a vacuum as in vacuum molding.

Calender a machine that converts liquid plastic into sheets.

Catalyst a chemical agent that assists a chemical reaction without entering into the reaction itself.

Cellophane the original man-made plastic wrap.

Celluloid the original man-made plastic, made of wood pulp and nitric acid; also called nitro-cellulose.

Cellulosics the group of plastics made from nitro-cellulose.

Cloisonné a method for decorating bronze adapted to plastics. The design is laid out with wire, usually brass, copper, or aluminum. Liquid plastic is poured into the individual compartments. When set, the surface is polished level with the wire.

Cohesives solvents that dissolve the surfaces of pieces to be bonded together.

Compression molding process in which the melted plastic is forced into the shape of the mold halves, as they are brought together.

Dip molding a solid mold is dipped into liquid plastic. The coating when removed is the finished product.

Drape mold a convex mold over which a heated sheet of plastic is laid to take the form of the mold.

Extrusion a molding process in which liquid plastic is forced through a die in continuous length.

Fabricating to make an article from separate pieces of material welded or cemented together.

Foam forming beads of polystyrene fill the mold cavity. When the mold is heated the beads expand and fuse together.

Gel coat the first coat of liquid plastic in fiber glass lamination.

Granules small pieces of plastics materials fed into machines to produce finished products.

Injection a molding process in which plastic pellets are melted and the liquid forced into the mold cavity by pressure of a ram.

Laminates plastics materials built up by layers, such as fiberglass.

Molecule the smallest particle of a material that has all of the qualities of the material.

Monomers a term for the smaller plastics molecules.

Polymerization a chemical process in the manufacture of plastics in which small molecules are joined together to form larger ones.

Polymers the general term for the larger molecules of plastic materials.

Slush casting the use of a heated mold into which liquid plastic is poured. When poured out, a layer adheres to the cavity forming the products.

Synthetics man-made materials.

Thermal welding the bonding of plastics by softening the surfaces with heat. They are pressed together and cooled.

Thermoplastics a group of plastics, which soften with heat.

Thermosets a group of plastics, which harden when heated.

Transfer a molding process in which the liquid plastic is forced into the mold cavity by a plunger.

Vacuum molding the use of the pressure of the atmosphere to form heated sheet plastic.

Leather 9
Technology

Leather was among the first materials used by man. Its first use was in the form of animal skins rather than leather as we know it today. Early man used sharp edged shells or stone knives to skin the animals. Without a knowledge of tanning, his skins were hard and stiff when dried. Eventually he discovered that chewing and beating them with clubs made them softer.

One of the most significant of the many early inventions was the horse collar. Until this, the full pulling power of a horse could not be obtained. Leather was used for these as well as for saddles. Such leather was possible; we can call it harness leather because of the discovery of tanning. This is the process for converting hides and skins to leather. It was probably first used in that part of the Middle East where Bible history was made. The early Jews and Egyptians are known to have tanned leather. With this development leather found many uses. It was made into suits of armor, sandals, leggings, and for covering canoes. It was even used as money.

Fig. 9–1. Shoes around the world. These types are in use today.

Caribou boot, or Muckluck. Eskimo

Woven bark shoe. Russia, Bulgaria

Carved wooden sabot. Holland, Belgium

Wood bukia with leather or plastic toe. Philippines

Knob sandal. Malaysia

Goatskin "mule." Turkey, Pakistan

Cowhide huarache. Mexico

American Indian moccasin

LEATHER SHOES

When we think of leather it is often in connection with shoes. Shoes were a rather late development among leather products. Without leather foot protection the Greeks could not have conquered the Middle East, and the Romans Europe. Land travel was largely by foot in those days. The first shoe was probably a skin wrapped around the foot and tied. Eventually came the sandal and then the legging of straps for protection in battle. When the legging idea was eventually added to the sandal, a boot appeared. Boots were used for centuries before the shoe, or low boot, became popular. In America the first shoemaker was Thomas Beard, who arrived on the Mayflower in 1629. He set up his shop in Plymouth, Massachusetts, where the community paid him an annual salary of fifty pounds.

Fig. 9–2. Leather products are usually made from pieces sewed, laced, or cemented together. The parts of a baseball glove are shown here. How would you like to put this together?

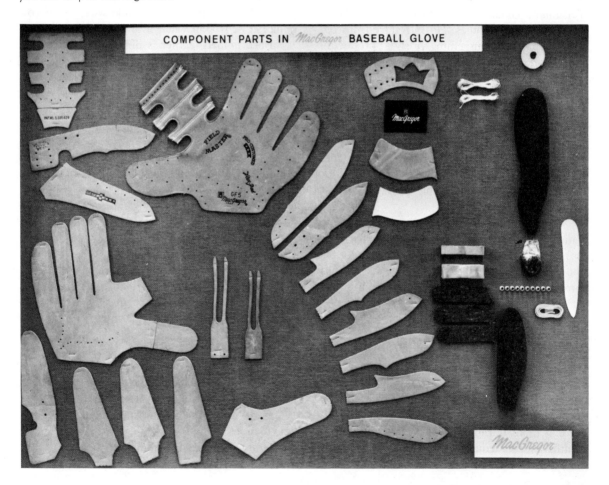

COMPONENT PARTS IN *MacGregor* BASEBALL GLOVE

The early shoemaker carved his own lasts of wood. A last is a mold for a foot over which the leather is fitted into a shoe. A child's shoe was usually made several sizes too large. In winter he wrapped his feet in woolen cloth to fill out the shoes and keep his feet warm. It was not until about 1800 that shoes were first made in pairs in this country. The idea of a left and a right shoe came from England.

The first shoe factory in the States is thought to have been set up at Lynn, Massachusetts, in 1636, by Philip Kirkland. Others were established at about the same time in other of the colonies. These early factories employed numbers of shoemakers. Each made the complete shoe. In 1750 a Welshman, Thomas Dagys, brought a revolutionary idea to the colonies. He introduced the concept of parts manufacture to shoe production. Each worker now made quantities of the same part. Others assembled them. It was not until a century later that the sewing machine invented by Elias Howe was used for sewing shoe uppers.

You and I can take shoes for granted. But to millions of people around the world they are highly prized. A pair of shoes for them is a symbol of status and refinement. It is a great thrill for such boys and girls to be able to have a pair of shoes to wear to school.

Plastics and rubber have replaced leather for many uses. Leather is still preferred for fine shoes, cases, upholstery, and certain items of clothing. The advantage of leather for shoes is its ability to "breathe." It is porous enough to allow the penetration of air but dense enough to shed water.

About Leather

Leather tanning began with the discovery that hides could be treated with a solution of tree bark and water. It was the tannic acid in the bark that reacted with the gelatin of the skin to toughen and make it more durable. The process usually took months and sometimes years. Today the tanning process is completed in weeks. Tanning is now a chemical process. Different kinds of leather require different chemical treatment. The first steps in leather manufacture include cleaning and dehairing the skins. Then the natural oils are dissolved out and replaced with the chemicals. Finally the skin is stretched, dried, and sometimes mechanically worked to soften it. Following this it may be grained and given surface texture and then dyed. Some cowhide is split after tanning. Run through a splitting machine, two or more full size skins are cut as layers from the original.

Kinds of Leather. Leather is typed by its source as well as by the tanning process. There is cowhide and horsehide, and calf, sheep, lamb, goat, kid, pig, snake, and alligator skins for example. It is usually called a hide when from cow or horse. It is skin when it is from other sources.

Bark tanned leather is different from chrome tanned. The latter is denser, more durable, and more nearly waterproof. Bark tanned leather is very porous and absorptive of water. For this reason we can shape, tool, and carve it readily. Following is a list of the common leathers and leather classifications.

Alligator: genuine alligator skin is from the underside of the animal. It has bold, checked grain pattern.

Alligator calf: this is calfskin grained to simulate alligator.

Back: a side usually of cowhide with the belly cut off; choice quality leather.

Belly: leather from the underside of the animal. Generally lower grade than backs.

Calf: skin of a calf; fine quality and expensive. When chrome tanned it is used for dress shoes and fine purses, wallets.

Chamois: split from the flesh side of heavy sheepskin.

Chrome tanned: leather tanned with chromium salts. Very durable and water resistant.

Cordovan: horsehide, commonly used in man's heavy shoes.

Grain leather: the grain side of leather is the hair side.

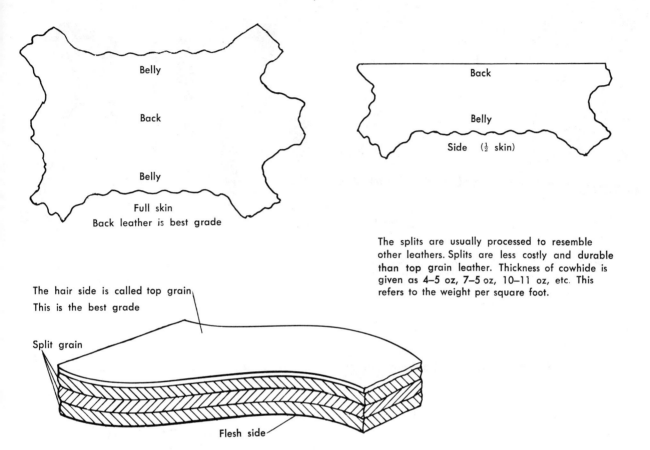

Belly

Back

Belly

Full skin
Back leather is best grade

Back

Belly

Side   (½ skin)

The splits are usually processed to resemble other leathers. Splits are less costly and durable than top grain leather. Thickness of cowhide is given as 4–5 oz, 7–5 oz, 10–11 oz, etc. This refers to the weight per square foot.

The hair side is called top grain. This is the best grade

Split grain

Flesh side

Fig. 9–3. Cowhide leather grades.

Cowhide: bark tanned cowhide is the favorite for tooling and carving. Chrome tanned is used for shoe soles, harness, safety belts for linemen and lumberjacks and other heavy duty articles. Bark tanned is also known as tooling cowhide. It is available in weights of 2 to 10 ounces per square foot.

Full grain means the full leather, only the hair has been removed.

Top grain is the hair side layer after the flesh side layers have been split off.

Kid: fine quality leather from young goats.

Kip: one side of a large calf skin.

Morocco: goatskin originally from Morocco.

Pecca pig: simulated pigskin made from lamb.

Pig: pigskin, usually soft and pliable with distinct hair follicles.

Plivers: the grain side leather split from sheepskin.

Side: one half hide or skin.

Skivers: very thin top grain split from sheepskin. Used for cemented linings.

Split: the under layer when split from the grain side. Often called *split grain.*

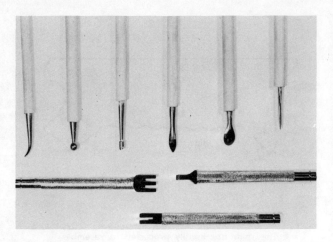

Fig. 9–4. Some typical modeling tools are shown above. They are used for leather tooling. The thonging chisels below are used to punch holes for flat lacing.

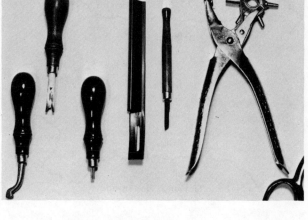

Fig. 9–5. From left to right these tools are: edge creaser, edge beveler, background stippler, skiver knife, mat knife, and hollow tube punch.

Steerhide: lighter than full grain cowhide but heavier than kip or calfskin.

Suede: a finish given to leather on which a nap has been raised by machine. Usually made from sheepskin.

LEATHER TOOLS

The following are essential hand tools in a leather craftsman's kit.

Tracer: This is for tracing designs onto leather and for fine detailing in leather tooling.

Modeler: Any one of several different tools used for decorative leather tooling.

Mallet: A mallet or soft-faced hammer is used for stamping and installing rivets and snaps. A ball peen or claw hammer is never used.

Leather stamps: Punch type tools used to stamp decorative designs into leather.

Swivel knife: The cutting tool for leather carving. The blade can be turned in the handle while cutting.

Edge creaser: Rounded edges and corners are made on thick leather with this tool. The tool is pressed and rubbed back and forth on half-dry leather.

Revolving punch: It punches several sizes of holes.

Thonging chisel: A punch type tool with cutting edges arranged as on a fork. It is used for making holes for flat lacing.

Mat knife: A heavy handled knife with a replacement blade, often of a razor type.

Straight edge: For marking and cutting straight lines a wood straight edge is recommended. Metal can stain the grain side of the leather.

Edge beveler: The cutting tool for beveling the edges of leather before cementing together.

Skiver knife: Skiving is a cutting process used to make leather thinner over a wide area. The knife has a long thin replaceable and very sharp blade.

Awl: A needlelike punch set in a handle. Used for making small holes and for tightening lace.

Fig. 9–6a. A close-up photo of the ends of several saddle stamps. There are many other styles. From left to right these are: smooth pear shader, ribbed pear shader, ribbed camouflage, plain camouflage, and beveler.

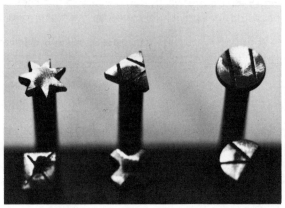

Fig. 9–6b. (top) You can make your own saddle stamps. This close-up photo shows the designs cut into the heads of 12d common nails. Only a file and a grinder are needed. Polish the faces so as not to scratch the leather. **Fig. 9–7.** (bottom) From left to right: a swivel knife, awl, lacing needle, and the last two are a snap fastening set.

Leather shears: A heavy shears with cutting edges especially ground to cut leather evenly. Ordinary shears leave a fuzzy edge.

Hole spacer: A revolving wheel with points on the rim. It spaces marks for single hole punching and for hand sewing.

Snap fastener: There are two parts to this tool designed for installing snap fasteners.

Eyelet setter: A punch with a conical point for spreading the end of an eyelet.

Miscellaneous tools and accessories: An assortment of other helps is needed: brushes, sponge, cutting board of soft wood, tooling board with a hard smooth surface.

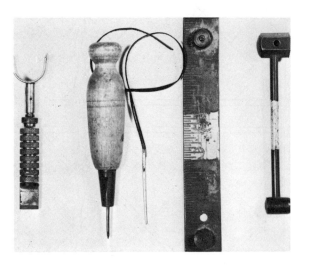

Design in Leather

Several principles and suggestions apply to the design of leather projects.

1. Determine the character or feeling of the piece. If it is to be rough, heavy, and primitive, it should be very simple in form and decoration. Tooling and carving may actually be out of place. If it is to be more refined in feeling, careful precise workmanship is in order.

Fig. 9–8. Decorative patterns with straight lines. Such simple ideas are appropriate on leather and clay. Use the tracer and a straight edge.

The spacing makes a difference

Cross hatching

Diamond

Woven pattern over and under

2. Let the form originate with the function. For example, if you are designing a wallet, decide first on what you expect it to contain. This will help to determine its size, shape, and construction.

3. Select the appropriate leather. Different kinds and weights of leather are suited to different purposes. For example, chrome tanned leather is better for moccasins than is bark tanned. Whatever the project, let it look like leather when it is completed.

4. Consider the texture and color. If you will use a simulated grain or dyed leather, it should be appropriate to the project.

5. The decorative design should be appropriate to the piece. Cowhide, for example, does not need to be covered with tooling or carving to be good or right. It can be left plain and be even more effective. Simple line designs can be in better taste than carved flowers and leaves. Beginners often tend to overdecorate leather.

6. Select an appropriate binding. Your choices may be from several stitches of hand lacing, hand sewing, riveting, or cementing. Because hand lacing is common does not mean that it is best for your project.

LEATHER PROJECT FABRICATION

To fabricate a leather project calls for a pattern. When several parts are required, cut and fit full-size paper pieces together. This will help you check out the idea. It enables you to make accurate allowances for bends and folds. Once you have these parts properly shaped, decide on any surface decoration. Cut out the pieces. Lay the patterns on the grain side of the selected leather. Arrange them to cut most economically. Trace around each with the tracing tool and cut them out. Use either the mat knife or the leather shears. To cut with the knife, place the leather on the cutting board. The knife should cut clear through the leather in one stroke.

Edge Beveling. When two or more pieces of leather make too much thickness for lacing or

Fig. 9–9. The edge beveler actually cuts a chamfer on the edge of heavy leather. An edge creaser rounds the edge without cutting.

sewing, the edges may be beveled. Use the special edge beveler knife.

Cementing. The parts are held in place for lacing or sewing by rubber cement. Coat each surface and let it dry. Press them together for a flexible joint.

Trimming. Final trimming of the edges if needed is done before marking or punching holes for lacing.

Lacing. Decide on lacing or sewing. If lacing, select the appropriate stitch. Mark a light line with the tracer about 1/8″ in from the edge. For lacing, the holes should be punched with either the thonging chisel or the revolving punch. The choice depends on the lace selected.

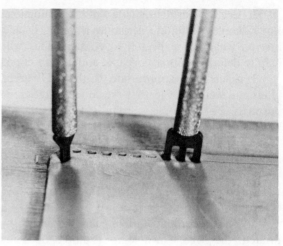

Fig. 9–10. (top) The skiver tapers the edge. This is usually done on the opposite sides of two pieces of leather to be cemented together.

Fig. 9–11. (top) Holes for flat and beveled edge lace are punched with thonging chisels. The single prong is used at corners. Place the last prong of the three- or four-pronged chisel in the last hole punched to get even spacing.

Fig. 9–12. (left) The hollow tube punch is used to punch holes for heavy lace. It is also used to install snap fasteners and other fittings.

Leather Tooling. Tooling is a process of modeling a decorative design into the surface of leather without cutting it. It can be used on any weight of leather. The idea is to press part of the detail firmly down. This leaves some parts raised and gives a low relief effect. Thick leather can be depressed deeper than thin so the modeling is more distinct.

Follow these steps:

1. Moisten the leather. A quick dip in and out of a pan of water is usually enough. Let it soak through for a few minutes. Blot up any excess water. When the leather is properly moist a deep impression can be easily made. It tends to remain.

If too dry, the leather resists the tool. When too wet, water oozes out and the mark tends to float away.

2. Lay the leather with grain side up on the tooling board, and trace the decorative pattern. Do not press hard enough to cut through the tracing paper. Never use carbon paper. The ink is permanent.

3. Retrace the outline, using the same tracing tool. Press firmly and slide the tool back and forth in the grooves. This burnishes and darkens the lines. A design done only in lines can be very appropriate and attractive.

4. Press down the background areas with the spoon end of the tool or with another modeler. Go over the entire area. When the leather is about half-dry repeat this. Rub and press to sharpen up the detail.

Fig. 9-13. (bottom) Using the tracer tool to deepen a traced design in the leather.

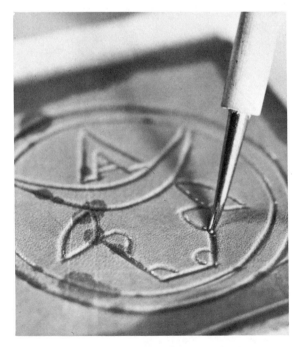

Fig. 9-14. (top) The spoon-tipped modeler is used to sharpen up detail in tooling. It presses down the background.

Leather Stamping. Damp leather can be decorated with stamped designs. These are available in many shapes and sizes. You can make your own if you wish. The stamps can be carved in the ends of hardwood blocks. To experiment, get a stick of hard maple or cherry about ⅜″ to ½″ square. Cut it into three-inch lengths. With a fine file or sharp knife, carve designs into one end of each piece. Place the leather on the tooling board. Set the stamp and rap it sharply with the mallet. The impressions can be repeated as for an all-over pattern. They can be used as borders. When the leather is dry, trim the edges to fit the pattern.

Leather Carving. In this process, the grain side of the leather is cut but not as in carving wood

or clay. Here are the steps. Experiment with designs, cutting, and stamping on scrap leather. Once done on your project they are permanent.

1. Trace the design on the damp leather as for tooling.

2. Make the cuts. With the swivel knife cut through the top layer of leather as you retrace the outline.

3. Depress the background of the design with appropriate stamps and mallet. This leaves some parts raised to give a relief effect.

4. Sharpen up the modeled detail with a small bevel stamp or modeling tool.

Lacing Stitches. There are many different lacing stitch materials. Some are of leather and some are plastic. They come in various widths and colors. The lace you select should suit the stitch to be used.

Sewing. For some articles sewing is more suitable than lacing. Light weight leather can often

Fig. 9–15. (top) Saddle stamps make interesting permanent impressions in damp leather. Cowhide is preferred. Your problem is to compose good patterns appropriate to your project. A soft-faced mallet is used to drive the stamps.

Fig. 9–16. (bottom) Some designs for carving or tooling.

be sewn on a home sewing machine. Heavy leather can be sewn by hand. This is slow but strong and attractive. Use the kind of thread with which the shoe repairman usually sews on soles. It is waxed linen.

Finishing. The natural color of leather is often preferred for the project. To protect it, apply several coats of natural shoe wax or a special leather finish.

To Dye Leather without streaking is tricky. There are special finishes, which include both color and wax, that go on smoothly. Dye penetrates the fibers of the leather. It may be applied by spray, sponge, brush, or swab. Several coats are usually needed. Apply them in different directions. Always do this on bone dry leather. Neat's-foot oil is used on leather to soften it and make it more water resistant. It soaks in readily and is especially useful on boots, saddles, and bridles.

Cleaning Leather. Natural leather should be cleaned before finishing. An oxalic acid solution is used. Dissolve one teaspoon of oxalic acid crystals in a pint of warm water. Use a sponge to apply it, rubbing gently. When the leather is clean, rinse off the acid with a clean sponge.

Findings. These are devices such as key holders, snaps, eyelets, buckles, rings, and purse latches used on leather projects.

Fig. 9–17. The carving cut with the swivel knife.

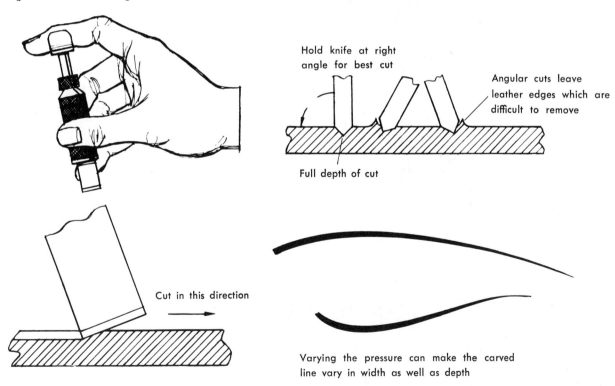

Hold knife at right angle for best cut

Angular cuts leave leather edges which are difficult to remove

Full depth of cut

Cut in this direction

Varying the pressure can make the carved line vary in width as well as depth

Fig. 9-18. Hand lacing stitches for bevel edge lace.

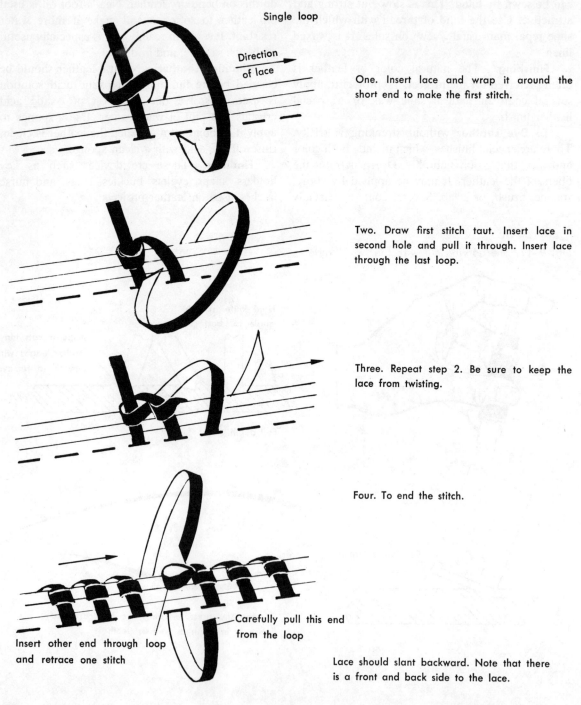

Single loop

Direction of lace

One. Insert lace and wrap it around the short end to make the first stitch.

Two. Draw first stitch taut. Insert lace in second hole and pull it through. Insert lace through the last loop.

Three. Repeat step 2. Be sure to keep the lace from twisting.

Four. To end the stitch.

Carefully pull this end from the loop

Insert other end through loop and retrace one stitch

Lace should slant backward. Note that there is a front and back side to the lace.

Single whip

Pull free end down into joint and clip off

Single cross
Lace every other hole in one direction
Then reverse and lace in the other holes

Double cross
One. Insert lace through first hole and go over the edge
Two. Through third hole
Three. Cross lace over edge and through second hole
Four. Cross over edge and through fourth hole, and repeat

Around a corner
One. Cut off the corner
Two. Take two stitches in the corner hole

Fig. 9–18a. Hand sewing stitches for leather thong and linen cobbler's thread.

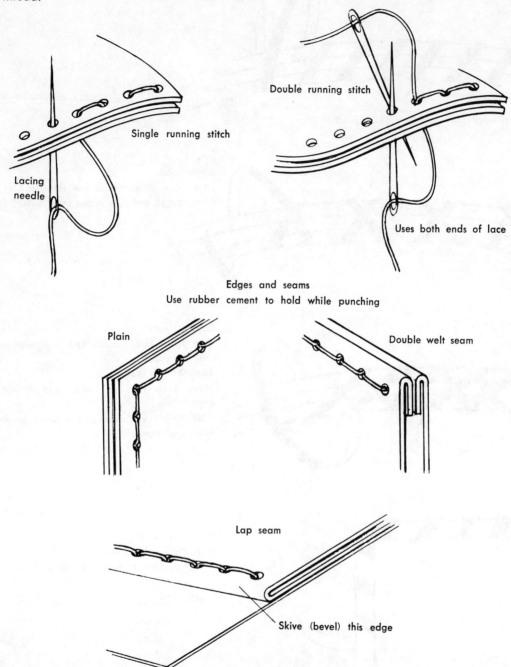

Single running stitch

Lacing needle

Double running stitch

Uses both ends of lace

Edges and seams
Use rubber cement to hold while punching

Plain

Double welt seam

Lap seam

Skive (bevel) this edge

Fig. 9–19. Installing snap fasteners.

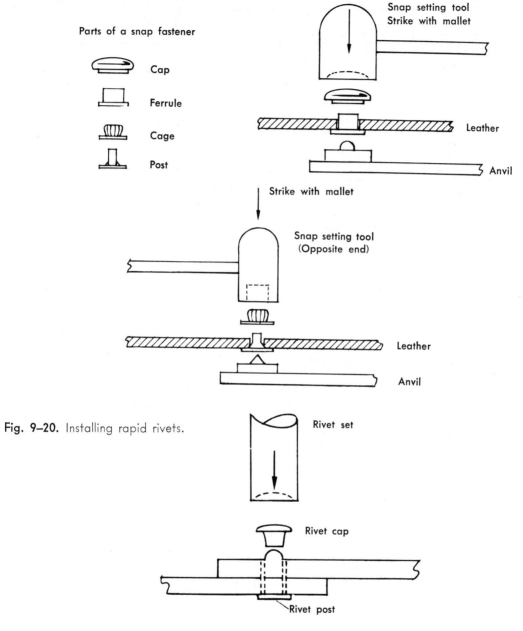

Parts of a snap fastener

Cap

Ferrule

Cage

Post

Snap setting tool
Strike with mallet

Leather

Anvil

Strike with mallet

Snap setting tool
(Opposite end)

Leather

Anvil

Fig. 9–20. Installing rapid rivets.

Rivet set

Rivet cap

Rivet post

Punch hole slightly larger than rivet post.
Press cap into place.
Place rivet set on cap and strike sharply with a mallet.

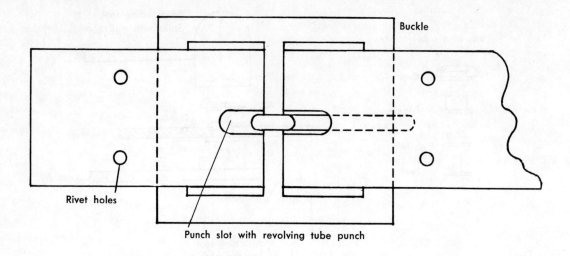

Buckle

Rivet holes

Punch slot with revolving tube punch

Fig. 9–21. Installing a buckle.

LEATHER AND OCCUPATIONS

Leather products manufacturing is one of the smaller industries in this country. It is small as compared to metals, plastics, and graphic arts, for example. Shoe production consumes most of the leather used. Leather has had stiff competition from other materials such as plastics, rubber, and paper. Increasing amounts of leather and leather goods are being imported. This affects the occupational opportunities here.

There are two main areas of occupations in leather manufacturing. The first is the production of leather itself, or tanning. The second is the manufacture of products from leather. There was a time when the tannery was a dark, foul-smelling place to work. Today it is much different. Machines do most of the work. Cleanliness and ventilation have lessened the odor problem. The future of the industry will depend on how leather can become more competitive with other materials. On-the-job training is the practice in tanneries and leather goods manufacture. Positions in management, chemistry, and product design generally call for college preparation.

At present leather is enjoying a new acceptance. It is an "in" material among teenagers and young adults. Handmade coats, jackets, blouses, bags, belts, shoes, boots, sandals, and moccasins are in demand. This return to handcrafted leather goods is interesting. It may not be a fad. A century ago all leather goods were handmade in industry. Can you explain this preference by teenagers? We find that handmade goods are a new experience for this group. They have been brought up with the machine-made and want a change. A similar feeling is being expressed for handmade pottery. A handmade pot is prized by its creator and admired by his peers. We suggest that you keep watch on such developments in this country. They may be the forerunners of a great movement. It could be a trend to personal craftsmanship. In this the individual finds satisfaction and fulfillment in creating things by hand in materials.

LEATHER AND ENVIRONMENT

Leather may be considered a byproduct of the meat industry. Raw hides are salted and stored for a time. This permits decomposition of the remaining flesh. Consequently a hide cellar can be pretty smelly. Tanneries were once very odorous because of the green hides going through the aging or curing process. Much of this kind of air pollution has been eliminated from modern tanneries. Discharge of water wastes has been another form of pollution.

With much of our leather being imported the expansion of tannery operations in our country has been limited. Once small tanneries were a common industry in each city. While tannery wastes are decreasing in our country they may be increasing in others. Should we be concerned about pollution problems in other countries? Why not think about this and discuss it in class?

Leather products are considered biodegradable. To the environmentalist this is desirable. Perhaps you have observed that leather deteriorates with age and dampness. This makes it very different from plastics and rubber.

LEATHER AND RECREATION

Leatherwork has been a popular recreational activity ever since the period of the Thirties. These were times when jobs were few and income scarce throughout the country. Crafts centers sponsored by the Federal government were set up in many cities. The object was to train people to work with various materials and to produce saleable items. Leather work was commonly included. Today, as we mentioned earlier, hand leatherworking is enjoying a big boom. There are leathercraft supply stores in cities as well as mail order supply houses.

Part of the appeal of leather work as a hobby is the actual work by hand. This · is the way with leather. There are no machines to do it. Compare this with working with woods and metals. Doing something solely by hand gives one a good feeling. This is much a part of leatherwork. If you want to go all the way, you can tan the leather from the raw skin. (Will your mother let you use the kitchen sink or the bathtub?) The tools needed for leatherwork are few and not expensive. They can be kept in a small toolbox and conveniently stored. Actually you could make most of them in industrial arts.

Genuine leather has a strong appeal over imitations and substitutes. Some suppliers sell scrap leather. These are less expensive. We recommend them for beginners. We suggest that you create your own designs. This includes the article itself as well as any decoration for it. And we remind you that a fine piece of leather work can be spoiled by the decoration applied. Be critical of your efforts. Study the principles of good design and make many drawings. Leather is well adapted to "way out" designing. It is just as suited to the traditional. So, have fun.

TO EXPLORE AND DISCOVER

1. There are two kinds of standard constructions used in sewn shoes: The Goodyear welt and the stitchdown. Find out the difference. Make a cross-sectional drawing of them.
2. Why doesn't chrome tanned leather tool or carve? Prove it.
3. What is rawhide boot lace made of?
4. What is there about leather that makes an old shoe more comfortable than a new one?
5. Find out how to braid four strands of leather lace. Make both round and square braids.
6. How are saddles formed of leather?
7. How are catcher's mitts formed? You may have to write a manufacturer for this information.
8. Visit a harness maker if there is one near you. How does he make harness and bridles to fit? Do you know that the horse population in this country is at an all-time high?
9. What were the so-called wineskins as used in Biblical times?
10. In how many thicknesses is tooling cowhide available? How does thickness relate to use?

FOR GROUP ACTION

1. Design and make a prototype belt including a cast bronze buckle. This should be a neat design that will have sales appeal. Take orders and mass produce them.
2. Design and make a prototype suede vest for students in your industrial arts club. Design a tooled emblem to be sewed on. From this vest make paper patterns for different sizes. Take orders and mass produce them.
3. Design and mass produce an industrial arts souvenir. For example, this could be a keyholder, pendant, notebook cover or other simple products. Can you

make a die to cut them out so that they will be uniform? How about the insignia cut into a wood block? This could be a sort of embossing plate for impressing the design in damp leather. A hand printing press or a vise could be used.

4. Organize a neighborhood shoe collection project. Collect used but good shoes and deliver or send them to an orphanage.

5. Conduct classes in leatherworking in a rest home, hospital, orphanage, or elementary school class.

FOR RESEARCH AND DEVELOPMENT

1. Compare the water absorbency and wet tensile (stretch) strength of bark-tanned and chrome tanned leather of the same thickness and width.

2. Compare the adhesive qualities of different surface finishes on tooling cowhide. For example, try lacquer, enamel, acrylic plastic. Which is more flexible? Do they bond or do they peel?

3. Suppose you want to make a quantity of cowhide tiles to cement to a table top. They must be of uniform size. Could you design and make a stamping die to cut them? Hint: the end grain of a block of hard wood makes an excellent surface on which to use such a die. It can be driven through the leather and slightly into the wood to produce a clean cut.

4. Have you ever thought of designing and making a pair of your own shoes? Work out the top, front, side, and rear view drawings full size. Perhaps you can use your feet as the lasts.

5. Make samples of ten different stitches for use with leather lacing. Can at least one of these be your own invention?

FOR MORE IDEAS AND INFORMATION

Books

1. Aller, Doris. *Leather Craft Book*. Menlo Park, Cal.: Lane Publishing Co., 1957.
2. Klingensmith, W. P. *Leatherwork*. Milwaukee: Bruce Publishing Co., 1958.
3. McCoy, R. A. *Basic Leathercraft*. Austin, Tex.: The Steck Co., 1961.
4. Stohlman, A., A. D. Patten, J. A. Wilson, *Leatherwork Manual*. Ft. Worth, Tex.: Tandy Leather Co., 1961.
5. Zimmerman, Fred W. *Leathercraft*. So. Holland, Ill.: Goodheart-Willcox, 1961.

Teacher's Packet

School Program Materials (booklet, transparencies, charts). Tandy Leather Company, Ft. Worth, Texas 76101

Booklets

1. *Adventures in Leather.* Tandy Leather Co., Advertising Dept., 1001 Foch St., Ft. Worth, Texas 76101
2. *Basics for Arts and Crafts.* Best Foods, Division of Corn Products Co., Home Service Dept., 717 Fifth Ave., New York, N.Y. 10017
3. *The History of Shoes.* Kiwi Polish Co., 2 High St., Pottstown, Penna. 19465
4. *Leather in Our Lives; Teen Tips on How to Buy Shoes.* Leather Industries of America, 411 Fifth Ave., New York, N.Y. 10016
5. *Let's Look at Leather.* Wolverine World Wide, Advertising Dept., 9341 Courtland Drive, Rockford, Mich. 49341
6. *More and Better Shoes for America.* U.S.M. Corporation, Advertising Dept., 140 Federal St., Boston, Mass. 02110
7. *The Story of Leather and the Techniques of Tanning.* Eberle Tanning Co., Westfield, Penna. 16950
8. Student Kit (booklets on leather and shoes). Brown Shoe Co., Public Relations Dept., 8300 Maryland Ave., St. Louis, Mo. 63105

GLOSSARY—LEATHER

Alligator grain An embossed pattern, usually on calfskin, resembling alligator. The same method is used for lizard and snake grains.

Awl A needlelike tool for punching small holes.

Back The center section of a hide of leather, from the upper back of the animal.

Bark tanning A process used for tanning leather, originally using a solution made from tree bark and water that provided tannic acid. Vegetable tanning is the modern term for essentially the same process.

Belly The leather from the under- or bellyside of the animal.

Binding post A twopiece screwtype fastener used with keyholders and albums.

Buckskin Leather made from deer and elk skins; it is very soft and pliable.

Capeskin A superior grade of sheepskin leather made from the skins of South African sheep.

Chamois A split leather made from the flesh side of sheepskin.

Chrome tan A process for tanning leather using chromium salts; produces a nearly waterproof leather.

Carving A decorative process in which outlines for designs in heavy cowhide are cut partially through the leather. Detail is added by stamping.

Cordovan Leather made from horsehide, very tough and durable.

Drive punch A hollow tube punch driven through the leather with a mallet.

Edge beveler A hand tool for chamfering the edge of heavy leather.

Edge creaser A hand tool for rounding the edge of heavy leather, adding a border.

Eyelet A metal reenforcement for small holes in leather.

Emboss The process of raising certain parts of a decorative design by depressing the surrounding area.

Fid An awl used for stretching holes when lacing.

Florentine lace A wide thin leather lace.

Grommet A large eyelet for reenforcing holes.

Hide The leather made from the skins of large animals, as cows, horses.

Kip Leather made from the skins of the very young of large animals.

Modeler A hand tool for impressing designs in leather; available in many shapes.

Morocco Goatskin leather used chiefly for linings not also available for tooling.

Oxalic acid A chemical cleaner for leather. One tablespoon of the crystals is dissolved in one quart of water. The solution is sponged on and rinsed off.

Peccary Leather made from the skin of the wild boar found in Central America.

Pigtex Sheepskin grained to resemble pigskin.

Plivers The grain side of split sheepskin.

Revolving punch A hand operated tool for punching holes of different sizes.

Skin A term usually applied to the leather made from the pelts of small animals; see Hide.

Skive To taper the edge of leather by shaving off thin layers.

Skiver Split sheepskin used for lining. Also the tool for skiving leather.

Snap A fastening device for holding two pieces of leather temporarily.

Split leather The thickness of the leather is cut into two or more layers. The under layers are the "splits." The hair side is the top grain.

Spots Ornaments usually of polished brass, sometimes jewelled; sometimes used on belts, dog collars.

Stamps A punchtype tool used with a mallet to impress a design in thick leather.

Stipple To stamp or press backgrounds in leather when tooling or carving. The stippled surface is usually rough textured.

Suede A soft, fine nap on leather. A type of leather having this nap; usually of sheepskin leather.

Swivel knife A hand tool for cutting through the top layer of the leather as a part of the carving process.

Tan The process of converting hides and skins into leather.

Thonging chisel A chisel for cutting slots in leather for lacing.

Tooling A hand process for decorating the surface of leather. Parts of the design are raised as backgrounds are depressed. The leather is not cut as in carving.

Top grain The split from the hair side of the leather. It is generally the most expensive and durable of the splits.

Vegetable tan A process for tanning leather in which the leather remains absorbent of water, as in tooling and carving leathers.

Measurement Aids

FORMULAS

Circles
Diameter $= 2$ R (Radius)
Circumference $= 3\frac{1}{7}$ (or 3.1416) $\times$ D (Diameter)
Area $= 3\frac{1}{7} \times R^2$ (or R $\times$ R)

Areas
Square $= L \times L$ (Length)
Rectangle $= L \times W$ (Length $\times$ Width)
Triangle $= \dfrac{\text{Altitude} \times \text{Base}}{2}$

Volumes
Cone $= \dfrac{R^2 \times 3.1416 \times H \text{ (Altitude)}}{3}$
Cube and rectangular box $= L \times W \times H$ (Height)
Cylindrical tank $=$ Area of bottom $\times$ Height
Sphere $= \dfrac{R^3 \times 3.1416 \times 4}{3}$

WEIGHTS

Troy Weight
(Used in weighing gold, silver, jewels)
24 grains (gr.) $= 1$ pennyweight (pwt)
20 pwt $= 1$ ounce (oz.)
12 oz $= 1$ pound

Avoirdupois Weight (U.S. System)
16 oz $= 1$ pound (lb)
100 lb $= 1$ hundredweight (cwt)
2000 lb $= 1$ short ton
2240 lb $= 1$ long ton
1 lb Avoirdupois $= 7000$ grains Troy

MEASURES

Linear Measure
12 inches (in.) $= 1$ foot (ft)
3 ft $= 1$ yard (yd)
5½ yds $= 1$ rod
5280 ft $= 1$ statute mile

Mariners' Measure
6 ft $= 1$ fathom
6,076.10 ft $= 1$ nautical mile (1 knot $= 1$ nautical mile per hour)

Square Measure
144 square inches (sq in.) $= 1$ square foot
9 sq ft $= 1$ sq yard
43560 sq ft $= 1$ acre
640 acres $= 1$ sq mile

Circular Measure
60 seconds $= 1$ minute
60 min $= 1$ degree
90 degrees $= 1$ quadrant
360 degrees $=$ a circle

Cubic Measure
1728 cubic inches $= 1$ cubic foot
27 cu ft $= 1$ cu yd
231 cu in. $= 1$ gallon

U.S. Fluid Measures
1 fluid ounce (oz) $= 2$ tablespoons (T)—measuring type spoons
1 tablespoon $= 3$ teaspoons (t)
16 tablespoons $= 1$ cup (c) measuring cup holds 8 ounces
4 gills $= 1$ pint
1 pint (pt) $= 16$ ounces
2 pints $= 1$ quart
4 quarts $= 1$ gallon
1 cubic foot of water weighs 62.42 pounds
1 gallon of water weighs 8.3453 lbs at 60° F
1 cubic foot of water $= 7.481$ gal
1 gallon of water $= 231$ cubic inches

METRIC EQUIVALENTS

Linear Measure
1 centimeter (cm) $= 0.3937$ in.
1 inch $= 2.54$ cm
1 meter $= 39.37$ in.
1 mile $= 1.6093$ kilometers (km)

Square Measure

1 sq meter = 1.196 sq yd
1 hectare = 2.47 acre

Volumes

1 cu cm = 0.061 cu in.
1 cu in. = 16.39 cu cm
1 liter liquid = 1.0567 quart

Weights

1 gram = 0.03527 oz
1 oz = 28.35 gr
1 kilogram = 2.2046 lbs
1 metric ton = 2200 lbs

FRACTIONS OF AN INCH IN EQUIVALENTS

Fraction	Decimal	Millimeters (1/1000 meter)
1/32	0.0313	0.794
1/16	0.0625	1.588
1/8	0.1250	3.175
3/16	0.1875	4.763
1/4	0.2500	6.350
5/16	0.3125	7.938
3/8	0.3750	9.525
1/2	0.5000	12.700
5/8	0.6250	15.875
3/4	0.7500	19.050
1	1.0000	25.400

WEIGHTS OF COMMON MATERIALS

		Lbs. per Square Foot
Wood		
	Apple	41–52
	Balsa	7–9
	Basswood	20–37
	Birch	32–48
	Ebony	69–83
	Hickory	37–58
	Lignum Vitae	73–83
	Mahogany	41
	Maple	39–47
	White pine	22–31
	Yellow pine	23–37
	Black walnut	40–63
Others		
	Acrylic plastic	74
	Brass	511–543
	Brick	87–137
	Clay	112–162
	Concrete	145
	Cork	14–16
	Glass	150–370
	Gold	1207
	Iron	439–492
	Lead	687
	Silver	662
	Snow	5–12
	Steel	489
	Vinyl plastic	87

Project Ideas

YOU AND YOUR PROJECTS

The project in industrial arts is that activity on which you spend most of your time. It may be a two-dimensional or a three-dimensional thing, using materials. It may be a search, a study for the solution of a problem, which when completed is not a take-home thing. In either case the project is an idea developed into reality. It is an idea, which "turns you on." It is composed of several problems all of which must be effectively solved. It challenges you to do your very best. (Any project that doesn't is a waste of time, money, and talent.) Your very best shows up in the quality of the idea you propose. It appears in the orderly way you go about to develop it. It shows off in the finished work as the example of the best thinking and craftsmanship you are capable of at the time. When you care enough to do your very best, you give yourself a full opportunity to discover and develop your interests and talents.

To Copy or Create?

Should you copy ideas or should you dream up your own? Every time you begin a new project you must answer this question. We assume that you can imagine and create because every student can. We know that creativity is the highest level of one's intellect. It needs to be used to be developed. In general we encourage you to use your own ideas. But we know, too, that there are times when you must be well-prepared in a subject before you can do this. What about clay and electronics, for example? With clay you can do your own thing sometimes in the first project. With electricity you do best to master what is first necessary before you experiment on your own. How can you explain this difference?

Copying doesn't give you a true design experience; this is an experience in original thinking. You can reproduce antiques but this is neither creating nor designing. Suppose you assume that in material things there is always a better way possible. The problem is to find it. You can study any industrial arts project or industrial product and come up with suggestions for improvement.

Sometimes it is more difficult than others and sometimes an engineering background is necessary. The simple round coffee cup after centuries of use is still round because no one has been able to come up with a better idea. Why is a round cup better than square, oval, or other shape? Nevertheless one of the reasons for which our civilization has become great is that its founders were searching for a better way of life and we have kept on searching. In industry no product is good for all time. It is assumed that there is always a better idea. This gives the young engineer and designer a goal and a challenge. This is why they are hired.

THE PROJECT PROCESS

A project that is worth doing is worth doing some thinking about. To cut down on the time you may spend on spinning your wheels, study the following steps in our "project process."

1. Identify the Project

What is it you wish to do?

What is it to do?

What are the problems to be solved in it?

What are the specifications?

Does it really turn you on?

2. Start the Idea Train

Where can you get ideas about such projects?

Books, magazines, catalogs, museums, movies, stores, factories?

Who can you discuss this with?

Teachers, classmates, parents, engineers, designers? How about a brainstorming session in class?

Has it been done before?

If so, where, by whom, when?

Can it be improved upon?

Can you improve it enough to be worthwhile?

3. Organize the Research

Which idea will you select to develop?

Does it need further study? Consultation?

What possible solutions are there to it?

Which seems to be best in terms of your abilities and the available materials, tools, machines?

Do you anticipate any really difficult problems? How can they be solved?

4. Design and Develop the Idea

Sketch, draw, and make models of the idea.

Select the appropriate materials and learn how to use them.

What tools and machines will you learn to use?

Construct the project.

5. Evaluate Your Efforts

Is the project as functional as desired?

Was there a good selection of materials?

Is it economical of construction?

How about safety?

What will it do for the environment?

Is it æsthetically good?

6. The Refinement

What can be done to simplify the idea and the construction?

Can it use less material?

Can it be improved or should you start over?

Where does it need redesign?

How can I know when it is as good as I can make it?

The Ideas Section

The following section is full of ideas for projects. They are only ideas. None has all of the construction answers. We have done this on purpose. When you use one of them you can develop your own version. Then you can say that what you have produced is your own. One of the ideas may suggest something quite different from the original. This is good. It means that your imagination is in top gear. What finally happens to the project is up to you.

SAFETY SENSE

Fitted into this book in many places are blocks of Safety Sense. These are guides to help you work safely and accurately. Safety, you see, is an attitude, a way of thinking. Carelessness, too, is a way of thinking. Most of us have to work at it if we will think and work safely. Here is a list of questions to think about and discuss in class. Perhaps such discussions will help you to draw up a class safety code.

1. Is there a law in your state requiring the wearing of eye protection devices in the industrial arts laboratory? Why is there such a law? Don't people value their eyesight?

2. What causes pain from an injury? What if we had no sensations of pain? What causes the blackening under the nail of a bruised finger? How does a splinter cause infection? Could you consult a doctor about these?

3. Why should there be a "speed limit" on pedestrians in your laboratory?

4. Do safety rules interfere with your freedom? Does an injured finger interfere with freedom?

5. Why should you know a machine well before using it? Machines make sounds. Do any of these suggest trouble or danger?

6. Are there safety and fire hazards in your laboratory? What are they? Are you one?

7. Is it "chicken" to act safely?

8. Why are machines not stupid? (How come only people are?)

9. Is there a difference between the safe and the proper way of operating a machine?

10. What should you do first in case you injure yourself? In case someone else in your class is injured?

If you need some safety signs for your lab, try these or dream up some of your own:

SAFETY IS SENSIBLE; BE SAFETY SMART; BE SAFETY SURE; BE SAFETY SHARP; SAFETY SENSE IS GOOD SENSE.

THE EYE (FINGER) YOU SAVE MAY BE YOUR OWN.

Our Best Wishes

You now have our best wishes for a great experience in industrial arts. Want to send us a picture of a project that you have completed? We'd surely like to see it. Meantime we hope that both you and your teacher are having a "ball" in industrial arts. We think it has to be fun to be good.

Sincerely,

Delmar W. Olson

ABCDEF

THIS IS THE UPPER CASE OF THE ALPHABET.

GHIJKL

MNOPQR

STUVW

XYZ

A GOOD ALPHABET FOR CUT-OUTS

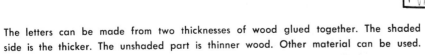

The letters can be made from two thicknesses of wood glued together. The shaded side is the thicker. The unshaded part is thinner wood. Other material can be used.

abedefg

THIS IS THE LOWER CASE OF THE ALPHABET.

hijklmno

pqrstuv

wxyz

1234 56

7890

AN ALPHABET FOR POSTERS AND SIGNS

ABCDEF
GHIJKLM
NOPQR
STUVW
XYZ1234
567890

SOME POSTER IDEAS

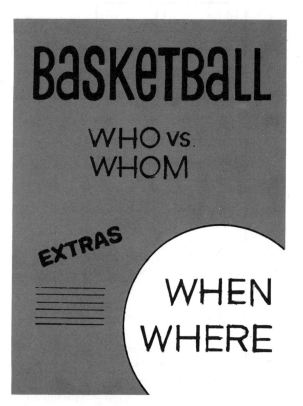

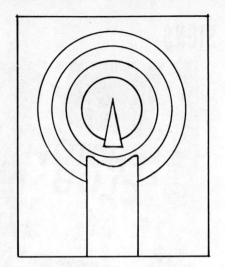

GREETING CARD IDEAS

For silkscreen and block printing.

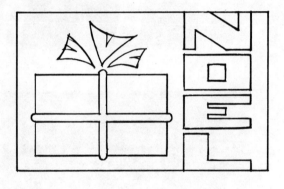

DESIGNING AN INSIGNIA

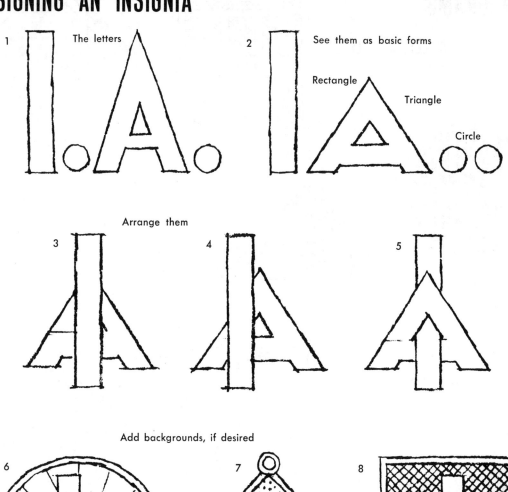

1   The letters

2   See them as basic forms

Rectangle

Triangle

Circle

Arrange them

3

4

5

Add backgrounds, if desired

6

In a wheel

These could enamel on copper, ceramic, inlaid wood, cast plastic, leather.

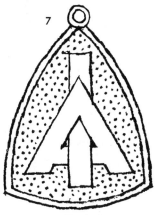

7

Pendant

8

Plaque or tile

MONOGRAMS

GOTHIC LETTER STYLE

ROMAN LETTER STYLE

These monograms illustrate the principle: *Make the design fit the space.* These ideas are suitable for leather, woodcarving, clay. Note the effects of different background textures.

SOME IDEAS FOR LEATHER

A wallet of separate pieces

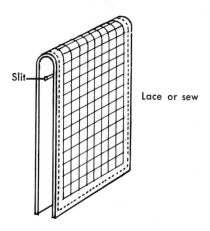

Slit

Lace or sew

Pocket notebook cover

Use lightweight lining. Cement edges only with rubber cement

or, folded

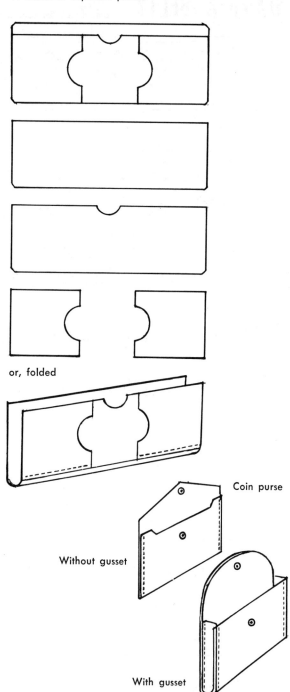

Photo album cover

Coin purse

Without gusset

With gusset

WANT A BELT?

Some ideas for tooled line designs.

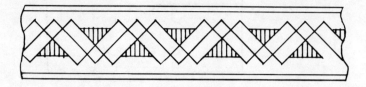

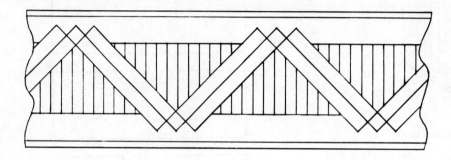

BUCKLES

Cast or cut from heavy brass.

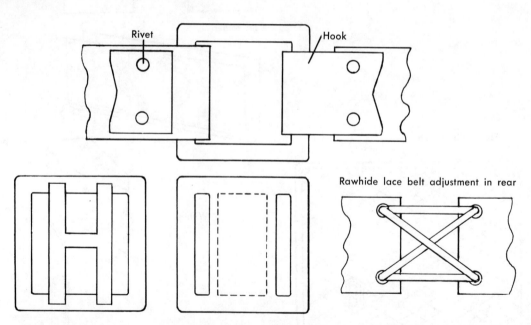

Overlay an initial

Heavy copper with ceramic enameled center

Rawhide lace belt adjustment in rear

STAMPS YOU CAN MAKE

For leather select simple designs that can be made with a file or jackknife in the end of a stick of hardwood.

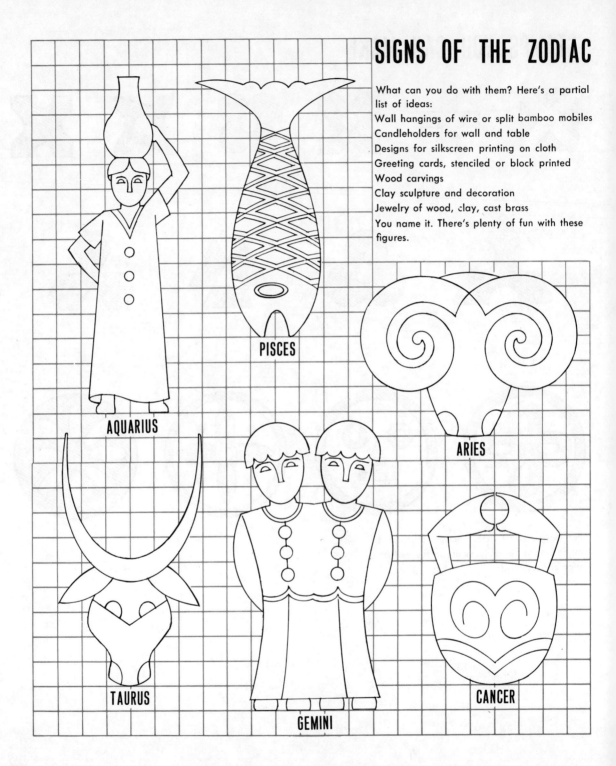

SIGNS OF THE ZODIAC

What can you do with them? Here's a partial list of ideas:

Wall hangings of wire or split bamboo mobiles
Candleholders for wall and table
Designs for silkscreen printing on cloth
Greeting cards, stenciled or block printed
Wood carvings
Clay sculpture and decoration
Jewelry of wood, clay, cast brass
You name it. There's plenty of fun with these figures.

PISCES

AQUARIUS

ARIES

TAURUS

GEMINI

CANCER

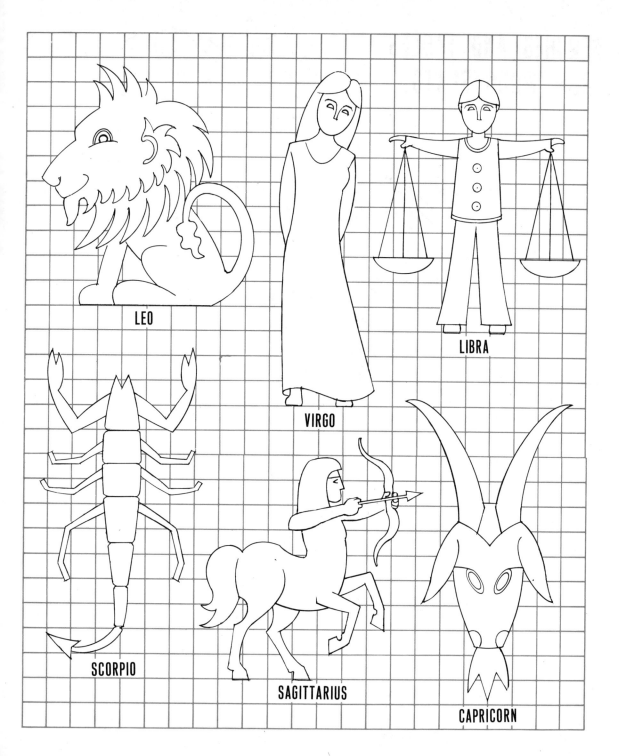

LEO

VIRGO

LIBRA

SCORPIO

SAGITTARIUS

CAPRICORN

BENCHES AND TABLES
FROM WOOD SLATS

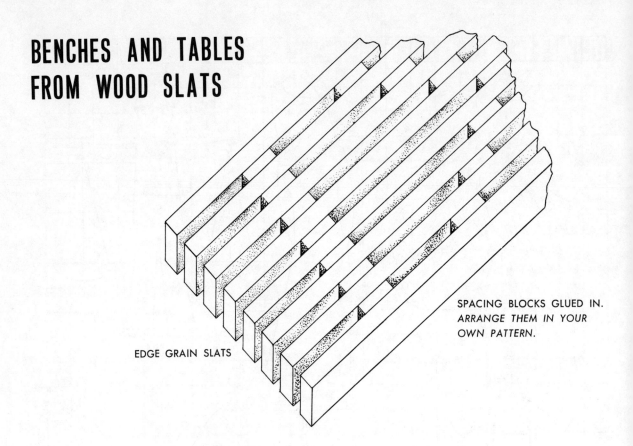

SPACING BLOCKS GLUED IN.
*ARRANGE THEM IN YOUR
OWN PATTERN.*

EDGE GRAIN SLATS

THE ENDS NEED NOT BE EVEN. TRY THEM AT DIFFERENT LENGTHS.

THE SLATS MAY BE SQUARES OR FLATS.

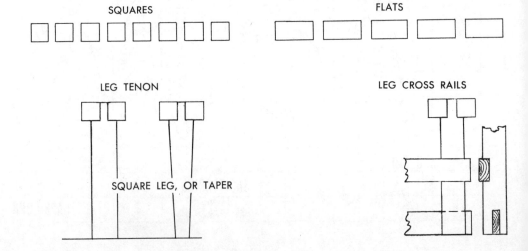

SQUARES

FLATS

LEG TENON

LEG CROSS RAILS

SQUARE LEG, OR TAPER

BOOK RACK FOR YOUR DESK TOP

Books are easily accessible and are up off the desk.

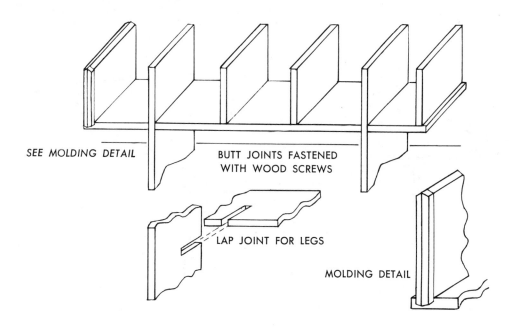

SEE MOLDING DETAIL

BUTT JOINTS FASTENED WITH WOOD SCREWS

LAP JOINT FOR LEGS

MOLDING DETAIL

SPACE AGE BOOK RACK

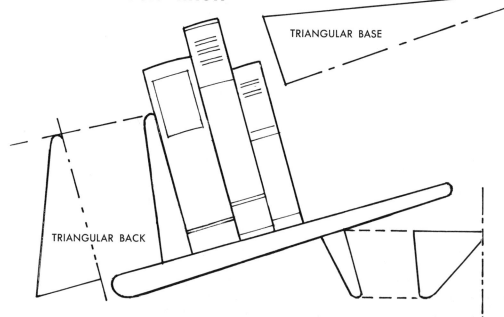

TRIANGULAR BASE

TRIANGULAR BACK

BIRD CAFETERIA

Make it small or large with generous eaves to keep out rain.

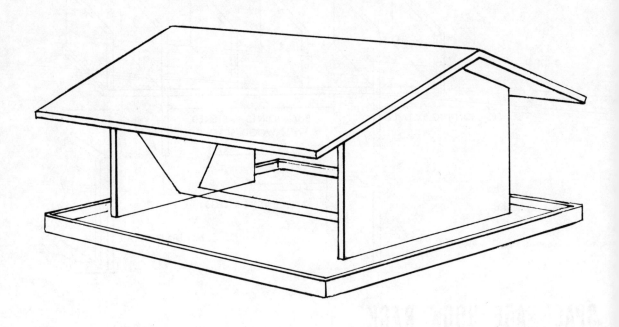

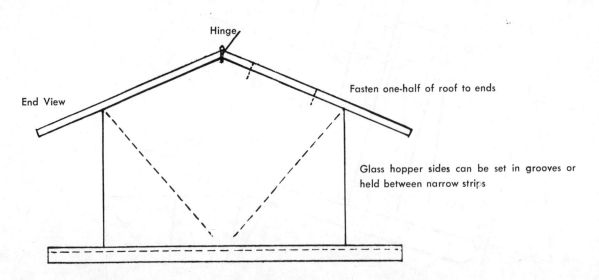

Hinge

End View

Fasten one-half of roof to ends

Glass hopper sides can be set in grooves or held between narrow strips

IDEAS FOR WOOD TURNED BOWLS

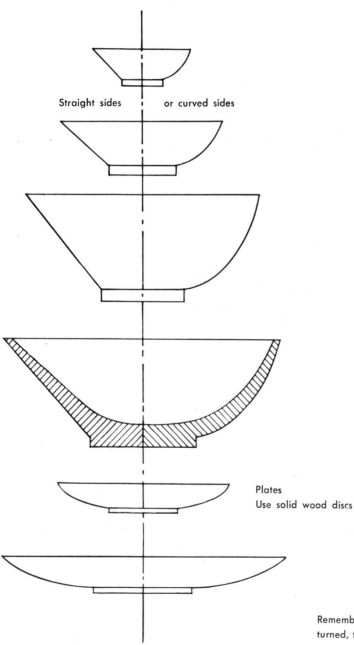

Straight sides | or curved sides

Plates
Use solid wood discs

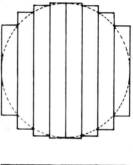

Stock glued on for a bowl.
Alternate the rings in the end grain.
This minimizes warping.

Remember: the larger the diameter being turned, the lower the lathe RPM's!

BOOK ENDS FROM YOUR INITIAL

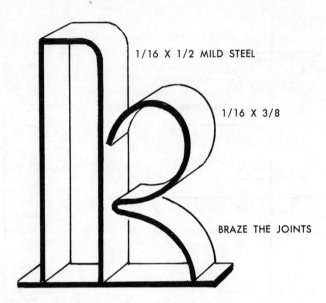

1/16 X 1/2 MILD STEEL

1/16 X 3/8

BRAZE THE JOINTS

The cut-out alphabet can be used with metals and plastics.

LAMP BASES

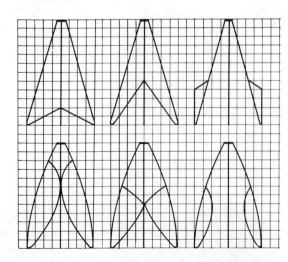

MADE OF 3 PIECES OF 1/4" BLACK IRON ROD AND A 3/8" WASHER. BRAZE THE 3 LEGS TO THE WASHER. MOUNT THE SOCKET ON THE WASHER WITH 1/8" PIPE NIPPLE AND LOCK NUT.

LAMP BASES 24"–38" TALL

Make the pedestal of turned wood, with the bases of clay or wood—round, square, or triangular.

CARVED OR INLAID DECORATION

A SCULPTURED BASE

THIS IS FOR LAMP CORD AND THE 1/8" PIPE NIPPLE TO HOLD THE SOCKET

THE TURNING STOCK:
*MAKE PLOW CUT 3/16" X 3/8"
ON CENTER LINES
BEFORE GLUING*

BASE DECORATION CARVED OR OF TOOLED LEATHER

CANDLE HOLDERS WALL TYPE

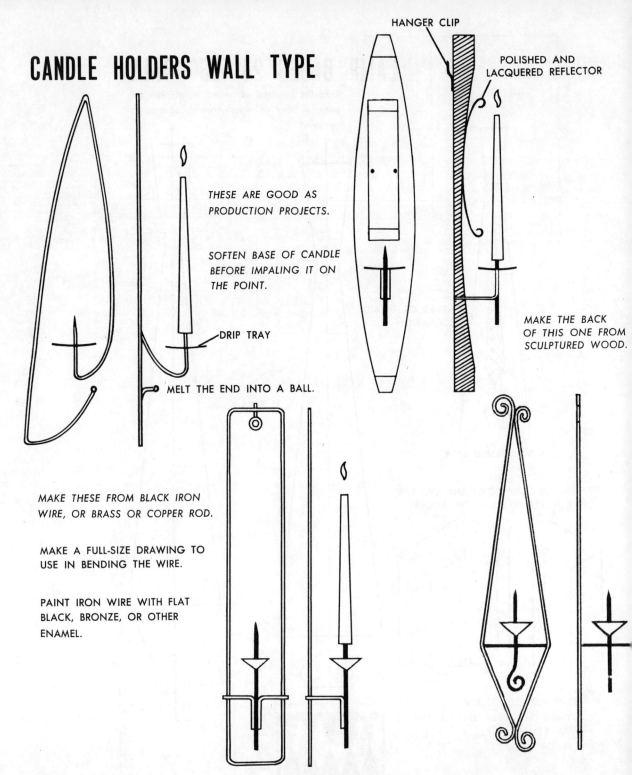

HANGER CLIP

POLISHED AND
LACQUERED REFLECTOR

*THESE ARE GOOD AS
PRODUCTION PROJECTS.*

*SOFTEN BASE OF CANDLE
BEFORE IMPALING IT ON
THE POINT.*

DRIP TRAY

*MAKE THE BACK
OF THIS ONE FROM
SCULPTURED WOOD.*

MELT THE END INTO A BALL.

*MAKE THESE FROM BLACK IRON
WIRE, OR BRASS OR COPPER ROD.*

*MAKE A FULL-SIZE DRAWING TO
USE IN BENDING THE WIRE.*

*PAINT IRON WIRE WITH FLAT
BLACK, BRONZE, OR OTHER
ENAMEL.*

WOOD TURNED CANDLE HOLDERS

These are designed as intersecting cones. Try other diameters and heights.

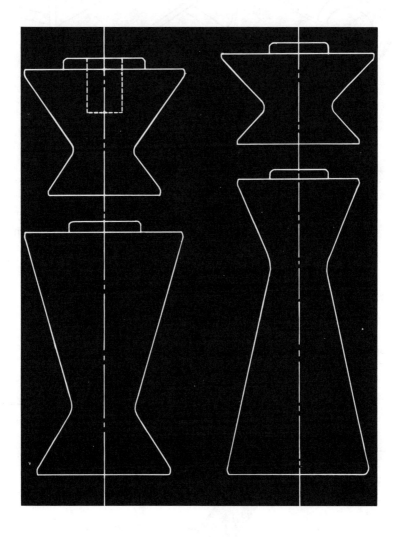

WANT TO BUILD A CAR?

Here's a simple idea that's easy to build if you can weld.

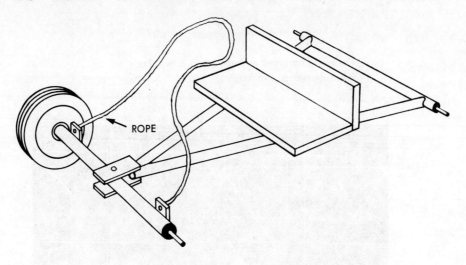

ROPE

A MORE ADVANCED AXLE AND SPINDLE

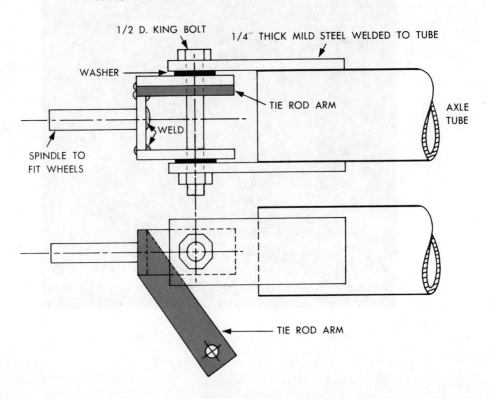

1/2 D. KING BOLT

1/4" THICK MILD STEEL WELDED TO TUBE

WASHER

TIE ROD ARM

AXLE TUBE

WELD

SPINDLE TO FIT WHEELS

TIE ROD ARM

An all-welded axle

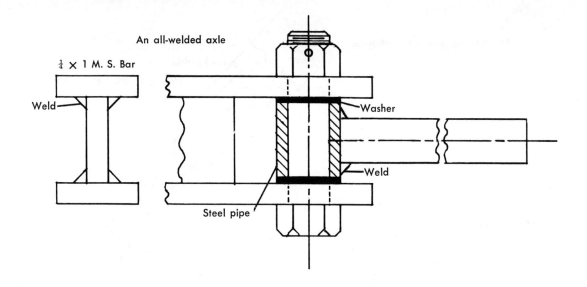

$\frac{1}{4} \times 1$ M. S. Bar

Weld

Washer

Weld

Steel pipe

A coil spring suspension for each wheel

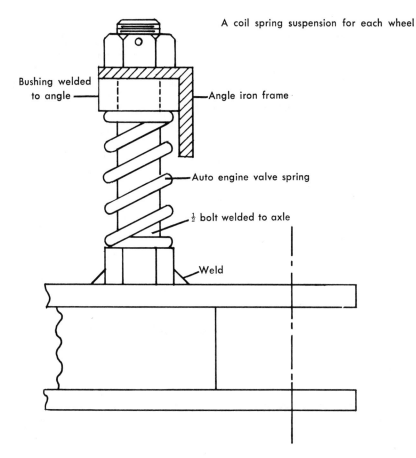

Bushing welded to angle

Angle iron frame

Auto engine valve spring

$\frac{1}{2}$ bolt welded to axle

Weld

An Alternate Steering Mechanism

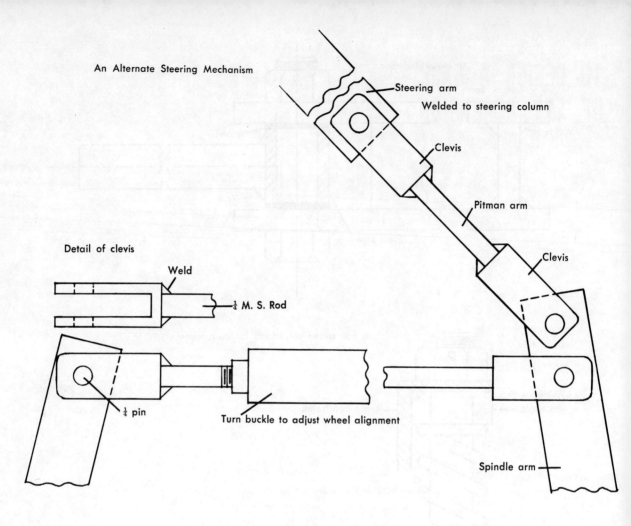

Steering arm

Welded to steering column

Clevis

Pitman arm

Clevis

Detail of clevis

Weld

$\frac{1}{4}$ M. S. Rod

$\frac{1}{4}$ pin

Turn buckle to adjust wheel alignment

Spindle arm

A brake system

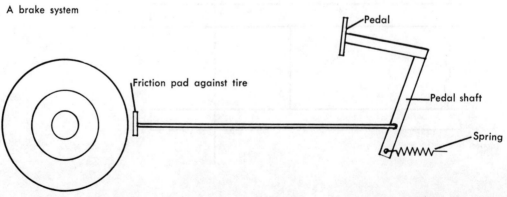

Friction pad against tire

Pedal

Pedal shaft

Spring

TO DESIGN A CASE
OF ANY DIMENSION

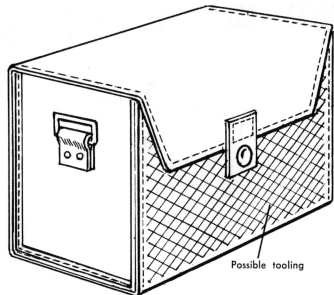

Sew or lace

Possible tooling

A typical pattern

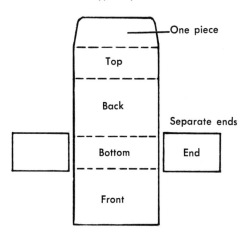

One piece

Top

Back

Separate ends

Bottom

End

Front

Construction detail

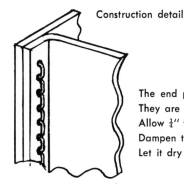

The end pieces are set in
They are called "gussets"
Allow ¼" for the fold
Dampen the gusset and bend the edges
Let it dry before cementing into place

Design tips:

1. Decide on the inside dimensions.
2. Lay it out full size on heavy paper and cut it out.
3. Fold it up and lightly cement the edges.
4. Make any desired changes. You now have the pattern.
5. Lay out any decoration on the pattern.
6. Trace the pattern on the leather. The larger the case, the heavier the leather up to 7–8 oz. cowhide.

7. Cut it out, moisten, tool, and let it dry in a folded position.
8. Insert gussets, punch, and sew or lace.
9. Add buckle, strap hangers.
10. Make a carrying strap to suit.
11. Fit a piece of ⅜" hardboard for the inside bottom.
12. Apply an appropriate leather finish.
13. On large cases it is a good idea to line the lid. Cement a light leather to the inside after step 7.
14. Accept compliments.

A POCKET CASE

This is a good design problem for your first work with leather. A pocket case can be made for different uses. You could make one to hold pliers, screw driver and other small tools. It could fit a hip pocket.

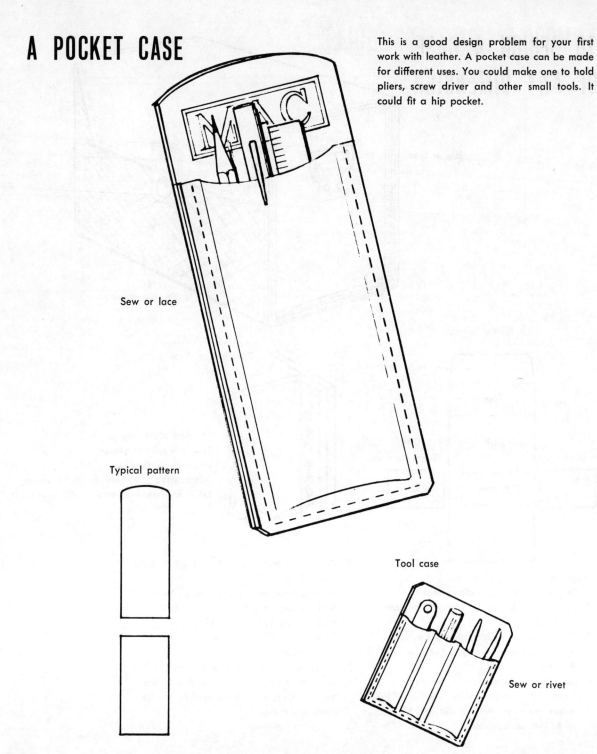

Sew or lace

Typical pattern

Tool case

Sew or rivet

A MOBILE HOME FOR DOLLS

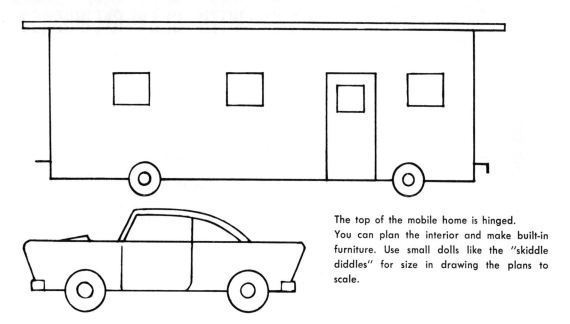

The top of the mobile home is hinged.
You can plan the interior and make built-in
furniture. Use small dolls like the "skiddle
diddles" for size in drawing the plans to
scale.

BIKE TRAILER HEAVY DUTY

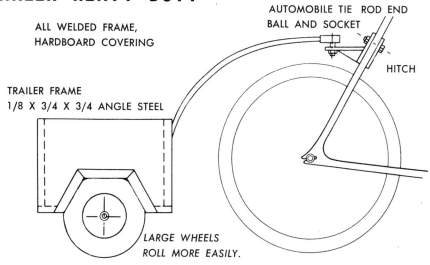

ALL WELDED FRAME,
HARDBOARD COVERING

AUTOMOBILE TIE ROD END
BALL AND SOCKET

HITCH

TRAILER FRAME
1/8 X 3/4 X 3/4 ANGLE STEEL

LARGE WHEELS
ROLL MORE EASILY.

WHAT CAN YOU DREAM UP WITH CLAY CYLINDERS?

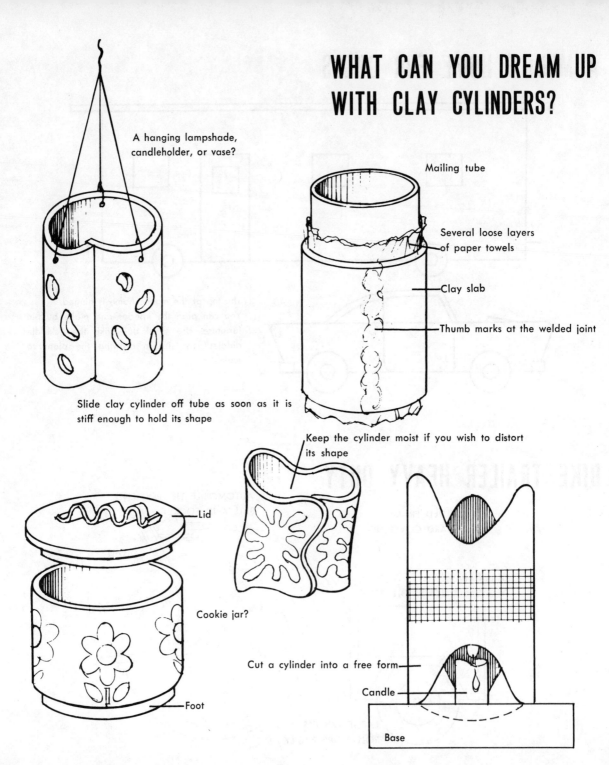

A hanging lampshade, candleholder, or vase?

Slide clay cylinder off tube as soon as it is stiff enough to hold its shape

Mailing tube

Several loose layers of paper towels

Clay slab

Thumb marks at the welded joint

Keep the cylinder moist if you wish to distort its shape

Lid

Cookie jar?

Foot

Cut a cylinder into a free form

Candle

Base

IDEAS FOR FREE-FORM COFFEE TABLES

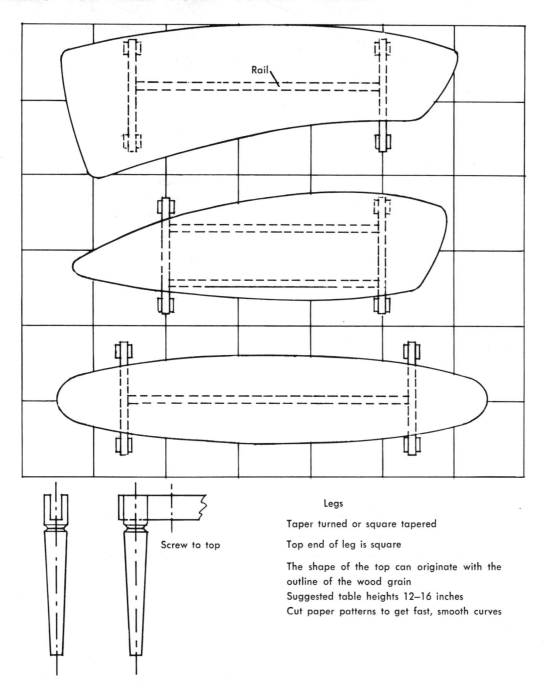

Rail

Legs

Taper turned or square tapered

Top end of leg is square

The shape of the top can originate with the
outline of the wood grain
Suggested table heights 12–16 inches
Cut paper patterns to get fast, smooth curves

Screw to top

A CONCRETE BENCH

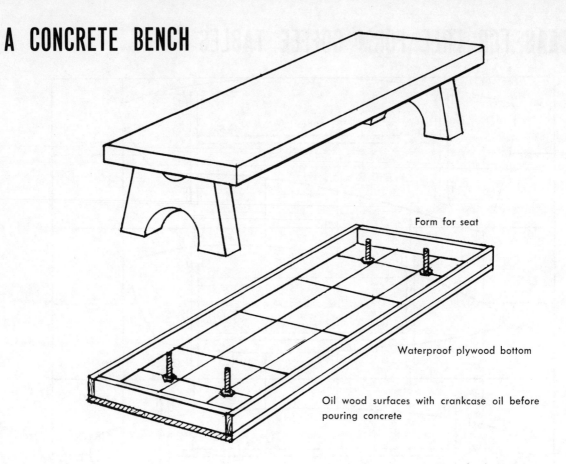

Form for seat

Waterproof plywood bottom

Oil wood surfaces with crankcase oil before pouring concrete

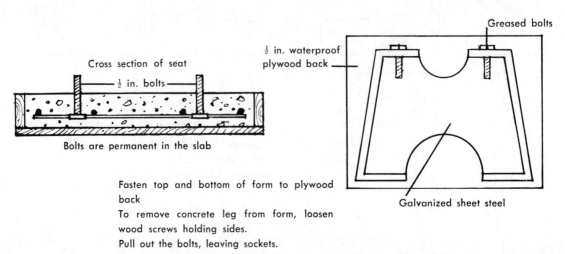

Form for legs

Greased bolts

$\frac{1}{2}$ in. waterproof plywood back

Cross section of seat

$\frac{1}{2}$ in. bolts

Bolts are permanent in the slab

Galvanized sheet steel

Fasten top and bottom of form to plywood back

To remove concrete leg from form, loosen wood screws holding sides.

Pull out the bolts, leaving sockets.

CORNER SHELVES

Some shelves

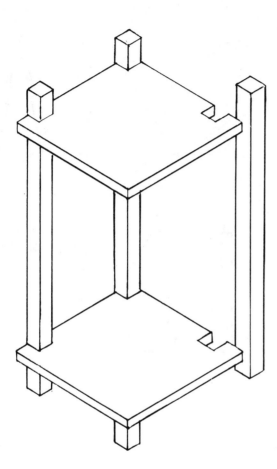

The basic idea

Changing the shape of the shelves adds interest.

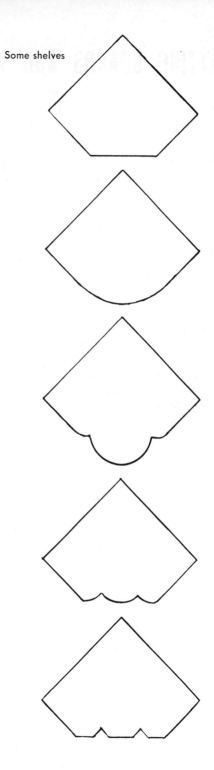

CUTTING BOARDS AND WALL HANGINGS

THIN   THICK

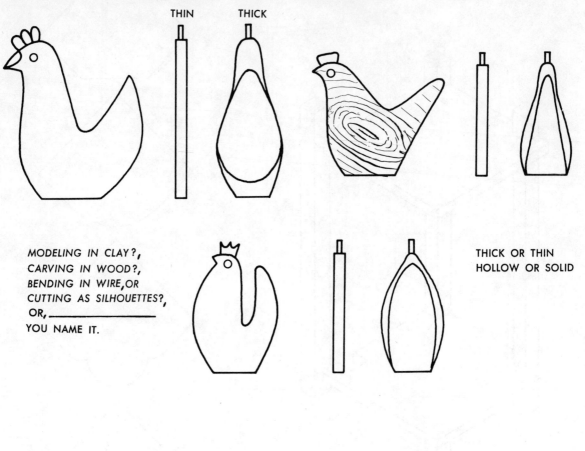

MODELING IN CLAY?,
CARVING IN WOOD?,
BENDING IN WIRE, OR
CUTTING AS SILHOUETTES?,
OR, _____
YOU NAME IT.

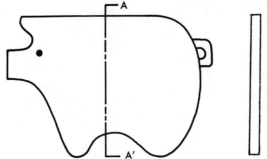

THICK OR THIN
HOLLOW OR SOLID

SECTION AA'

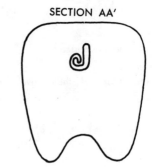

NEED A STUDY DESK?

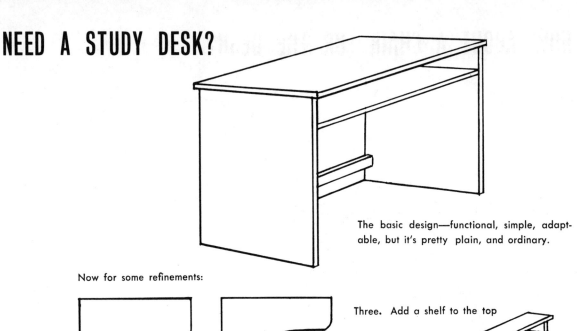

The basic design—functional, simple, adaptable, but it's pretty plain, and ordinary.

Now for some refinements:

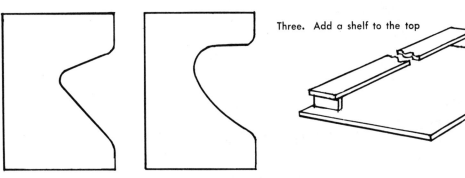

Three. Add a shelf to the top

One. Lighten the ends

Four Add paneling to the rack

Rabbet

Two. Add drawers

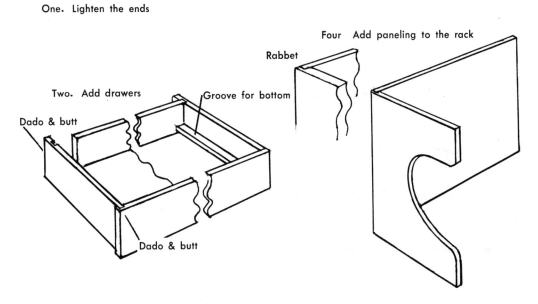

Groove for bottom

Dado & butt

Dado & butt

HOW ABOUT A CHAIR FOR THE DESK?

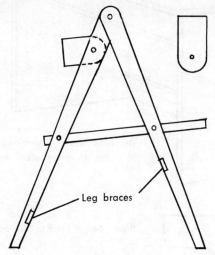

Back is of 14 G. sheet steel.
Cement on foam padding and cover with leather or plastic fabric.

Leg braces

Make seat of solid wood or plywood. Cover with foam padding.
Fasten with R. H. wood screws

FOOTSTOOLS   WROUGHT IRON FRAMES AND UPHOLSTERED PADS

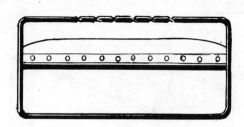

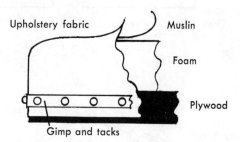

Upholstery fabric   Muslin

Foam

Plywood

Gimp and tacks

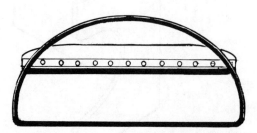

Note: Weld two crossbraces connecting the sides.
Fasten top to these with wood screws.
What would be the best size of the steel?
$\frac{3}{8}''$ sq. ? $\frac{1}{8}'' \times \frac{3}{4}''$? $\frac{3}{16}'' \times \frac{3}{4}''$?

2-DIMENSIONAL DESIGNS FOR DECORATION

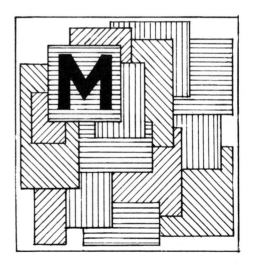

These examples show how the space can be filled easily. Make the design fit the space.
These are adaptable to leather, ceramics, printing.

DRAFTING MACHINES

You can make these—they can speed up your drawing.

A PARALLEL RULE

1

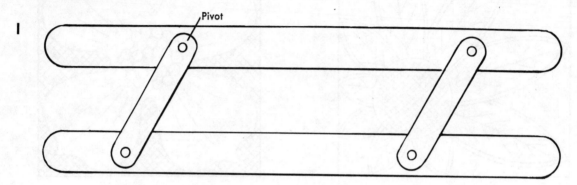

Pivot

A PARALLEL RULE WITH GREATER CAPACITY

2

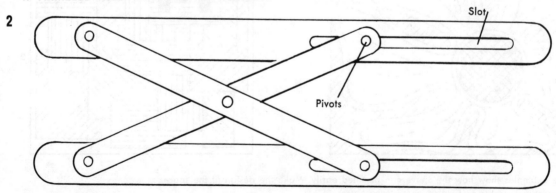

Slot

Pivots

A PARALLEL RULE FOR DRAWING BOARD OR CHALKBOARD

3

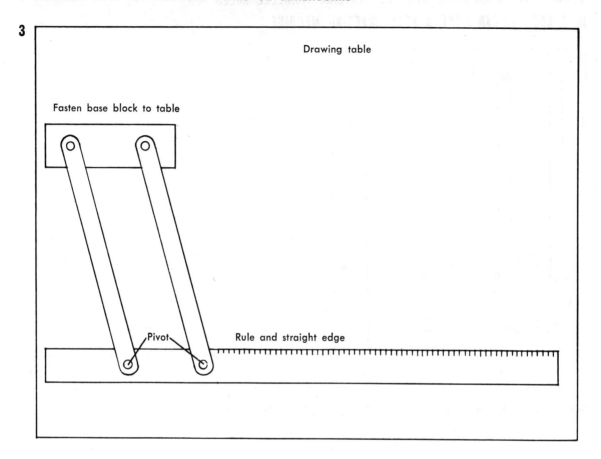

Drawing table

Fasten base block to table

Pivot

Rule and straight edge

This can be mounted at the end of the board
for vertical lines

COMBINE THE HORIZONTAL AND THE VERTICAL PARALLEL RULES
IN 3 AND YOU CAN MAKE A REAL DRAFTING MACHINE.

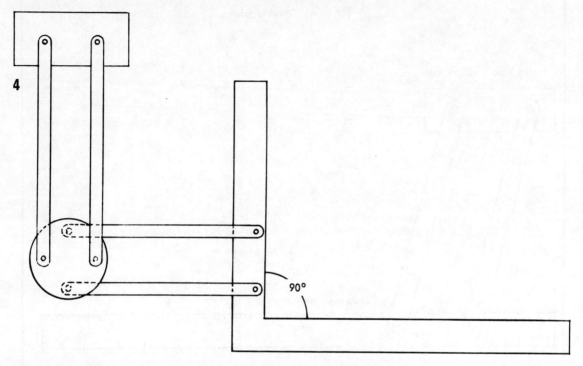

4

90°

THE PANTOGRAPH CAN ENLARGE OR REDUCE A DRAWING.

5

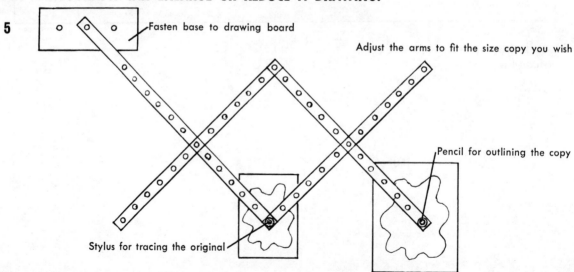

Fasten base to drawing board

Adjust the arms to fit the size copy you wish

Pencil for outlining the copy

Stylus for tracing the original

XYLOPHONE

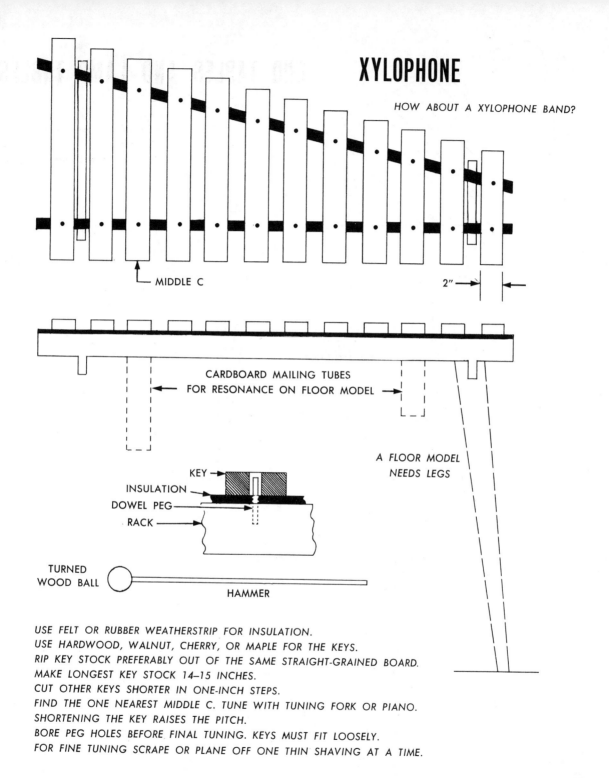

HOW ABOUT A XYLOPHONE BAND?

MIDDLE C

2"

CARDBOARD MAILING TUBES
FOR RESONANCE ON FLOOR MODEL

A FLOOR MODEL
NEEDS LEGS

KEY
INSULATION
DOWEL PEG
RACK

TURNED
WOOD BALL

HAMMER

USE FELT OR RUBBER WEATHERSTRIP FOR INSULATION.
USE HARDWOOD, WALNUT, CHERRY, OR MAPLE FOR THE KEYS.
RIP KEY STOCK PREFERABLY OUT OF THE SAME STRAIGHT-GRAINED BOARD.
MAKE LONGEST KEY STOCK 14–15 INCHES.
CUT OTHER KEYS SHORTER IN ONE-INCH STEPS.
FIND THE ONE NEAREST MIDDLE C. TUNE WITH TUNING FORK OR PIANO.
SHORTENING THE KEY RAISES THE PITCH.
BORE PEG HOLES BEFORE FINAL TUNING. KEYS MUST FIT LOOSELY.
FOR FINE TUNING SCRAPE OR PLANE OFF ONE THIN SHAVING AT A TIME.

END TABLES AND LAMP TABLES

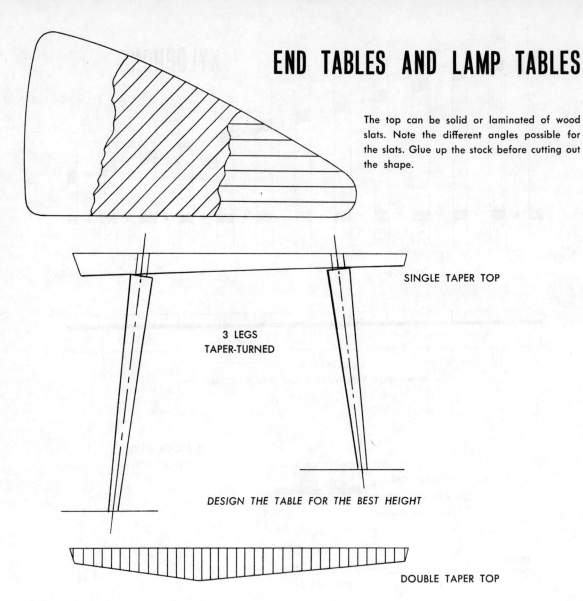

The top can be solid or laminated of wood slats. Note the different angles possible for the slats. Glue up the stock before cutting out the shape.

SINGLE TAPER TOP

3 LEGS
TAPER-TURNED

DESIGN THE TABLE FOR THE BEST HEIGHT

DOUBLE TAPER TOP

TRY THE SAME IDEA WITH A ROUND TOP

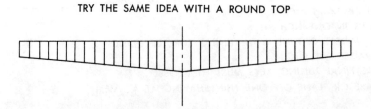

FISH FORMS

FISH COME IN MANY SHAPES AND SIZES

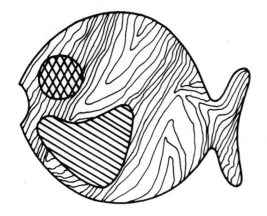

ADAPTED TO MANY USES SUCH AS:

GLASS PLAQUES AND MOSAIC CERAMIC TILES
CARVED WOODEN TRAYS
INLAID CUTTING BOARDS
DECORATIVE WIRE FIGURES

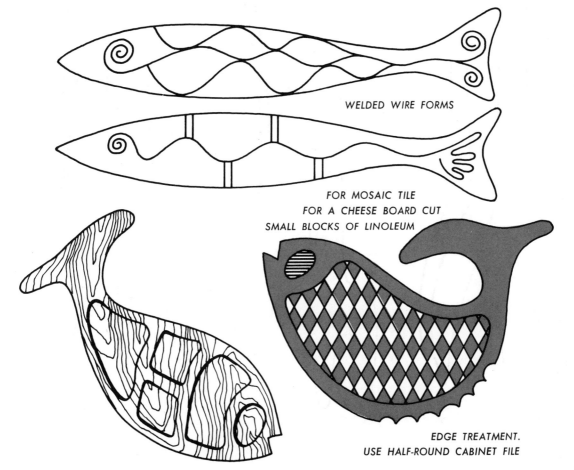

WELDED WIRE FORMS

FOR MOSAIC TILE
FOR A CHEESE BOARD CUT
SMALL BLOCKS OF LINOLEUM

EDGE TREATMENT.
USE HALF-ROUND CABINET FILE

431
Project Ideas

NEED A SERVING FORK? Try sheet plastic, hard wood, or stainless steel.

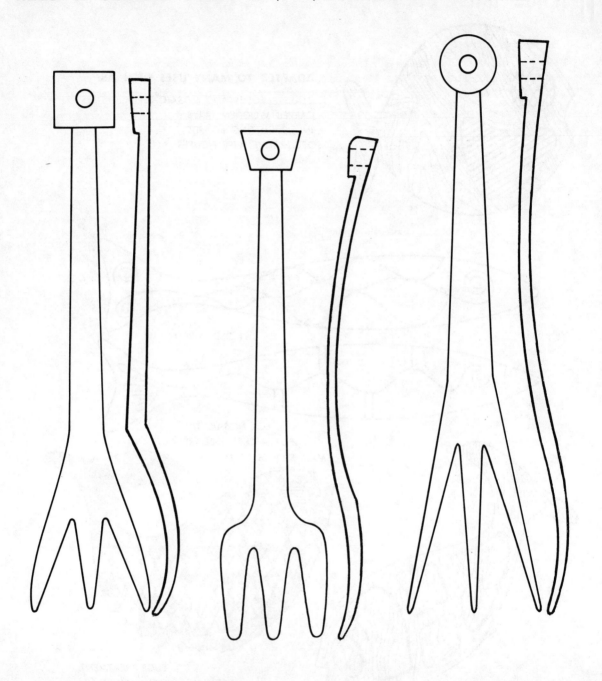

A GLASS RACK OF WOOD

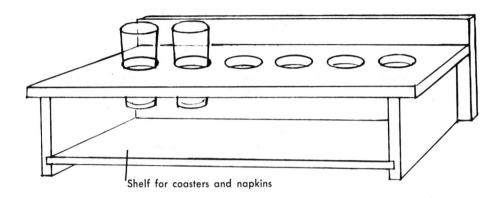

Shelf for coasters and napkins

This can be mounted on the wall, back of a
door, or it can be a tote rack to carry.
To design it, start with the glasses to be used.
Or,
A formed glass rack of acrylic plastic sheet,
aluminum sheet, or laminated veneer.

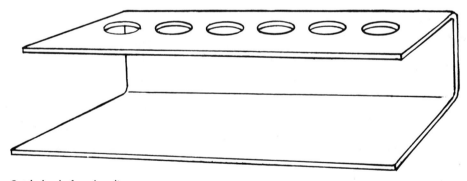

Cut holes before bending

GAVEL

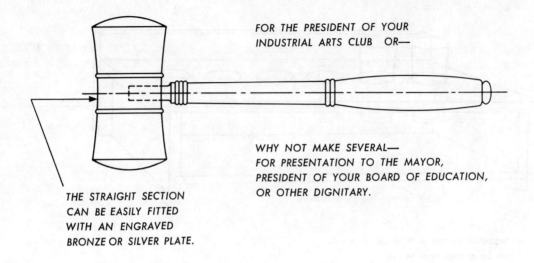

FOR THE PRESIDENT OF YOUR
INDUSTRIAL ARTS CLUB OR—

*THE STRAIGHT SECTION
CAN BE EASILY FITTED
WITH AN ENGRAVED
BRONZE OR SILVER PLATE.*

WHY NOT MAKE SEVERAL—
FOR PRESENTATION TO THE MAYOR,
PRESIDENT OF YOUR BOARD OF EDUCATION,
OR OTHER DIGNITARY.

GLASS BOTTLE AND JUG CUTTER

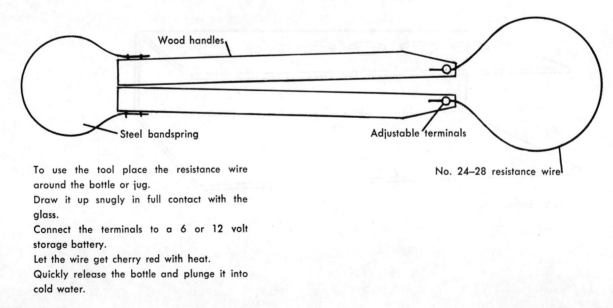

Wood handles

Steel bandspring

Adjustable terminals

No. 24–28 resistance wire

To use the tool place the resistance wire around the bottle or jug.

Draw it up snugly in full contact with the glass.

Connect the terminals to a 6 or 12 volt storage battery.

Let the wire get cherry red with heat.

Quickly release the bottle and plunge it into cold water.

GRINDING AND POLISHING HEAD

DETAILS OF ASSEMBLY

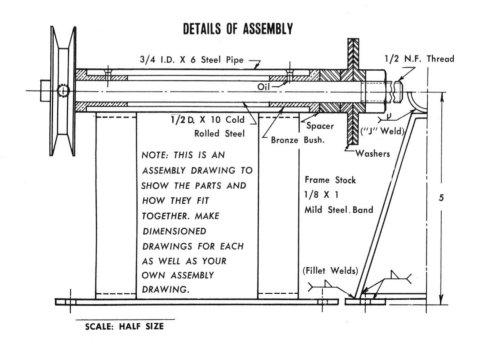

3/4 I.D. X 6 Steel Pipe

Oil

1/2 N.F. Thread

1/2 D. X 10 Cold Rolled Steel

Spacer

Bronze Bush.

("J" Weld)

Washers

Frame Stock
1/8 X 1
Mild Steel Band

5

NOTE: THIS IS AN ASSEMBLY DRAWING TO SHOW THE PARTS AND HOW THEY FIT TOGETHER. MAKE DIMENSIONED DRAWINGS FOR EACH AS WELL AS YOUR OWN ASSEMBLY DRAWING.

(Fillet Welds)

SCALE: HALF SIZE

The grinding and polishing head can become a headstock for a wood lathe for your home workshop. Develop a prototype. Get any bugs out of it. It might be a good item for a mass production project.

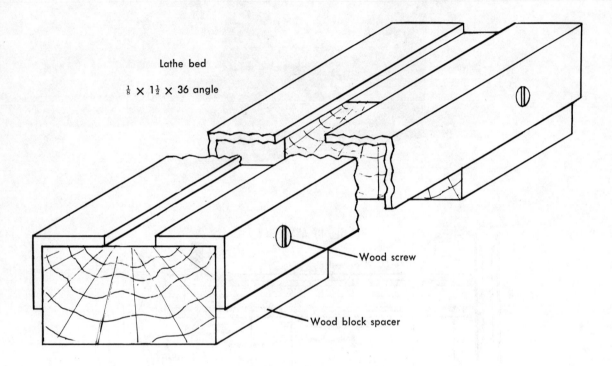

Lathe bed

$\frac{1}{8}$ × 1$\frac{1}{2}$ × 36 angle

Wood screw

Wood block spacer

Lathe tail stock

Same design and overall dimensions as in the
grinding and polishing head

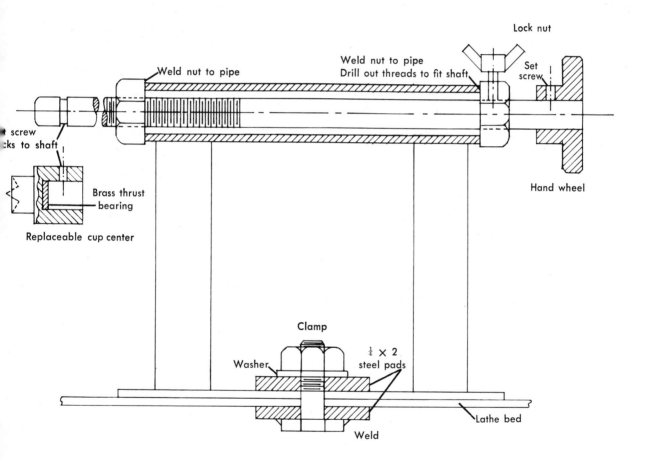

Lock nut

Weld nut to pipe

Weld nut to pipe
Drill out threads to fit shaft

Set
screw

screw
cks to shaft

Brass thrust
bearing

Replaceable cup center

Hand wheel

Clamp

Washer

$\frac{1}{4} \times 2$
steel pads

Lathe bed

Weld

Details of tool rest for wood lathe

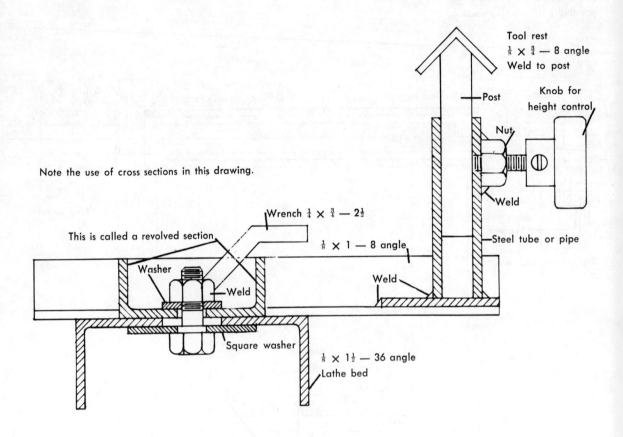

Note the use of cross sections in this drawing.

Tool rest
$\frac{1}{8} \times \frac{3}{4}$ — 8 angle
Weld to post

Post

Knob for
height control

Nut

Weld

Wrench $\frac{1}{4} \times \frac{3}{4}$ — 2$\frac{1}{2}$

$\frac{1}{8} \times 1$ — 8 angle

This is called a revolved section

Washer

Steel tube or pipe

Weld

Weld

Square washer

$\frac{1}{8} \times 1\frac{1}{2}$ — 36 angle

Lathe bed

WALL SHELF

Wrought iron brackets.

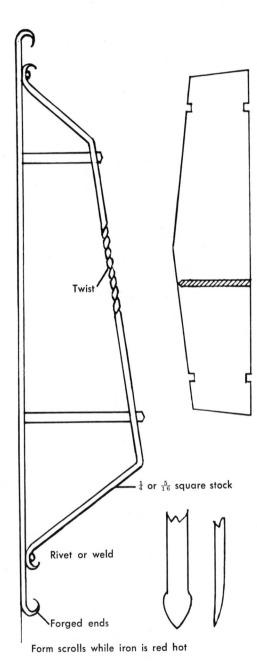

Twist

¼ or 5/16 square stock

Rivet or weld

Forged ends

Form scrolls while iron is red hot

Shelf plan

Chamfer ends and front

Design tip:
First determine distance between shelves.
Draw the bracket full-size as a pattern for shaping.
An overall height of 18–30 inches is recommended.

Cold formed ends using flat band,
⅛ × ½ rear and ⅛ × ⅜ front

To twist square stock grip it horizontally in a vise at the point where the twist begins. Use a crescent or monkey wrench to twist.

CAMPING STOVE—PORTABLE, FOR CHARCOAL

USE WELDED OR RIVETED CONSTRUCTION

SUGGESTED OVERALL DIMENSIONS
2-MAN 10 X 18   4-MAN 12 X 20

DETAIL OF
RIVETED CORNER

0.125 SOFT ALUMINUM SHEET.
SOLDER CORNERS.

GRIDDLE

RODS—1/8 STEEL, WELDING

GRILL

BOX—20–24G.
BLACK STEEL
SHEET

1/8 X 1/2 BAND-
IRON FRAME

BOX

DRAWER FOR
CHARCOAL—20–22 G.
BLACK STEEL
SHEET

LID

CARRY HANDLES
SERVE AS
LEGS—5/16 ROD.

A HOT DOG ROASTER

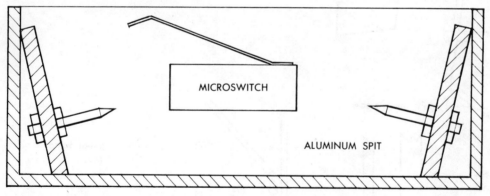

MICROSWITCH

ALUMINUM SPIT

Designed and made by a student. The box is
of sheet plastic. Shutting the clear plastic lid
closes the microswitches to complete the circuit. As indicated in the wiring diagram, one
switch is installed in each wire.

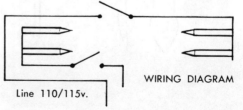

Line 110/115v.

WIRING DIAGRAM

JIGSAW

JIGSAW MECHANISM

DETAIL OF SAW

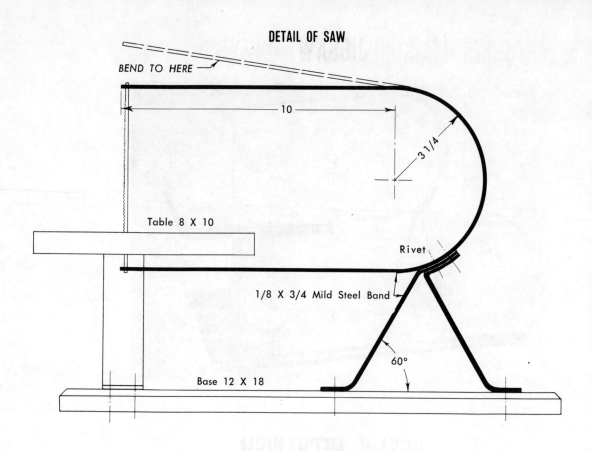

BEND TO HERE

10

3 1/4

Table 8 X 10

Rivet

1/8 X 3/4 Mild Steel Band

60°

Base 12 X 18

DETAIL OF MECHANISM

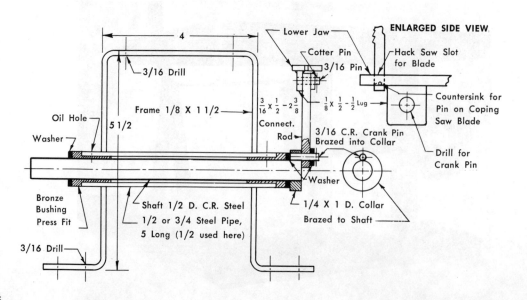

ENLARGED SIDE VIEW

Lower Jaw

Cotter Pin

3/16 Pin

Hack Saw Slot for Blade

4

3/16 Drill

Frame 1/8 X 1 1/2

$\frac{3}{16} \times \frac{1}{2} - 2\frac{3}{8}$

Connect. Rod

$\frac{1}{8} \times \frac{1}{2} - \frac{1}{2}$ Lug

Countersink for Pin on Coping Saw Blade

Oil Hole

5 1/2

Washer

3/16 C.R. Crank Pin Brazed into Collar

Drill for Crank Pin

Bronze Bushing Press Fit

Washer

Shaft 1/2 D. C.R. Steel

1/4 X 1 D. Collar Brazed to Shaft

1/2 or 3/4 Steel Pipe, 5 Long (1/2 used here)

3/16 Drill

A MATCHING MACHINE FOR TEACHERS

A suggestion: Put U.S. weights and measures on one set of tags and the metric equivalents on the other. Mix up the order, of course. What other uses could this teaching machine have?

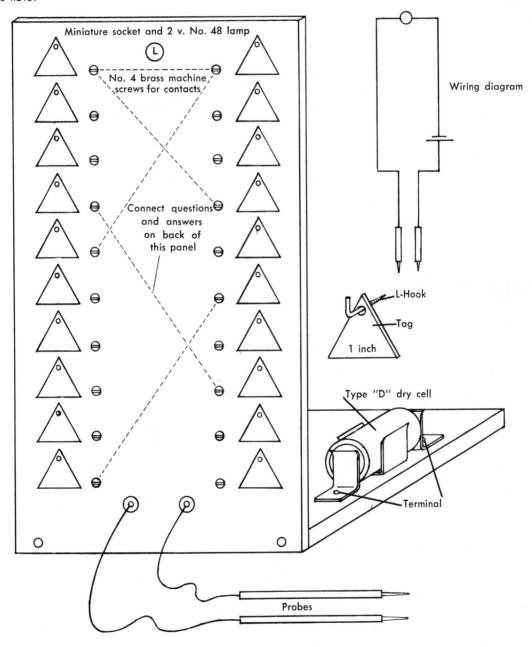

Miniature socket and 2 v. No. 48 lamp

No. 4 brass machine screws for contacts

Connect questions and answers on back of this panel

Wiring diagram

L-Hook

Tag

1 inch

Type "D" dry cell

Terminal

Probes

MOLDS *A SIMPLE MOLD FOR FORMING ACRYLIC PLASTIC SHEET.*

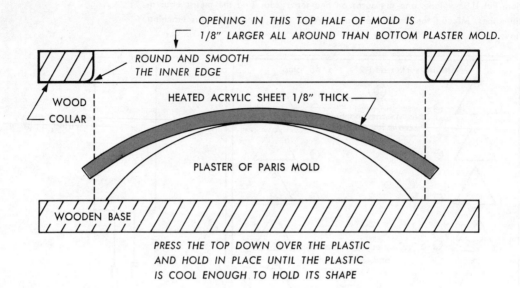

OPENING IN THIS TOP HALF OF MOLD IS
1/8" LARGER ALL AROUND THAN BOTTOM PLASTER MOLD.

ROUND AND SMOOTH
THE INNER EDGE

WOOD
COLLAR

HEATED ACRYLIC SHEET 1/8" THICK

PLASTER OF PARIS MOLD

WOODEN BASE

PRESS THE TOP DOWN OVER THE PLASTIC
AND HOLD IN PLACE UNTIL THE PLASTIC
IS COOL ENOUGH TO HOLD ITS SHAPE

BISQUE MOLD *FOR FORMING SHEET GLASS.*

ADD CONSIDERABLE SILICA SAND TO
THE CLAY TO MAKE IT MORE REFRACTORY
(RESISTANT TO THERMAL SHOCK)

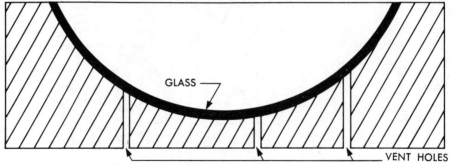

GLASS

VENT HOLES

CUT WINDOW GLASS TO DESIRED SHAPE.
PLACE OVER MOLD CAVITY. INSERT IN KILN AND
SLOWLY RAISE TEMPERATURE TO THE POINT WHERE THE
GLASS SAGS INTO THE CAVITY AND THE EDGES LOSE
THEIR SHARPNESS. COOL SLOWLY. THE GLASS CAN BE
DECORATED WITH CERAMIC DECORATING COLORS OR
CERAMIC ENAMEL BEFORE HEATING.

LAMPS USING UNIT DESIGN

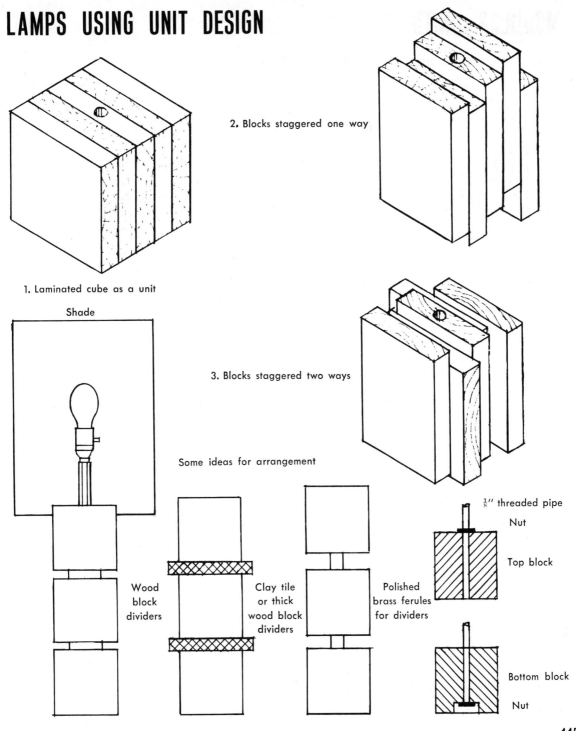

1. Laminated cube as a unit

2. Blocks staggered one way

3. Blocks staggered two ways

Some ideas for arrangement

Shade

Wood block dividers

Clay tile or thick wood block dividers

Polished brass ferules for dividers

$\frac{1}{8}''$ threaded pipe

Nut

Top block

Bottom block

Nut

MODULAR UNITS

Make shelves for books, knickknacks, room dividers.

SCREW HOLES ARE COUNTERBORED AND PLUGGED

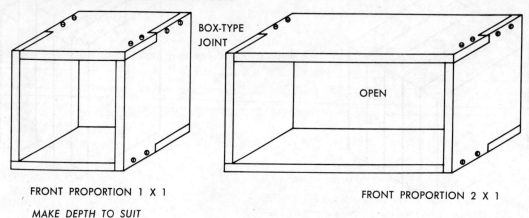

BOX-TYPE
JOINT

OPEN

FRONT PROPORTION 1 X 1

MAKE DEPTH TO SUIT

FRONT PROPORTION 2 X 1

*MODULAR CONSTRUCTION IS BASED
ON STANDARD SIZE UNITS, SUCH
AS CEMENT BLOCKS AND BRICKS.
THIS PERMITS MANY DIFFERENT .
ARRANGEMENTS. THE BOX-TYPE JOINT
IS FASTENED WITH WOOD SCREWS.*

A BASE FOR STACKING THE UNITS

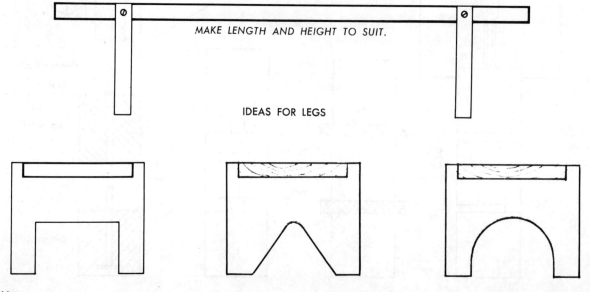

MAKE LENGTH AND HEIGHT TO SUIT.

IDEAS FOR LEGS

MODULAR UNITS
SOME COMBINATIONS

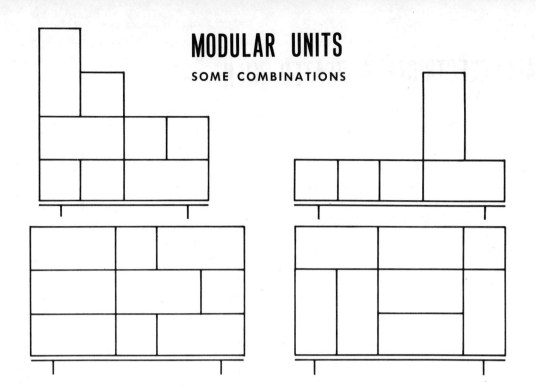

IDEAS FOR BRASS OR IRON ROD

RECORD CADDY

DESIGN IT TO FIT
YOUR RECORDS

LOOP
FOR
HANGING

HAND GRIP OF LEATHER, REED, PLASTIC, ETC.

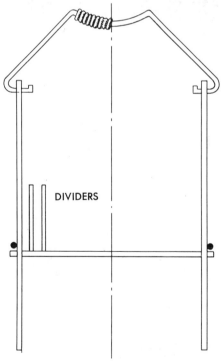

DIVIDERS

SMALL ELECTRICALLY HEATED KILN

For enameling, glaze tests, or heat treating.
Building the kiln. The element: 30.5 ft. No. 18
chromel "A." Wind a tight coil on a $\frac{3}{8}$ in.
steel rod. This plugs into 110/120 v. The case:
Use 26 G. black iron sheet and sheet metal
screws.

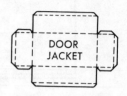

TOP and BOTTOM

Fold

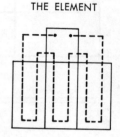

DOOR JACKET

SIDES

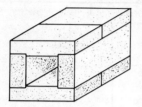

Standard insulating brick,
2000° or 2300°.
Rabbet 1/4".
Back is 1/2 brick.

Saw bricks with hacksaw.
Cut grooves for the
Element with wood chisel.

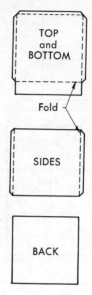

BACK

THE ELEMENT

AN ELECTRIC MOTOR

You can build this electric motor.

ARMATURE SUPPORT (2)

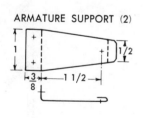

MAGNET TEST

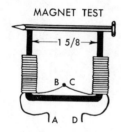

1 5/8

B C

A D

ARMATURE CORE

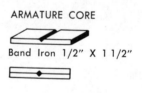

Band Iron 1/2" X 1 1/2"

SERIES HOOK-UP

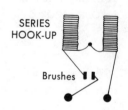

Brushes

ARMATURE ASSEMBLY

Plastic Rod
1/4 D.—5/8" long

Shaft, 8d finishing nail

PARALLEL OR SHUNT HOOK-UP

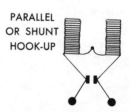

A PICTORIAL SKETCH IN ISOMETRIC

It shows the components and the wiring of the
motor on page 448.

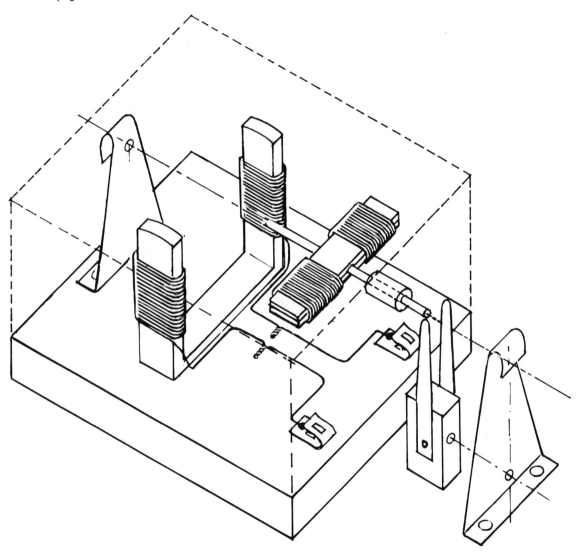

A SYNCHRONOUS MOTOR

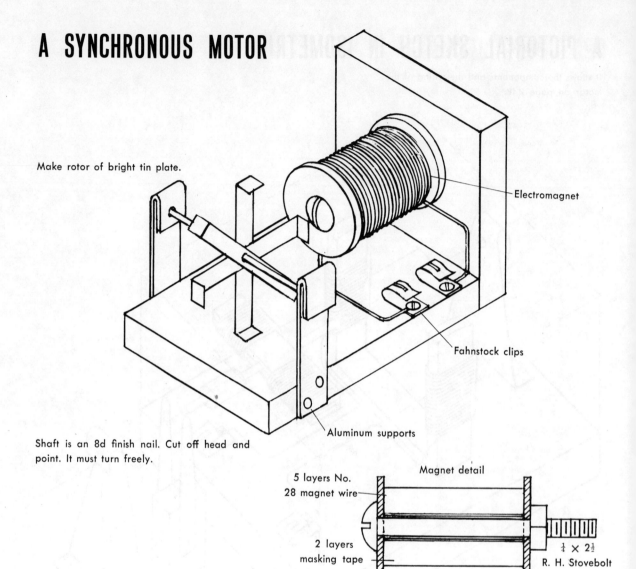

Make rotor of bright tin plate.

Electromagnet

Fahnstock clips

Aluminum supports

Shaft is an 8d finish nail. Cut off head and point. It must turn freely.

Magnet detail

5 layers No. 28 magnet wire

2 layers masking tape

Plastic or fiber washer

¼ × 2½

R. H. Stovebolt

The synchronous motor is commonly found in electric clocks. Since 60 cycle alternating current changes direction 120 times per second, the polarity of the magnet changes accordingly. The magnet attracts each arm of the rotor as the polarity changes. This operates on a toy transformer. Why won't it run on direct current? You may have to spin the rotor to start it.

OPAQUE PROJECTOR

Screen magnification up to 7X is obtained with this easy-to-build opaque projector. It will project any kind of opaque copy, such as photos, halftones, and comics. The screen image is erect and is black-and-white or full color—just like the copy.

Two 100-watt lamps provide a fair degree of screen brightness, although it must be remembered all opaque projectors need at least semi-darkness for good visibility. A good projection lens like the one specified will cost several dollars, but you can make a less expensive job by using the alternate simple lens duplet costing less than a dollar. The barrel for this lens is a small tin can, the kind used for frozen fruit juice.

The construction may be simplified by using the split tin can as reflectors, as in drawing below. However, the actual plan uses these only as a kind of heat shield, the actual reflectors being bright tin or foil on asbestos paper, fitted as shown.

COMPACT TABLE MODEL COVERS
4 X 4" COPY AT 3X TO 7X BLOW-UP

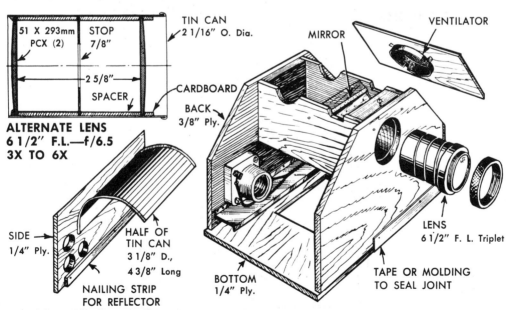

51 X 293mm PCX (2) STOP 7/8"

2 5/8"

SPACER

TIN CAN
2 1/16" O. Dia.

CARDBOARD

**ALTERNATE LENS
6 1/2" F.L.—f/6.5
3X TO 6X**

SIDE
1/4" Ply.

HALF OF TIN CAN
3 1/8" D.,
4 3/8" Long

NAILING STRIP
FOR REFLECTOR

BACK
3/8" Ply.

MIRROR

VENTILATOR

BOTTOM
1/4" Ply.

LENS
6 1/2" F. L. Triplet

TAPE OR MOLDING
TO SEAL JOINT

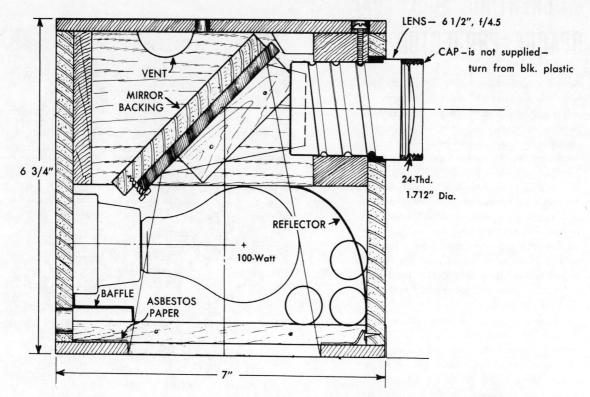

VENT

MIRROR BACKING

LENS— 6 1/2", f/4.5

CAP—is not supplied— turn from blk. plastic

REFLECTOR

100-Watt

24-Thd. 1.712" Dia.

6 3/4"

BAFFLE

ASBESTOS PAPER

7"

1/2 FULL SIZE

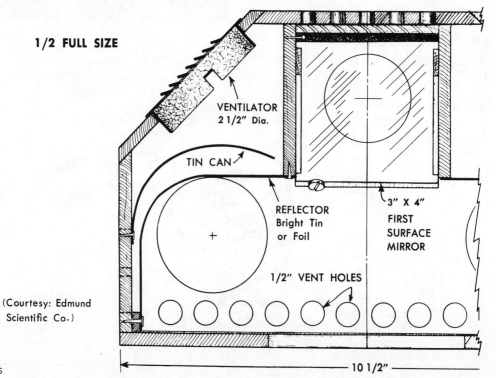

VENTILATOR 2 1/2" Dia.

TIN CAN

REFLECTOR Bright Tin or Foil

3" X 4" FIRST SURFACE MIRROR

1/2" VENT HOLES

(Courtesy: Edmund Scientific Co.)

10 1/2"

CREATING SCULPTURE

Suppose you wanted to create sculpture on the wood lathe?

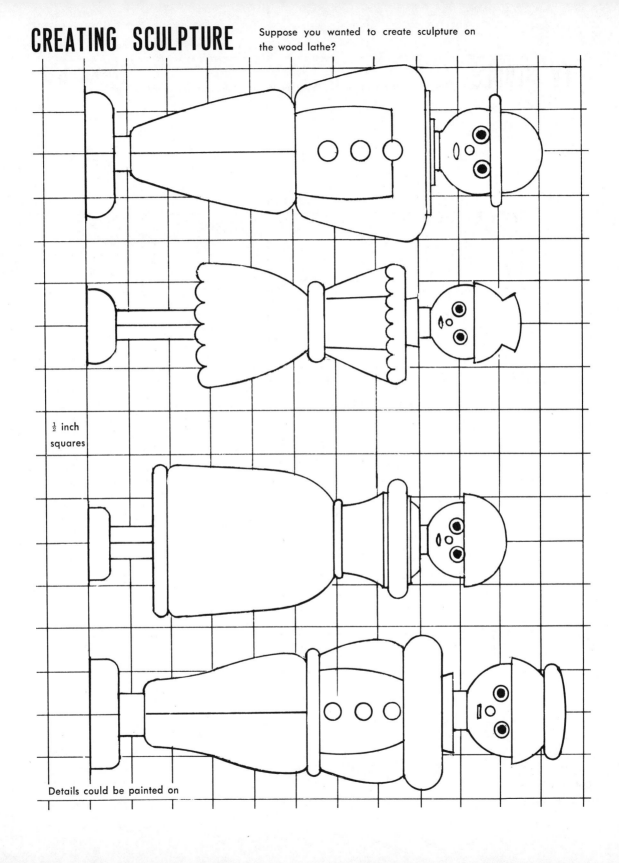

½ inch squares

Details could be painted on

TV STOOLS

Turned on the wood lathe, the three-legged stool can take many forms.

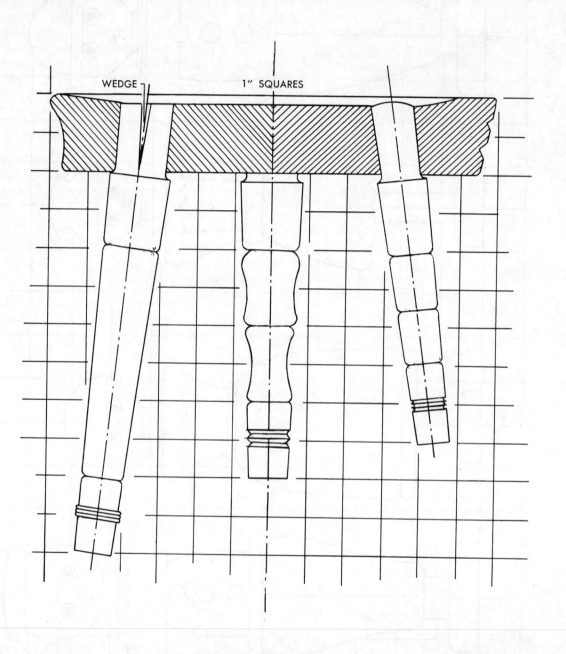

WEDGE

1" SQUARES

TRAYS

For snacks and serving. Items can be used with fiber glass or ceramics.

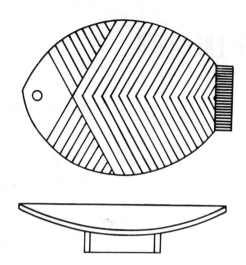

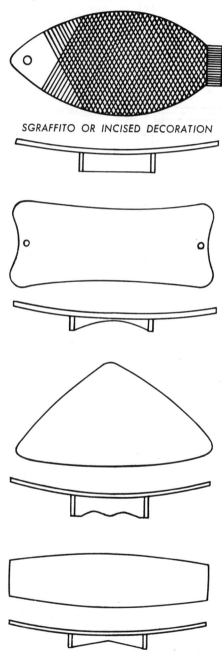

SGRAFFITO OR INCISED DECORATION

TO MAKE THE MOLD:

(1) THE CLAY MODEL, UPSIDE DOWN

(2) POUR POTTERY PLASTER OVER THE MODEL.

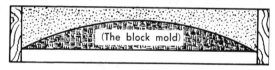

(The block mold)

(3) REMOVE CLAY AS SOON AS PLASTER HARDENS. COAT INSIDE WITH SOAPSUDS (NOT DETERGENT). POUR THE CAVITY FULL OF PLASTER. WHEN HARD REMOVE THIS AND LET IT GET BONE DRY.

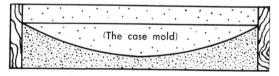

(The case mold)

(4) LAY SLAB OF CLAY OVER THE MOLD AND SMOOTH WITH A DAMP SPONGE.

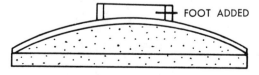

← FOOT ADDED

NOTE: SOAP THE BLOCK MOLD WELL BEFORE FILLING WITH PLASTER.

PHOTO ENLARGER

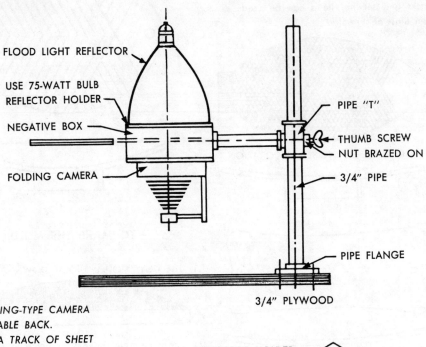

FLOOD LIGHT REFLECTOR

USE 75-WATT BULB
REFLECTOR HOLDER

NEGATIVE BOX

FOLDING CAMERA

PIPE "T"

THUMB SCREW
NUT BRAZED ON

3/4" PIPE

PIPE FLANGE

3/4" PLYWOOD

NOTE:
USE A FOCUSING-TYPE CAMERA
WITH REMOVABLE BACK.
MAKE CAMERA TRACK OF SHEET
METAL TO FIT CAMERA.
USE GROUND GLASS DIFFUSION
IF OPAL GLASS IS NOT AVAILABLE.

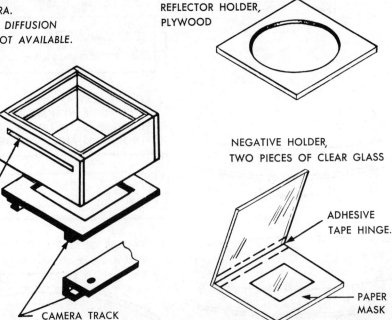

REFLECTOR HOLDER,
PLYWOOD

NEGATIVE BOX

RABBET FOR OPAL
NEGATIVE HOLDER

MAKE SLOT FOR
GLASS DIFFUSER.

CAMERA TRACK

NEGATIVE HOLDER,
TWO PIECES OF CLEAR GLASS

ADHESIVE
TAPE HINGE.

PAPER
MASK

DOUBLE EASEL

Want to make something useful for a teacher in your elementary school? The children may need an easel.

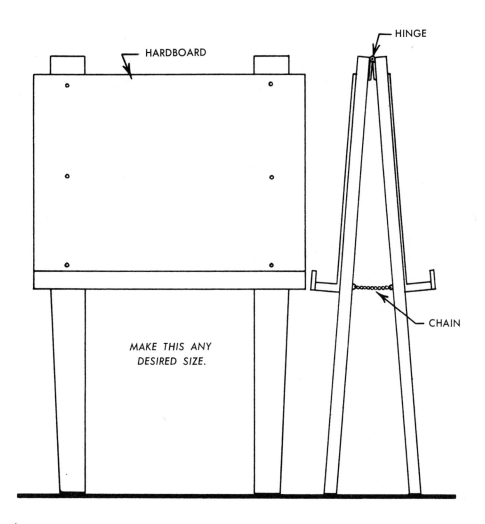

HARDBOARD

HINGE

CHAIN

MAKE THIS ANY DESIRED SIZE.

TO USE AS A CHALKBOARD APPLY 2 COATS OF CHALKBOARD PAINT TO ONE SIDE. LEAVE OTHER SIDE PLAIN TO USE WITH PAPER OR CANVAS AS A PAINTING EASEL.

SCRAPBOOK OR PHOTO ALBUM

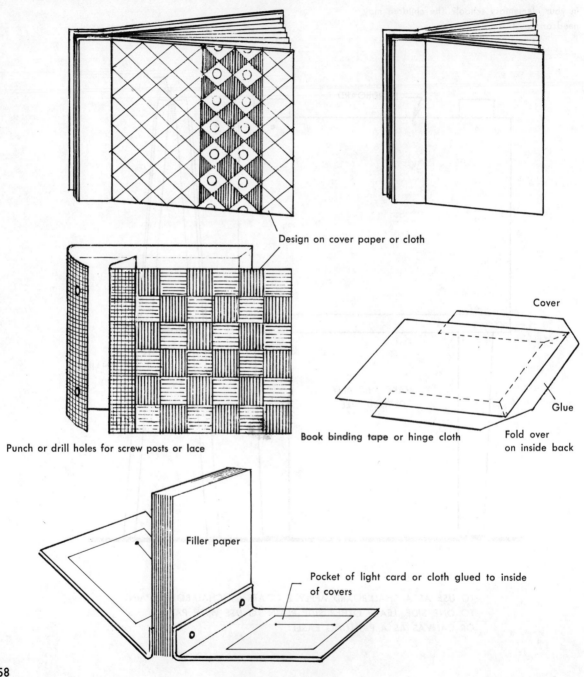

Design on cover paper or cloth

Punch or drill holes for screw posts or lace

Book binding tape or hinge cloth

Cover

Glue

Fold over
on inside back

Filler paper

Pocket of light card or cloth glued to inside
of covers

TOYS FOR SMALL FRY

A TRAIN

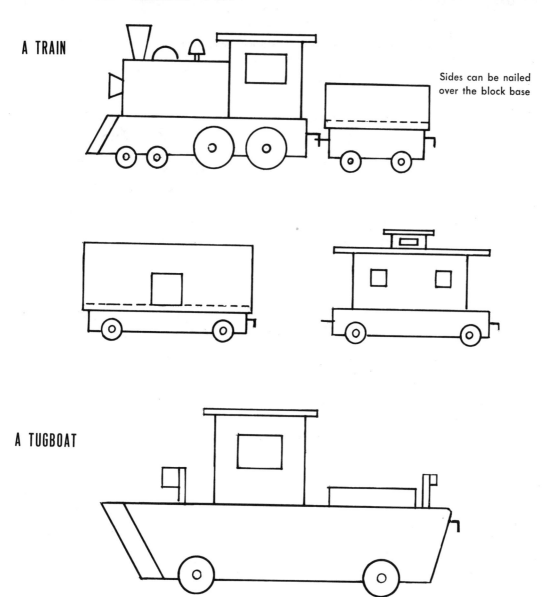

Sides can be nailed over the block base

A TUGBOAT

Construction tips: For simplest construction use solid wood blocks for the parts. Glue them together. For wheels, turn a length of wood to the desired diameter on the wood lathe. Slice off wheels. Or cut out discs from a board, fasten to a face plate with a center screw and turn to shape.

A "ROCK" TESTER

Can you hold your hand steady as a rock? Try the probe in each hole. When it lights the light, you've missed. This demonstrates the simplest circuit.

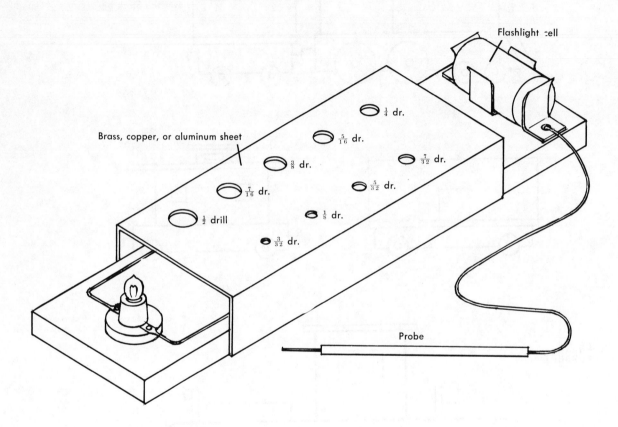

Flashlight cell

$\frac{1}{4}$ dr.

$\frac{5}{16}$ dr.

Brass, copper, or aluminum sheet

$\frac{3}{8}$ dr.

$\frac{7}{32}$ dr.

$\frac{7}{16}$ dr.

$\frac{5}{32}$ dr.

$\frac{1}{2}$ drill

$\frac{1}{8}$ dr.

$\frac{3}{32}$ dr.

Probe

A CRYSTAL RADIO

Diagram for a crystal radio that can pull in stations 25 to 50 miles away. For greater selectivity, tap the secondary at 30, 50, 70, 90, 110 turns, as shown in diagram at right. Connect these to the points on a selector switch. Then connect a fixed condenser across the phones. You may find parts in an old radio.

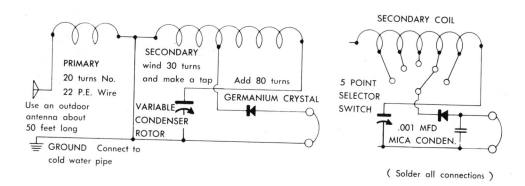

A SIMPLE RADIO

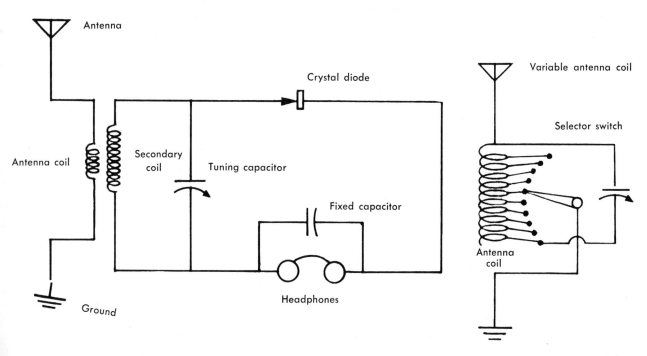

TELEGRAPH SET

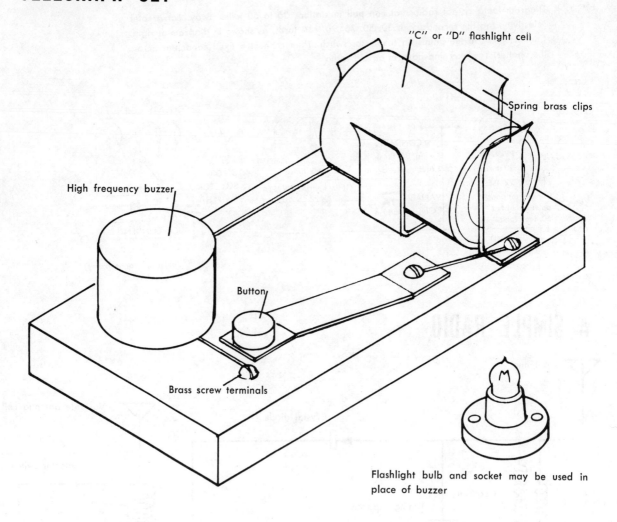

"C" or "D" flashlight cell

Spring brass clips

High frequency buzzer

Button

Brass screw terminals

Flashlight bulb and socket may be used in place of buzzer

The buzzer or light socket may be purchased from an electronics supply company. The buzzer emits a high-pitched "squeal."

In practicing the international code hold the dash, or "dit," three times as long as the dot, or "dah." The interval between dahs and dits should be equal to three dots. The interval between letters should equal three dots. Between words, leave an interval equal to five dots.

A TELEGRAPH SOUNDER

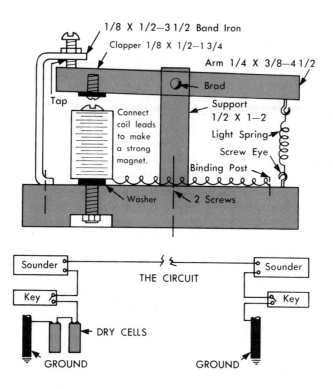

1/8 X 1/2—3 1/2 Band Iron
Clapper 1/8 X 1/2—1 3/4
Arm 1/4 X 3/8—4 1/2
Brad
Tap
Connect coil leads to make a strong magnet.
Support 1/2 X 1—2
Light Spring
Screw Eye
Binding Post
Washer
2 Screws

Sounder
THE CIRCUIT
Sounder
Key
Key
DRY CELLS
GROUND
GROUND

You can make this easily. It produces a click signal like the old-fashioned telegraph. To signal your partner, tap your key, then close it and wait for the reply. Keep the key closed when receiving. Two electromagnets are required. Core is $\frac{3}{16}$ X 2 in. stovebolt. Wrap with 2 layers of heavy paper and model airplane cement. Make washers of cardboard. Fill the spool with No. 26 E.S.C.C. wire wound evenly. The key is a simple switch. Use thin brass for the arm and a large button for the finger rest.

A CONTACT PRINTER

You can make this yourself. Follow the details in the diagram.

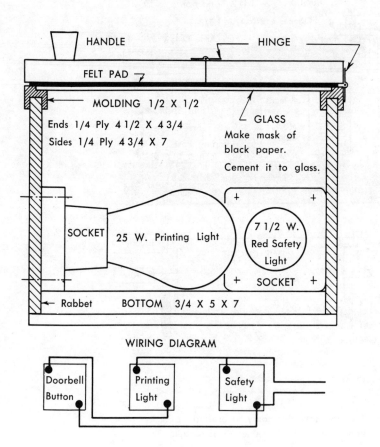

HANDLE

HINGE

FELT PAD

MOLDING 1/2 X 1/2

Ends 1/4 Ply 4 1/2 X 4 3/4

Sides 1/4 Ply 4 3/4 X 7

GLASS

Make mask of black paper.

Cement it to glass.

SOCKET   25 W. Printing Light

7 1/2 W. Red Safety Light

SOCKET

← Rabbet   BOTTOM 3/4 X 5 X 7

WIRING DIAGRAM

Doorbell Button

Printing Light

Safety Light

illustration credits

ABC Television Network, Figs. 5-32, 5-32a. Airco, Fig. 4-91. Alcoa, Figs. 4-13, 4-14, 4-21. Aluminum Co. of America, Figs. 4-12, 4-49. America House, Fig. 7-33. American Electric Power System, Fig. 5-21. American Forest Industries, Fig. 3-3. American Plywood Assn., Figs. 3-4, 3-5. American Type Founders, Fig. 6-10. Bd. of Educ., City of New York, Figs. 6-28, 6-31, 6-31a, 6-66, 7-34, 7-35. Behr-Manning, Div. of Norton Co., Fig. 3-44. Bethlehem Steel Corp., Figs. 4-16, 4-17, 4-18, 4-19. Black & Decker, Figs. 3-45a, 3-45b, 3-64. Brunswick Corp.—MacGregor, Fig. 9-2. Bulova Watch Co., Figs. 5-47, 5-48. Bureau of Mines, U.S. Dept. of the Interior, Figs. 4-2, 4-5. Fred S. Carver, Inc., Fig. 8-23. Chandler & Price Co., Figs. 6-15, 6-16, 6-18. Christian Science Publishing Society, Figs. 6-14, 6-19. Colonial Williamsburg Photograph, Figs. 6-1, 6-2. Copper Development Assn., Inc., Fig. 4-1. Corning Glass Works, Figs. 7-39, 7-44, 7-46, 7-46a, 7-47. Di-Acro Metalworking Equipment, Div. of Houdaille Industries, Inc., Figs. 8-17, 8-24, 8-25, 8-26, 8-27, 8-28, 8-32. Do-All Co., Figs. 4-69, 4-70, 4-76. Eastman Kodak Co., Figs. 6-56, 6-57, 6-67. Edison Electric Institute, Fig. 5-1. Evinrude Motors, Figs. 4-65, 4-66, 4-67, 4-68. Ford Motor Co., Fig. 4-20. General Motors, Figs. 2-1, 2-26. General Motors Corp., Electro-Motive Div., Figs. 5-16, 5-17. Goodyear Tire & Rubber, Fig. 2-6. Graymark Enterprises, Inc., Figs. 5-26, 5-45a, 5-46. Heidelberg Eastern, Inc., Fig. 6-4. Hillerich & Bradsby Co., Figs. 3-77, 3-78. International Paper Co., Figs. 6-46, 6-47, 6-48, 6-49, 6-50. Kennecott Copper Corp., Fig. 4-15. Lincoln Electric Co., Fig. 4-94. Mergenthaler Linotype Co., Fig. 6-20. *The New York Times,* Figs. 6-23, 6-34, 6-39. Owens-Corning Fiberglas Co., Figs. 8-3, 8-6, 8-31, 8-39, 8-40, 8-41. Piper Aircraft Corp., Figs. 2-3, 8-36, 8-37, 8-38. *Popular Science Monthly,* Figs. 6-22, 6-24. Porcelain Enamel Institute, Fig. 7-58. Portland Cement Assn., Figs. 7-61, 7-62, 7-63, 7-64, 7-65, 7-68. Rand McNally & Co., Figs. 6-42, 6-43, 6-44. RCA, Fig. 5-39. Rebman Photo Service, Figs. 3-46, 3-47. Rockwell Mfg. Co., Figs. 3-65, 3-66. Rohm and Haas Co., Fig. 8-18. South Bend Lathe, Figs. 4-83, 4-85. Southern Forest Institute, Fig. 3-6. Southern Furniture Manufacturers Assn., Figs. 2-28, 2-29. Sperry Rand Corp., Fig. 5-31. *The Staten Island Advance,* Figs. 6-17, 6-33, 6-34a, 6-35, 6-36, 6-37, 6-38, 6-40, 6-41. Steuben Glass, Figs. 7-40, 7-41, 7-42, 7-43, 7-45, 7-48. Syracuse China Corp., Figs. 7-3, 7-5, 7-6, 7-7, 7-8, 7-9, 7-10. Tennessee Eastman Co., Fig. 8-5. TVA, Fig. 5-7. Union Carbide—Linde Division, Figs. 4-87, 4-88, 4-89, 4-90, 8-1, 8-2, 8-4, 8-16. U.S. Steel Corp., Figs. 4-10, 4-11. Vermont Log Buildings, Inc., Fig. 3-63. The Warner-Swasey Co., Fig. 4-71. Western Electric Co., Inc., Figs. 5-37, 5-38. Western Printing and Lithographing Co., Figs. 6-29, 6-30, 6-32. Western Wood Products Assn., Figs. 3-7, 3-10. Westinghouse, Figs. 5-15a, 5-15b. Weyerhaeuser & Co., Fig. 6-45 Kay V. Wiest, Institute of American Indian Arts, Figs. 7-1, 7-2. Woodlam, Inc., Fig. 3-41.

Index